MW01517986

ARSENIC

SOURCES, ENVIRONMENTAL IMPACT, TOXICITY AND HUMAN HEALTH - A MEDICAL GEOLOGY PERSPECTIVE

CHEMISTRY RESEARCH AND APPLICATIONS

Additional books in this series can be found on Nova's website
under the Series tab.

Additional e-books in this series can be found on Nova's website
under the e-book tab.

PUBLIC HEALTH IN THE 21ST CENTURY

Additional books in this series can be found on Nova's website
under the Series tab.

Additional e-books in this series can be found on Nova's website
under the e-book tab.

CHEMISTRY RESEARCH AND APPLICATIONS

ARSENIC

SOURCES, ENVIRONMENTAL IMPACT, TOXICITY AND HUMAN HEALTH - A MEDICAL GEOLOGY PERSPECTIVE

ANDREA MASOTTI
EDITOR

publishers

New York

NOTICE TO THE READER

The Publisher has taken reasonable care in the preparation of this book, but makes no expressed or implied warranty of any kind and assumes no responsibility for any errors or omissions. No liability is assumed for incidental or consequential damages in connection with or arising out of information contained in this book. The Publisher shall not be liable for any special, consequential, or exemplary damages resulting, in whole or in part, from the readers' use of, or reliance upon, this material. Any parts of this book based on government reports are so indicated and copyright is claimed for those parts to the extent applicable to compilations of such works.

Independent verification should be sought for any data, advice or recommendations contained in this book. In addition, no responsibility is assumed by the publisher for any injury and/or damage to persons or property arising from any methods, products, instructions, ideas or otherwise contained in this publication.

This publication is designed to provide accurate and authoritative information with regard to the subject matter covered herein. It is sold with the clear understanding that the Publisher is not engaged in rendering legal or any other professional services. If legal or any other expert assistance is required, the services of a competent person should be sought. FROM A DECLARATION OF PARTICIPANTS JOINTLY ADOPTED BY A COMMITTEE OF THE AMERICAN BAR ASSOCIATION AND A COMMITTEE OF PUBLISHERS.

Additional color graphics may be available in the e-book version of this book.

Library of Congress Cataloging-in-Publication Data

Arsenic : sources, environmental impact, toxicity and human health : a medical geology perspective / editors, Andrea Masotti.
    p. cm.
 Includes index.
 ISBN: 978-1-62081-320-1 (hardcover)
 1. Arsenic. 2. Arsenic--Health aspects. 3. Arsenic--Environmental aspects. 4. Toxicology. I. Masotti, Andrea.
 QD181.A7A77 2012
 553.4'7--dc23
                        2012007377

Published by Nova Science Publishers, Inc. †*New York*

CONTENTS

PREFACE

"Omnia venenum sunt: nec sine veneno quicquam existit. Dosis sola facit, ut
venenum non fit" Paracelsus (1493–1541)

"All substances are poisons; there is none which is not a poison. The right dose
differentiates a poison and a remedy"

This sentence has been formulated by Paracelsus (Philippus Aureolus Theophrastus
Bombastus von Hohenheim) many centuries ago, but has become the fundamental principle
of toxicology: the negative biological effects in the organism increase (up to death), as the
concentration of a certain element increases. But some chemical elements are essential for life
and even the decreasing concentrations of essential elements can result in deleterious health
effects. So the Paracelsus' sentence should be read in this double meaning: too much or too
little can be 'a poison'. All of the elements that influence human health and constitute all
living creatures are found in Nature. Therefore, we have to look at the environment under
different eyes, the eyes of a new discipline: the medical geology. Medical geology is the
science dealing with the relationship between natural geological factors and health in humans

and animals. Medical geology tries also to understand the influence of ordinary environmental factors on the geographical distribution of such health problems. This novel science requires the interdisciplinary contribution of various research fields (i.e., geoscientists, clinicians and public health researchers) to understand, mitigate, and resolve emerging problems. It appears clear that medical geology, focusing on the impacts of geologic materials (rocks, minerals, and water) and processes (volcanic eruptions, earthquakes, and dust), thus 'the natural environment', on animal and human health, has to be considered complementary to environmental medicine. In fact, most of the elements the human body takes mainly come from air, water, soil and food. We inhale atmospheric dusts and gases, and the water we drink travels through rocks and soils as part of the hydrological cycle and by breaking down rocks, forms the soils on which crops are grown and animals raised. For all of these reasons, human beings and human health are directly linked to geology.

People in many developed and developing countries is becoming aware that geologic processes and human activities can have a crucial impact on human health by redistributing minerals and elements contained within, from one site to another, displaying all their harmful potential. Many issues and safety concerns are found around the globe and several billions of people can be at risk. Arsenic in drinking water is one among the several examples of international health safety concerns, with Bangladesh and West Bengal, India the most exposed regions characterized by high arsenic levels in water and soils. Many other countries are suffering from arsenic exposure and millions of people are at risk of arsenic poisoning.

This book, focused on describing arsenic sources, environmental distribution and human health issues, is aimed at illustrating only a little part of the medical geology approach. I hope that the content could help to understand and integrate such two different fields and their complex relationships. Outstanding scientists from many disciplines all around the globe have contributed to this initiative with their enthusiasm and expertise, different scientific cultures, different terminologies and different ways of thinking, but jointly leading to the birth of this brief compendium.

SECTIONAL PLAN

There are three main sections in the book: i) Sources and Distribution of Arsenic in the Environment (Chapters 1-5), ii) Pathology and Toxicology (Chapters 6-11) and iii) Techniques and Tools for Arsenic removal (Chapters 12-17).

Chapter 1 gives a general overview of Arsenic sources and its distribution in the Environment, with insights of its mechanism of action. Chapter 2 deals with mobility, bioavailability and toxicity of arsenic in the aquatic environment that depends by its physical distribution and chemical speciation in rocks and soils. Speciation of arsenic in aquatic and terrestrial environments is also of utmost importance in terms of environmental management and remediation and Chapter 3 considers the effects of natural organic matter, pH, Eh, adsorption, desorption, biological transformations and key inorganic interactions on arsenic speciation. The mobility and stabilization of arsenic in contaminated soil is described in Chapter 4 and the uses of sludge-based stabilization techniques are suggested. Finally,

Chapter 5 discusses the bioaccumulation, biotransformation, and trophic transfer of arsenic in the aquatic food chains in relation to its ecotoxicological risks in the freshwater environment.

The toxicological concerns and effects on human health are clearly illustrated in Chapter 6 where chronic arsenic toxicity (arsenicosis), one of the major environmental health hazards, is reviewed. This issue does not concern only India and Bangladesh but also China, Chile, Argentina, Mexico and United States. Europe is not spared either by this problem. Occurrence of arsenic in water and food, especially in rice-based foods for infants, poses several questions about food safety and outlines the importance of the axis soil-food-health. Chapter 7 describes this aspect and the need to find practical guidelines to minimize damage especially to infants. Arsenic contamination in drinking water has implications also on pregnancy and Chapter 8 reports an overview of Hungarian findings. Contaminated food and water are not the only routes of arsenic poisoning. A voluntary practice such as geophagy, adopted in various countries like Ghana for cultural or spiritual beliefs (Chapter 9), consists in eating soil, clay or chalk to obtain essential elements, and represents another way of poisoning. Chronic arsenic poisoning causes several subtle damages such as those induced to the central nervous system: from peripheral neuropathy to cognitive dysfunctions, decrease of attention, alteration of verbal comprehension and long-term memory loss in children and adolescents (Chapter 10). Moreover, if it has been clearly demonstrated the existence of a link between human health and arsenic contamination and intake, little is known about the effects of arsenic and its metabolites when interact with other metals, such as iron, in the environment. This is the subject in Chapter 11, which emphasizes also the need to find novel strategies and innovative purification systems for arsenic removal.

Therefore, the last section describes some of the current and most innovative technologies to remove arsenic from groundwater. The biological approach is one of the most economical approaches, is eco-friendly as no chemicals are used and the sorbent materials are continuously produced in situ. Two biological filtration techniques have been analyzed, the first one, described in Chapter 12, is linked to iron and/or manganese removal and the second one, sulfate-reducing bacteria and titanium dioxide, used for groundwater and industrial wastewater applications is described in Chapter 13. More conventional complexation-ultrafiltration processes, also called 'polymer-assisted' or 'polymer-enhanced' ultrafiltration, are described in Chapter 14. Natural and modified adsorbents together with a compilation of up to date adsorption techniques are reviewed in Chapter 15. In this chapter, an analysis of the availability and cost of adsorbents and their absorption capacity from waters/wastewaters is also illustrated. Chapter 16 describes fundamental concepts associated with the removal of soluble arsenic species using continuous systems based on the Zero-Valent iron technology. Chapter 17 deals with a study about the adsorption process and the employment of mixtures of different iron and silica oxides to immobilize the arsenic-contaminated adsorbents in vitreous matrices, finally reconverting these products in useful decorative ceramics without safety concerns.

Finally, I would like to express my gratitude to all the forty-nine authors who wrote chapters and contributed to the success of this initiative. I also thank the reviewers who helped authors to improve their chapters with their helpful suggestions and comments.

A special thank goes to Dr. Ornella Walker Barbarito for the careful revision of many parts of the book and for continuous suggestions. Many thanks to Dr. Carra Feagaiga from the Department of Acquisitions of Nova Science Publishers for having assisted me during all the development phases of the book.

Last, but not least I would like to dedicate this book to my wife and my children, for the time spent to prepare this book that I unduly subtracted to them.

CONTRIBUTORS

Emmanuel Arhin
University of Leicester, Department of Geology, University Road, LE1 7RH, UK.

Elisabetta Bemporad
Italian Workers' Compensation Authority (INAIL), Department of Production Plants and Anthropic Settlements (DIPIA) ex National Institute for Occupational Safety and Prevention (ISPESL)- Rome - Italy

Mátyás Borsányi
National Institute of Environmental Health, Gyáli út 2-6, Budapest, Hungary

F. Burló
Universidad Miguel Hernández. Departamento Tecnología Agroalimentaria. Ctra. de Beniel, km 3.2. 03312-Orihuela, Alicante, Spain

A. A. Carbonell-Barrachina
Universidad Miguel Hernández. Departamento Tecnología Agroalimentaria. Ctra. de Beniel, km 3.2. 03312-Orihuela, Alicante, Spain

Barbara Casentini
Istituto di Ricerca sulle Acque (IRSA)/Water Research Institute, Consiglio Nazionale delle Ricerche (CNR) / National Research Council, Via Salaria km 29,300 C.P. 10, 00015 Monterotondo (RM), Italy

C. Castaño-Iglesias
Universidad Miguel Hernández
Departamento de Farmacología, Pediatría y Química Orgánica. Crta. Nacional N-332, s/n. 03350-Sant Joan, Alicante, Spain

Victor Raj Mohan Chandrasekaran
Department of Environmental and Occupational Health, College of Medicine, National Cheng Kung University, 138 Sheng-Li road, Tainan, Taiwan

M. Ciopec
University "Politehnica" Timisoara, Faculty of Industrial Chemistry and Environmental Engineering, 2 Piata Victoriei, 300006 Timisoara, Romania

Mihály Csanády
National Institute of Environmental Health, Gyáli út 2-6, Budapest, Hungary

Fernando S. García Einschlag
Instituto de Investigaciones Fisicoquímicas Teóricas y Aplicadas (INIFTA), CCT-La Plata-CONICET, Departamento de Química, Facultad de Ciencias Exactas, Universidad Nacional de La Plata. 64 Diag. 113, CP (1900), Argentina

D. N. Guha Mazumder
Department of Medicine and
Gastroenterology, Institute of Post
Graduate Medical Education & Research,
Kolkata - 700 020, India

P. I. Haris
Faculty of Health and Life Sciences. De
Montfort University. Hawthorn Building.
The Gateway. Leicester, LEI 9BH, UK

Hiroshi Hasegawa
Institute of Science and Technology,
Kanazawa University, Kakuma,
Kanazawa 920-1161, Japan

Christel Hassler
Plant Functional Biology and Climate
Change Cluster, School of the
Environment, Faculty of Science,
University of Technology Sydney, P.O.
Box 123, Broadway, NSW 2007,
Australia

María E. Jiménez-Capdeville
Department of Biochemistry, Medicine
Faculty, University of San Luis Potosi,
Av. V. Carranza 2405, 78210, San Luis
Potosi, S.L.P., Mexico

Chuanyong Jing
State Key Laboratory of Environmental
Chemistry and Ecotoxicology, Research
Center for Eco-Environmental Sciences,
Chinese Academy of Sciences, Beijing
100085, China

Mihály Kádár
National Institute of Environmental
.Health, Gyáli út 2-6, Budapest, Hungary

R. Lazău
University "Politehnica" Timisoara,
Faculty of Industrial Chemistry and
Environmental Engineering, 2 Piata
Victoriei, 300006 Timisoara, Romania

Richard Lim
Centre for Environmental Sustainability,
School of the Environment, Faculty of
Science, University of Technology
Sydney, P.O. Box 123, Broadway, NSW
2007, Australia

Ming-Yie Liu
Department of Environmental and
Occupational Health, College of
Medicine, National Cheng Kung
University, 138 Sheng-Li road, Tainan,
Taiwan

Ting Luo
State Key Laboratory of Environmental
Chemistry and Ecotoxicology, Research
Center for Eco-Environmental Sciences,
Chinese Academy of Sciences, Beijing
100085, China

L. Lupa
University "Politehnica" Timisoara,
Faculty of Industrial Chemistry and
Environmental Engineering, 2 Piata
Victoriei, 300006 Timisoara, Romania

Clemens Marb
Bavarian Environment Agency, Josef-
Vogl-Technology Center, Am Mittleren
Moos, 46, 86167 Augsburg, Germany

Claudio Minoia
Laboratory of Environmental and
Toxicology Testing, "S. Maugeri"-
IRCCS, Pavia, Italy

S. Munera
Universidad Miguel Hernández.
Departamento Tecnología
Agroalimentaria. Ctra. de Beniel, km 3.2.
03312-Orihuela, Alicante, Spain

A. Negrea
University "Politehnica" Timisoara,
Faculty of Industrial Chemistry and
Environmental Engineering, 2 Piata
Victoriei, 300006 Timisoara, Romania

Yuanming Pan
Department of Geological Sciences,
University of Saskatchewan, Saskatoon,
SK S7N 5E2, Canada

Dionisios Panagiotaras
Department of Mechanical Engineering,
Technological Educational Institute
(T.E.I.) of Patras, 26 334, Patras, Greece

Georgios Panagopoulos
Department of Mechanical and Water
Resources Engineering, Technological-
Educational Institute (T.E.I.) of
Messolonghi, Nea Ktiria, 30 200,
Messolonghi, Greece

Dimitris Papoulis
Department of Geology, University of
Patras, 26504 Patras, Greece

Maurizio Pettine
Istituto di Ricerca sulle Acque
(IRSA)/Water Research Institute,
Consiglio Nazionale delle Ricerche
(CNR) / National Research Council, Via
Salaria km 29,300 C.P. 10, 00015
Monterotondo (RM), Italy

M. Azizur Rahman
Centre for Environmental Sustainability,
School of the Environment, Faculty of
Science, University of Technology
Sydney, P.O. Box 123, Broadway, NSW
2007, Australia

A. Ramírez-Gandolfo
Universidad Miguel Hernández.
Departamento Tecnología Agroalimentaria. Ctra. de Beniel, km 3.2.
03312-Orihuela, Alicante, Spain

Rosalva Ríos
Department of Biochemistry, Medicine
Faculty, University of San Luis Potosi,
Av. V. Carranza 2405, 78210, San Luis
Potosi, S.L.P., Mexico

Bernabé L. Rivas
Polymer Department, Faculty of
Chemistry, University of Concepción,
Casilla 160-C, Concepción, Chile

Peter Rudnai
National Institute of Environmental
Health, Gyáli út 2-6, Budapest, Hungary

Danladi Mahuta Sahabi
Department of Biochemistry, Usmanu
Danfodiyo University, P.M.B. 2346,
Sokoto, Nigeria

Julio Sánchez
Polymer Department, Faculty of
Chemistry, University of Concepción,
Casilla 160-C, Concepción, Chile

Reinhard Schliebs
Paul Flechsig Institute for Brain Research,
Medical Faculty, University of Leipzig,
Jahnallee 59, 04109, Leipzig, Germany

Virender K. Sharma
Department of Chemistry, Florida Institute
of Technology, 150 West University
Boulevard, Melbourne, Florida 32901,
USA

Mary Sohn
Department of Chemistry, Florida Institute
of Technology, 150 West University
Boulevard, Melbourne, Florida 32901,
USA

Elena Sturchio
Italian Workers' Compensation Authority
(INAIL), Department of Production Plants
and Anthropic Settlements (DIPIA) ex
National Institute for Occupational Safety
and Prevention (ISPESL)- Rome – Italy

Minoru Takeda
Division of Materials Science and
Chemical Engineering, Graduate School
of Engineering, Yokohama National
University, 79-5 Tokiwadai, Hodogaya,
Yokohama 240-8501, Japan

Juan M. Triszcz ˙
Laboratorio de Ingeniería Sanitaria (LIS),
Departamento de Hidráulica, Facultad de
Ingeniería, Universidad Nacional de La
Plata. 48 e/ 115 y 116. CP (1900),
Argentina

Harald Weigand
Technische Hochschule Mittelhessen -
University of Applied Sciences, KMUB,
Wiesenstr. 14, 35390 Giessen, Germany

Miriam Zanellato
Italian Workers' Compensation Authority
(INAIL), Department of Production Plants
and Anthropic Settlements (DIPIA) ex
National Institute for Occupational Safety
and Prevention (ISPESL)- Rome - Italy

Musah Saeed Zango
University for Development Studies,
Department of Earth Science, P.O. Box
24, Navrongo-Ghana.

Sergio Zarazúa
Department of Toxicology, Chemistry
Faculty, University of San Luis Potosi,
Manuel Nava 6, 78210, San Luis Potosi,
S.L.P., Mexico

SECTION I.
SOURCES, SPECIATION AND DISTRIBUTION

In: Arsenic
Editor: Andrea Masotti

ISBN: 978-1-62081-320-1
© 2013 Nova Science Publishers, Inc.

Chapter 1

ARSENIC: ENVIRONMENTAL CONTAMINATION AND EXPOSURE

*Elena Sturchio[1], Miriam Zanellato[1], Claudio Minoia[2]
and Elisabetta Bemporad[1]*
[1]Italian Workers' Compensation Authority (INAIL),
Department of Production Plants
and Anthropic Settlements (DIPIA) ex National Institute for Occupational Safety
and Prevention (ISPESL)- Rome – Italy
[2]Laboratory of Environmental and Toxicology Testing,
"S. Maugeri"-IRCCS, Pavia, Italy

ABSTRACT

Arsenic (As) is a ubiquitous metalloid, widely distributed in the environment through both natural and anthropogenic pathways. Exposure to arsenic typically results from either oral arsenic consumption through contaminated drinking water, soil, and food or arsenic inhalation in an industrial work setting. Arsenic pollution is often traced to mining or mining-related activities, copper smelting, coal burning, other combustion processes, and volcanic eruptions that bring arsenic into the environment. Anthropogenic sources of arsenic include its use in various commonly used arsenical herbicides, insecticides, rodenticides, wood preservatives, animal feeds, paints, dyes, and semiconductors. Inorganic or organic species of the metalloid arsenic occur in the environment. Arsenic is a contaminant of concern in soil and groundwater, often difficult to remove because of the varying toxicity and mobility of its compound. Restrictions on manufacture, marketing and use of arsenic, limiting or targeting guideline values for the various environmental compartment for waste and emissions in air and water are prescribed by European and/or national laws to limit the presence and the spread of arsenic in the environment. European data from the latest available validated version of the E-PRTR for the year 2009 concerning arsenic and its compounds, show that the highest releasing sector among the large air polluting sectors is the energy sector, while for water it is waste and wastewater management. Generally, the inorganic arsenic forms (iAs) are thought to be more toxic, both acute and chronic, than the organic ones (oAs), but both forms are toxic to humans. iAs is a highly toxic element and its presence in food composites is a matter of concern to the well being of both humans and animals. The main exposure route to iAs still remains dietary. Arsenic-contaminated groundwater is often used in agriculture to irrigate crops for food and animal consumption, which could potentially lead to arsenic entering into the human food chain. As, like other metals and metalloids, resides in the soil for long periods of time, where it can either be taken up by plants or washed down into the groundwater, and present a risk to human health. A range of mitigation methods, from agronomic measures (e.g. raised beds, soil amendments), or plant breeding to genetic modification, may be employed to reduce As uptake by food crops. Human exposure to food and drinking water As contamination is a real concern. Arsenic is not an essential element for plants and its over-concentration in the soil can generate toxicity phenomena. Its translocation from soil to plant constitutes one of the main human exposure ways. Many of the human health effects of arsenic have been established based on epidemiologic studies, which have shown a significant association between the consumption of arsenic through drinking water and cancers of the skin, lung, bladder, liver, and kidney, neurologic disease, cardiovascular disease, as well as other non-malignant diseases.

1. ARSENIC IN THE ENVIRONMENT

Arsenic in the environment originates from natural and anthropogenic sources. It is commonly bound to carbon, iron, oxygen, and sulfur, forming inorganic and organic arsenical species in different oxidation states.

Commission (EC) Regulation no.1272/2008 on classification, labelling and packaging of substances and mixtures (CLP Regulation) and the subsequent Adaptation to Technical Progress (Commission Regulation n.790/2009) classify arsenic and its compounds, with the exception of those specified elsewhere in Annex VI (Table 3.1), in the following categories:

- Acute Tox. 3* (H331 Toxic if inhaled, H301 Toxic if swallowed)
- Aquatic Acute 1 (H400 Very toxic to aquatic life)
- Aquatic Chronic 1 (H410 Very toxic to aquatic life with long lasting effects)
  (the reference * indicates minimum classification for a category)

The physicochemical properties of many arsenical species are important determinants of the potential toxic effects.

Arsenic, mainly in inorganic form, shows a ubiquitous diffusion in soil, air and water (Figure 1).

It mainly originates from geochemical sources and is present as compounds derived by oxidation of metallic sulphides (pyrite, chalcopyrite, arsenopirite). Arsenic is also present in many industrial raw materials, products, and wastes, thus resulting as a contaminant of concern in soil and groundwater at many remediation sites. Arsenic is considered a trace element among the most "mobile" both in the hydrosphere and in atmosphere. Elementary As is not soluble in water, but arsenic salts are more or less soluble according to the pH and to the ionic medium. Arsenic can exist in four oxidation states: - 3, 0, +3 and +5.

Under reducing conditions, As occurs primarily as arsenite (AsIII), while arsenate (AsV) is the more stable form in an oxidant environment (WHO, 2001).

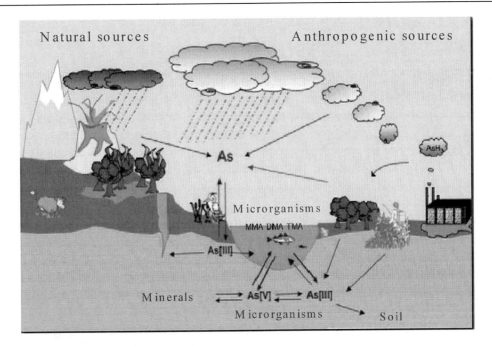

Figure 1. Inorganic arsenic diffusion in soil, air and water.

The inorganic species (arsenate plus arsenite) are the predominant forms of arsenic in natural water, where monomethylarsonic acid (MMA) can occur in lower concentration, as well as dimethylarsinic acid (DMA) and the monomethylated (MMAIII) and trivalent dimethylated (DMAIII) arsenical species. Further arsenical species found in sea water and in fresh water could represent up to 20% of total As (IARC, 2004).

In oceanic waters Arsenic is present in a concentration of 0.3 µg/L while in continental waters generally it has values between 0.05 and 1.00 µg/L. In natural waters the element is present as soluble oxyanion, arsenate V and arsenite III, in a concentration of 1-10 µg/L(WHO, 2001) in uncontaminated waters and of 100-5000 µg/L in contaminated waters (near mining zones).

In the terrestrial crust it is naturally present at relatively low concentrations (2 ppm).

Arsenic concentrations in soils can vary from 1 to 40 mg/kg (WHO, 2001), because of agricultural and industrial contaminations (employment of sewage sludge and use of cleansing and/or phosphate based fertilizers that contain appreciable quantities of this element). The volcanic activity and the microorganisms of soils release As into the air and its levels vary, with inferior values in rural areas (0.007-28 ng/m^3) and more elevated concentrations in urban areas (3-200 ng/m^3) (WHO, 2001).

Different studies performed in European Union States have underlined values of As in the air among 1-3 ng/m^3 in the urban areas and among 20-30 ng/m^3 in industrial areas (WHO, 2001).

The high mobility of As occurs because under oxidizing conditions the element is present in the oxidation state +5 and doesn't suffer remarkable trials of coprecipitation or adsorption with the exception of the coprecipitation induced by iron hydroxides. The environmental destiny of As has been studied extensively (Cullen and Reimer, 1989; Mandal and Suzuki, 2002; Nordstrom, 2002).

Arsenic toxicity has become a global concern owing to the ever-increasing contamination of water in many regions of the world.

In some areas of Argentina, Bangladesh, Chile, Hungary, India, Mexico, Romania, Taiwan, Vietnam, and in many parts of America (Smedley e Kinniburgh, 2002), arsenic concentrations are greater than 50 µg/L (Figure 2). In many parts of Argentina, in Japan, New Zealand, Chile, Iceland, France, USA, arsenic is present in thermal waters. In Ghana, Greece, Thailand and the USA problems related to the presence of arsenic exist in areas affected by mining activities. The presence of As groundwater is found in both oxidizing and reducing conditions, and in humid/temperate or dry climates. This, in particular, is the environmental situation in Bangladesh where many contaminated rural wells are used to irrigate rice fields. Many toxic species of As (III) are found in groundwater polluted areas (Mandal e Suzuki, 2002) both in Bangladesh and Bengal.

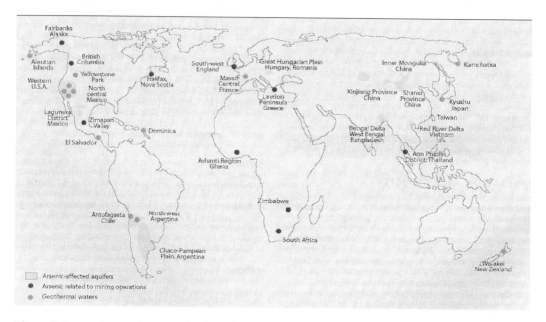

Figure 2. Inorganic arsenic contamination of water in many regions of the world.

The presence of arsenic in soil and water is noted in several Italian regions and is generally due to natural phenomena, even though there were clear cases of anthropogenic contamination.

In Italy abnormal concentrations of As have been found in different regions, such as Lombardy, Tuscany, Lazio, Sardinia, Campania and Trentino (Dall'Aglio, 1996) (Figure 3).

Legislative Decree no. 31/01 of December 2003, national transposition of the Council Directive 98/83/EC on the quality of water intended for human consumption established that arsenic is one of the critical parameters for the quality of drinking water, in accordance with the decisions of the WHO toxicity about many arsenical species in drinking water. Data show that levels of arsenic in Italy are frequently higher than 50 µg /L. Among the most polluted areas, the Piana di Scarlino, in Grosseto (Tuscany), presents a combination of natural and anthropogenic factors, which resulted, in some places, in an accumulation of arsenic in soils at concentrations of up to 1000 mg/kg.

In the area of Lazio, as concentrations ranging between 25-80 µg/L are higher than the limits set by law (10 µg/L)

Because arsenic readily changes valence state and reacts to form species with varying toxicity and mobility, effective treatment of arsenic can be difficult. Treatment can result in residuals that, under certain environmental conditions, become more toxic and mobile (U.S. EPA 2002).

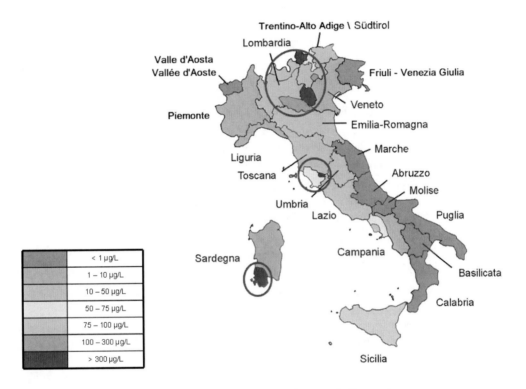

Figure 3. Inorganic arsenic concentration in drinking water in some Italian regions.

Criteria to Limit the Presence and the Spread of Arsenic in the Environment

Obviously limitations both on natural and anthropogenic sources should be in force.

The basic and main criterion is prevention.

To protect human health, thus also preserving the environment, restrictions on the manufacture, marketing and use have been in place for about thirty years; such restrictions are now consolidated in Annex VIII of the REACH (Regulation EC n.1907/2006):

Arsenic compound:

1) Shall not be placed on the market, or used, as substances or in mixtures where the substance or mixture is intended for use to prevent the fouling by micro-organisms, plants or animals of:

— the hulls of boats,

— cages, floats, nets and any other appliances or equipment used for fish or shellfish farming,
— any totally or partly submerged appliances or equipment.

2) Shall not be placed on the market, or used, as substances or in mixtures where the substance or mixture is intended for use in the treatment of industrial waters, irrespective of their use.

3) Shall not be used in the preservation of wood. Furthermore, wood so treated shall not be placed on the market.

4) By way of derogation from paragraph 3:

   a)   Relating to the substances and mixtures for the preservation of wood: these may only be used in industrial installations using vacuum or pressure to impregnate wood if they are solutions of inorganic compounds of the copper, chromium, arsenic (CCA) type C and if they are authorized in accordance with Article 5(1) of Directive 98/8/EC. Wood so treated shall not be placed on the market before fixation of the preservative is completed.

Wood treated with CCA solution in accordance with point (a) may be placed on the market for professional and industrial use provided that the structural integrity of the wood, required for human or livestock safety and skin contact by the general public, during its service life is unlikely:

   — as structural timber in public and agricultural buildings, office buildings, and industrial premises,
   — in bridges and bridgework,
   — as constructional timber in freshwater areas and brackish waters, for example jetties and bridges,
   — as noise barriers,
   — in avalanche control,
   — in highway safety fencing and barriers,
   — as debarked round conifer livestock fence posts,
   — in earth retaining structures,
   — as electric power transmission and telecommunications poles,
   — as underground railway sleepers.

   b)   Without prejudice to the application of other Community provisions on the classification, packaging and labelling of substances and mixtures, suppliers shall ensure before the placing on the market that all treated wood placed on the market is individually labelled 'For professional and industrial installation and use only, contains arsenic'. In addition, all wood placed on the market in packs shall also bear a label stating 'Wear gloves when handling this wood. Wear a dust mask and eye protection when cutting or otherwise crafting this wood. Waste from this wood shall be treated as hazardous by an authorized undertaking'.

c)   Treated wood referred to under point (a) shall not be used:

— in residential or domestic constructions, whatever the purpose,
— in any application where there is a risk of repeated skin contact,
— in marine waters,
— for agricultural purposes other than for livestock fence posts and structural uses in accordance with point (b),
— in any application where the treated wood may come into contact with intermediate or finished products intended for human and/or animal consumption.

5) Wood treated with arsenic compounds that was in use in the Community before 30 September 2007, or that was placed on the market in accordance with paragraph 4 may remain in place and continue to be used until it reaches the end of its service life.

6) Wood treated with CCA type C that was in use in the Community before 30 September 2007, or that was placed on the market in accordance with paragraph 4:

— may be used or reused subject to the conditions pertaining to its use listed under points 4(b), (c) and (d),
— may be placed on the market subject to the conditions pertaining to its use listed under points 4(b), (c) and (d).

7) Member States may allow wood treated with other types of CCA solutions that was in use in the Community before 30 September 2007:
— to be used or reused subject to the conditions pertaining to its use listed under points 4 (b), (c) and (d),
— to be placed on the market subject to the conditions pertaining to its use listed under points 4(b), (c) and (d).

Then, limit or target or guideline values are prescribed by European and/or national laws for the various environmental system compartments, making it possible to prevent both natural and anthropogenic sources.

Table 1 shows the values set by Europe and/or Italy. Medium for environmental contamination by arsenic are also wastes. European Council Decision 2033/33/EC establishes specific limit values for leachable arsenic concentration for waste acceptable in landfills. These limit values are reported in Table 2.

Italy also set rules specifying the types and quantities of waste that may be covered by an exemption from permit requirement and the method of treatment (recovery) to be used, as provided for in article 25 of Directive 2008/98/EC on waste.

Table 1. Limit or target values set by European and/or Italian regulations for the environmental compartments

ENVIRONMENTAL COMPARTMENT	INTENDED USE	LIMIT VALUE	UNIT	VALUE DETAILS	REGULATION
Air	-	6	ng/m^3	target value for the total content in the PM_{10} fraction averaged over a calendar year	Directive 2004/107/EC
Water	human consumption	10[1]	µg/l	chemical parameter not to be exceeded	Directive 98/83/EC
Inland surface waters	-	10	µg/l	environmental objectives for the annual average	Italian regulation in compliance with the directive 2000/60/EC
Transitional and coastal waters	-	5	µg/l		
Soil and subsoil	residential and green areas	20	mg/kg s.s.	contamination thresholds [2]	Italian Decree n.152/2006
	commercial and industrial areas	50	mg/kg s.s.		
Groundwater	-	10	µg/l		

(1) Member States may provide for derogations, limited to as short a period of time as possible, up to a maximum value to be determined by them (in Italy up to 50 µg/l), provided that no derogation constitutes a potential danger to human health and provided that the supply of water intended for human consumption in the area concerned cannot otherwise be maintained by any other reasonable means

(2) the contamination threshold for soil must be observed on sample without the fraction > 2 cm (to be discarded in field) and on the sieve fraction < 2 mm of the total dry mass, shell included.

Table 2. Limit values for leachable arsenic concentration for waste acceptable in landfills (European Council Decision 2033/33/EC)

CLASS OF LANDFILL	L/S = 2 l/kg	L/S = 10 l/kg	C_0 percolation test
	mg/kg dry substance	mg/kg dry substance	mg/l
for inert waste	0.1	0.5	0.06
for non-hazardous waste	0.4	2	0.3
for hazardous waste	6	25	3

The waste addressed by these rules, which could contain arsenic, referring to the European List of Waste in force, are:

- black liquor from wood processing and the production of panels and furniture, pulp, paper and cardboard 030199
- slag, dross and skimmings from primary and secondary production of lead 100401*, 100402*
- lead batteries and accumulators containing these batteries, included unsorted ones 160601*, 200133*
- zinc ferrites from zinc hydrometallurgy 110202*
- slags from primary and secondary production of zinc and lead 100501, 100401*
- solid salts, sludge and solutions containing high concentration of copper from the etching of printed circuit boards and solution processing 060313*, 060405*, 190205*
- shovelable sludge containing lead and sulphur from secondary production of lead by spent lead accumulators, primary production of zinc and lead oxides and production of lead accumulators 060405*, 100407*, 190205*
- acidic sludge from sulfonation plants and spent H_2SO_4 solutions from petrochemical and chemical industries 060101*
- spent activated carbon from craft and industrial production processes, incineration or pyrolysis of municipal and assimilated waste 060702*, 061302*, 190110*, 070109*, 070110*, 070209*, 070210*, 070309*, 070310*, 070409*, 070410*, 070509*, 070510*, 070609*, 070610*, 070709*, 070710*

The limits set for arsenic, to get the exemption from permit range from 0,00001% to 3%, depending on the origin, type and classification (that is arsenic speciation) of waste. The highest limits are reserved for spent activated carbons and waste from non-ferrous, particularly zinc and lead, thermal metallurgy. The recovery operations, which can be exempted by the permit, are also regulated for emission limits. Finally it should be highlighted that arsenic and its compounds are included in the list of polluting substances to be limited both in air and water as permit conditions for industrial activities specified in Annex I of the Directive 2010/75/EU on industrial emissions (Integrated Pollution Prevention and Control or IPPC). The IPPC EU Directive establishes a general framework for the control of the main industrial activities, giving priority to intervention at source, ensuring prudent management of natural resources and taking into account, when necessary, the economic situation and specific local characteristics of the place in which the industrial activity is taking place, to prevent, reduce and as far as possible eliminate pollution arising from industrial activities in compliance with the 'polluter pays' principle and the principle of pollution prevention. The industrial activities covered by this Directive from specified threshold values referring to production capacities or outputs include:

1) energy industries
2) production and processing of metals
3) mineral industry
4) chemical industry
5) waste management
6) other activities

The IPPC Directive also establishes specific limit values for waste incineration plants and particularly:

- an air emission limit value for the average total of nine heavy metals, including arsenic, over a sampling period of a minimum of 30 minutes and a maximum of 8 hours (0,5 mg/Nm^3);
- an emission limit value for discharges of waste water from the cleaning of waste gases (unfiltered samples) for arsenic and its compounds, expressed as arsenic (0,15 mg/L).

Table 3. Arsenic compounds and categories of interest for the "Seveso II" directive application

Named substances or category of substances and mixture	Qualifying quantities (t) for	
	Lower Tier Est.	Upper Tier Est.
Arsenic pentoxide, arsenic (V) acid and/or salts	1	2
Arsenic trioxide, arsenious (III) acid and/or salts	0.1	
Arsenic trihydride (arsine)	0.2	1
2. Toxic	50	200
9. Dangerous for the environment risk phrases: i) R50: 'Very toxic to aquatic organisms' (including R50/53)	100	200

Another regulation which could be useful to limit the spread of arsenic into the environment is the 96/82/EC or "Seveso II" directive on the control of major-accident hazards involving dangerous substances, which applies to the establishments where dangerous substances are present in quantities equal to or in excess of the qualifying quantities listed in Annex I, Parts 1 (named substances) and 2 (category of substances and mixtures not specifically named in part 1). The qualifying quantities for arsenic and its compounds are listed in Table 3. Finally, the census sources of arsenic were gathered to identify the crucial ones, where the attention to prevent and reduce arsenic in the environment should be focused.

The references used to gather data were:

- the first reconnaissance of the presence of hazardous substances in water bodies in Italy according to the 76/464/EC directive on pollution caused by certain dangerous substances discharged into the aquatic environment of the Community (APAT, 2003), which highlighted that arsenic in some Italian regions was the most detected pollutant but its average concentrations in the investigated water bodies were well below the Quality Standard of 10 µg/L with only few local exceptions (the highest values were detected in a river where mining wastes were poured in the past);

- the national database about air quality BRACE (ISPRA, 2010a), reporting few and localized data, with average values well below the Quality Standard of 6 ng/m^3;

- the National Air Emission CORINAIR Inventory (ISPRA, 2010b) revealing that the 84,5% of the arsenic emissions in air for the years 1990, 1995, 2000 and 2005 came from the Group 3 "Combustion in manufacturing industry". The description of the activities of the Group 3 emitting arsenic is given in Table 4.

Table 4. Description of the Activities of the EMEP/CORINAIR Group 3 emitting arsenic

SNAP	Name of SNAP/CORINAIR Activity
0301	Combustion in boilers, gas turbines and stationary engines
030101	Combustion plants > = 300 MW (boilers)
030102	Combustion plants > = 50 MW and < 300 MW (boilers)
030103	Combustion plants < 50 MW (boilers)
030104	Gas turbines
030105	Stationary engines
030106	Other stationary equipment
0302	Processes with or without contact
030203	Blast furnace cowpers
030204	Plaster furnaces
0303	Processes with contact
030301	Sinter and pelletizing plants (with the exception of 040209)
030302	Reheating furnaces steel and iron
030303	Grey iron foundries
030304	Primary lead production
030305	Primary zinc production
030308	Secondary zinc production
030309	Secondary copper production
030310	Secondary aluminum production
030311	Cement
030314	Flat glass
030315	Container glass
030316	Glass wool (except binding)
030317	Other glass (including special glass)

- the national transposition of the European Pollutant Release and Transfer Register (E-PRTR), replacing and improving upon the previous European Pollutant Emission Register (EPER), and particularly its last validated version, (INES, 2006) which highlighted that in Italy the plants subjected to the IPPC directive and to the duty to declare arsenic emissions (emissions in air > 20 kg/y and in water > 5 kg/y)

  o contribute poorly to the emissions to air and the highest emissions are from the combustion plants (IPPC activity 1.1), suggesting that at a national level the small industrial combustion sources control the overall emission;
  o emit into water consistently, but a national inventory is not available so that a comparison is impossible and the highest emission plants are included by IPPC activities 4.2 Production of inorganic chemicals, 5.3 Disposal of non-hazardous waste and particularly the physico-chemical treatment plants and 5.4 Landfills

The latest available validated version of the E-PRTR reports Arsenic and its compounds Releases for the year 2009 [E-PRTR, 2009). Excluding the accidental emissions, which are scarcely significant, the distribution of the arsenic emissions by environmental compartment is shown in Figure 4, while the distribution of the number of arsenic emitting facilities (Figure 5), and of the same arsenic emissions in air, in water and in soil by sector are shown respectively in Figure 6, 7, 8. It is evident that the emissions in soil are insignificant compared to water and air, and only the distribution of the emissions in water roughly reflects that of the number of facilities, except for the chemical industry.

European data confirm that the most polluting sector for air is the energy sector, especially the thermal power stations and other combustion installations aloof followed by the production and processing of metals, especially the production of non-ferrous crude metals from ore, concentrates or secondary raw materials, while for water the most polluting sector is the waste and wastewater management, particularly the urban wastewater treatment plants (responsible of the 22.5% of the total arsenic emissions), followed by the chemical industry, particularly the industrial scale production of basic inorganic chemicals.

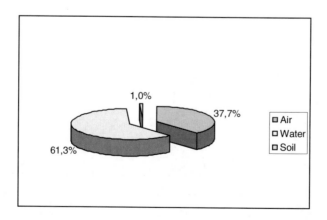

Figure 4. Distribution of arsenic emissions from large facilities for each environmental compartment.

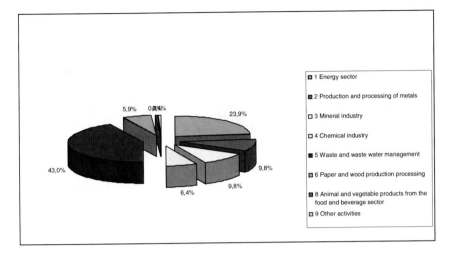

Figure 5. Distribution of the number of arsenic emitting large facilities by sector.

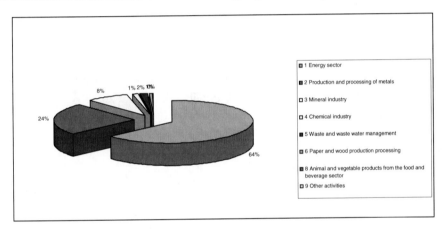

Figure 6. Distribution of arsenic emissions in air from large facilities by sector.

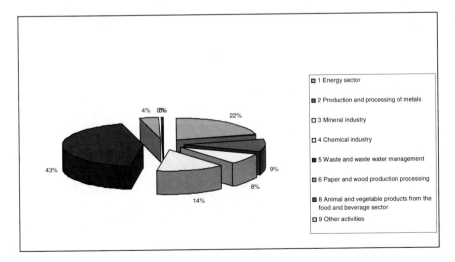

Figure 7. Distribution of arsenic emissions in water from large facilities by sector.

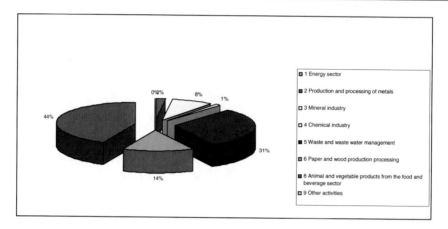

Figure 8. Distribution of arsenic emissions in soil from large facilities by sector.

For soil, 44% of the total emissions comes from the treatment and processing of the vegetables raw material, followed by the waste and waste water management, especially the urban wastewater treatment plants.

Finally, the urban wastewater treatment plants contribute to the 22.5% of the total arsenic emissions from large facilities in the environment.

Referring to the State, the 47% of the emissions in air by large plants are concentrated in Spain, Poland, Germany, France and Czech Republic, while Italy alone is responsible of the 25% of the emissions in water.

The significance of arsenic among the all heavy metals emitted by large plants in Europe is, however, very low if considered in absolute quantity, but becomes much higher, especially in water, if related to toxicity equivalents.

2. HUMAN EXPOSURE

Exposure to arsenic typically results from either oral arsenic consumption through contaminated drinking water, soil, and food, or arsenic inhalation in an industrial work setting (Figure 9).

Regarding the environmental exposure to arsenic in the population, it may occur close to smelters and settlements for production of energy (coal-burning energy producers). Exposure can also occur from skin contact with contaminated soil and less frequently with pesticides containing arsenic. Wood treated with chromated copper arsenate (CCA) as a preservative agent represents a source of exposure to arsenic. In the U.S. wood treated with CCA was widely used in many homes and playgrounds but has been phased out since 2003. Diet represents the main non-occupational source of exposure to arsenic in humans. Particularly the total contribution of inorganic As is about 17-24%. In most European and non-European countries human drinking water is the main exposure source of inorganic As. In this regard, it is important to stress that a study by the EPA showed that in the U.S. drinking water contained an As concentration of 2 µg/l. According to Karagas et al. (1998) the 12% of the surface water resources (regions of the north central of the United States) and the 12% of underground water supplies (western regions) have a superior element content of 20 µg/l.

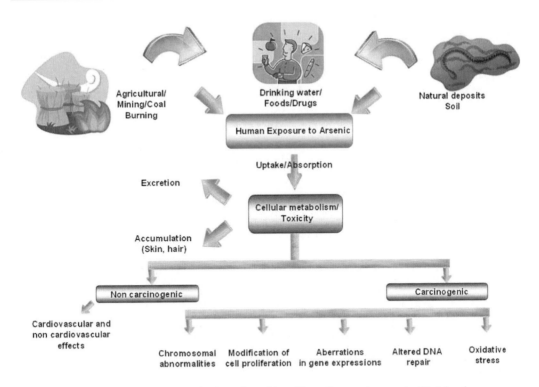

Figure 9. Human exposure to inorganic Arsenic and its effects. Inorganic arsenic (iAs) has been classified by IARC in group 1 as carcinogenic to human. Many of the human health effects of arsenic have shown a significant association between the consumption of arsenic through drinking water and cancers of the skin, lung, bladder, liver, and kidney, neurologic disease, cardiovascular disease, as well as other nonmalignant diseases.

On January 2001, the EPA decided to lower the standard of As in drinking water to 10µg/l, replacing the previous standard of 50µg/l (EPA, 2001). The new MCL (*Maximum Contaminant Level*) entered into force as of January 23rd 2006.

In Italy arsenic abnormal concentrations on water, marine and river sediments and on soils of Lazio and Campania were determined (Dall'Aglio, 1996). In hydrothermal conditions this element can be selectively mobilized reaching yields of the order of mg / l.

For inorganic arsenic in drinking water a PTWI (provisional tolerable weekly intake 0,015 mg / kg), has been established. A PTWI has not been established for food (WHO 1989).

Several studies (Vaessen and Van Ooik, 1989; Munoz, 2000; Storelli and Marcotrigiano, 2001; WHO, 2001; Fattorini, 2004; Argese, 2005; Hsiung and Huang, 2006; Schmeisser, 2006; Liu, 2007) have confirmed the presence of elevated As concentrations in seafood; in particular, the prevailing form has been found to be the organic As specie (arsenobetaina). (Mohri, 1990; Francesconi and Edmonds, 1997) Munoz et al. (2000) have found a content of inorganic As higher than 11% of total As, and its concentrations varied from 0.16 to 33.1 mg/Kg (Ghidini et al., 2000). Flat fishes (gudgeons, soles, codfishes) showed higher values than eels, mullets and striped sea breams.

It is important to stress that fish samples collected in brackish environments or near the mouths of rivers were characterized by a content of As lower than 1 mg/Kg. Different studies of total diet have confirmed meaningful variations of As content in fish products coming from different geographical areas (Basque Country: 0.004 mg/Kg; Catalonia: 0.002 mg/Kg;

Canada: 0.001 mg/Kg) and have correlated this comparison to the characteristics of the consumed fish. As regards shellfish, the total As values (2.3-149.2 mg/Kg) were higher than in mollusks (1.2-24.2 mg/Kg) while meaningful differences have not emerged on the content of inorganic As (shellfishes: 0.1-1.3 mg/Kg; mollusks: 0.1-1.6 mg/Kg).

Actually the influence of the consumption of crustacean e/o mollusks on the urinary levels of inorganic As, arsenobetaina, MMA (monomethyalsoric acid) and DMA (dimethylasr) appears controversial. Recently Soleo et al. (2008) have confirmed a statistically meaningful correlation among the presence in urine of inorganic As, arsenobetaina, MMA and DMA and the consumption of shellfishes and mollusks. Buchet et al. (1996) have determined an increase of DMA in urine without variations of the content of inorganic As and MMA. More recent studies (Heinrich-ramm 2001; Morton and Mason 2006) have confirmed the data of Buchet et al., so it can be hypothesized that the content of As in shellfishes and mollusks is tightly correlated with the level of sea pollution of this element.

According to Pearson et al. (2007) the consumption of American rice can significantly increase the urinary excretion of DMA, therefore confirming that other foods can represent a source of As for the general population. In fact groundwater contaminated by As is often used in agriculture to irrigate crops and for food and animal consumption. This could potentially bring arsenic into the human food chain (Beni et al., 2007). Arsenic can find its way into the grains of plants, such as rice and wheat, and into vegetables and fruit plants through irrigation with As-contaminated water.

According to Williams et al. rice has higher grain arsenic levels than other investigated cereals (wheat and barley) as it is much more efficient in accumulating arsenic from the soil (Williams et al., 2007b). Rice accumulates up to 2mg Kg-1 As in grain and to 92 mg kg -1 in straw.

It has been estimated that in Europe, in ethnic groups that are large consumers of rice, exposure to inorganic arsenic is about 1 µg/kg of body weight per day, while in the groups of large consumers of products based on algae it is about 4 µg/kg of body weight per day.

Different researchers share the belief that the prevailing contribution of inorganic As to the diet is due to the consumption of rice (Meacher, 2002; Meliker, 2006; Williams, 2007a). For this reason the As contamination of caryopses has been investigated extensively (Schoof, 1998; Heitkemper, 2001; Alam, 2003; Kohlmeyer, 2003; Meharg and Rahman, 2003; D'Amato, 2004; Mehari, 2004).

In further studies on this matter, Williams et al. (2005) have shown that inorganic As levels in rice varies according to its geographical origin, as the samples coming from Bangladesh (80±3%) and from India (81±4%) showed higher As values compared with those of European (64±1%) or American (42±5%) origin. Nevertheless investigations performed by other authors (Zavala and Dubury, 2008) have noticed that in caryopses total As concentration was higher in Europe and in America (0.16-0.19 mg/Kg) than in Asia and in South America (India and Pakistan: 0.051-0.053 mg/Kg; Bangladesh, Thailand and Venezuela: 0.062-0.089 mg/Kg). Total As content in foods made with rice varies between 0.14 and 0.28 mg/kg (Sun, 2008) and higher values were related to samples of cracker and blown rice. Instead, the As concentration in oil and vinegar had appreciably lower values (0.01-0.03 mg/l), probably because of the water dilution during the preparation phases.

To express a toxicological evaluation on the presence of As in rice samples it is necessary to effect the speciation of the different forms of the element, because the inorganic form is more toxic than the organic one (Petrick, 2000; Vahter and Concha, 2001). Such speciation has been conducted on rice samples (white and integral) in which the inorganic As and the DMA represented the principal components (Meharg, 2008a, b, c, Smith 2006; William, 2005 and 2006, 2007a). In further research performed by Signes-Pastor (2008) it has been demonstrated that the content of As in the rice of the western Bengal was higher than the As content found in other agricultural products and this confirmed the data obtained by Meharg et al. (2004). In this case the total arsenic concentration in caryopses was equal to 339. g/Kg (middle value). Relatively to vegetable products, As has been found in radishes (167µg/Kg), carrots (121µg/Kg), potatoes (80µg/Kg), cauliflowers (70µg/Kg), tomatoes (56µg/Kg) and white beans (41µg/Kg). It should be noted that, except for blown rice in which inorganic As concentration (40µg/Kg) was lower than DMA (80µg/Kg), in the other analyzed samples the prevailing form was AsIII. The recovery of elevated values of inorganic As in boiled rice (380µg/Kg) would be related instead to the high content of As in boiling water, as confirmed by previous studies (Devesi, 2001; Diaz, 1989; She, 1992). Evaluations performed on Bangladesh's population have confirmed that rice represents around 70% of the daily caloric intake (Signes, 2008) and, through the composites (flat ready to the consumption), it furnishes over 90% of total As intake (Roychowdhury, 2003). It is evident that in the absence of a varied diet, if the prevailing food is rice, it becomes the principal source of exposure to total As and to the inorganic form. A review of the current literature shows that in Bangladesh the consumption of drinking water contaminated by arsenic constitutes the most important way of food exposure (Mandal, 1998). Other studies support the idea that the introduction of rice and vegetables into the diet is a further exposure source. According to research performed by Roychowdhury et al. (2008) the presence of a positive correlation between the content of As in rice and in vegetables and the content of As in water used to irrigate the fields in the delta of the Ganges (Bengal) is evident. Elevated quantities of total As have been found in potatoes (335µg/Kg) and in samples of cumin (107µg/Kg) and curry (289µg/Kg) seeds. Besides, evaluations of rice and vegetables before and after cooking have confirmed a considerable increase of As concentrations (laughed: 239µg/Kg vs. 569µg/Kg; vegetables: 85µg/Kg vs. 369µg/Kg). A great variation has also been found among the inorganic forms of As, which was higher in rice (AsV/AsIII=3.2) than in vegetables (AsV/AsIII=2.3); during the interpretation of this data, consideration was given to possible reactions of oxidation during the preparation of the samples and the used procedures. As regards other food stuff (milk and its derivatives) a limited number of scientific studies is available. In a study performed in Turkey in 2008 (Ayar, 2008), As concentrations have been determined in samples of milk (0.02 mg/Kg), powdered milk (0.10 mg/Kg) and ice cream (0.09 mg/Kg). The values related to cheese have been between 0.02 and 0.07 mg/Kg, while a higher content of As has been determined in butter (0.15 mg/Kg) and in yogurt (0.15 mg/Kg). It should be stressed that the reported data are lower than the limits established by the *Codex Turkish Food* (CTF) and by the *European Communities* (0.1-1.0 mg/Kg) as regards the maximum contaminant levels in food stuff. Finally it is appropriate to signal an Italian publication, (Licata, 2004) in which the presence of As and other toxic metals has been evaluated in samples of milk coming from Calabria. 30% of the examined samples had a concentration between 0.24 and 0.68 mg/Kg, exceeding the data obtained by Cerutti et al. (1999) (0.03-0.06 mg/Kg), Alais et al. (2000) (0.04-0.10 mg/Kg) and Simsek et al. (2000) (0,0002-0,05 mg/Kg).

Recent studies focused their attention to the evaluation of arsenic effects on soil-plant system, performing genotoxicity and phytotoxicity tests on soil coupled with absorption spectroscopy techniques such as Magnetic Resonance Imaging (MRI) and Nuclear Magnetic Resonance (NMR) on vegetables. MRI performed in radish tubers grown in contaminated soil indicates that the morphology of the outermost cell layer varies with water As concentration, as a consequence of the physiological response of the hypocotyls to the arsenic presence, particularly in clay-loamy soil. On the contrary, in sandy soil the As leaching preserves the radish tuber by morphological modifications, even though the radish yield is lower than in clay-loamy ones. The larger value of soil exchange capacity in clay loamy soil determines a higher adsorption of As in the plant-system considered, probably due to the longer contact time between hypocotyls and soil solution containing arsenic. The different amount of organic matter and clay content in the two types of soil was correlated with the soil exchange capacity that regulates the release of toxic substances, so major attention should be paid to different aspects, such as different agronomic practices of edible plants. (Sturchio et al., 2011)

An emerging problem related to the arsenic content in rice regards its wide use as baby food, because the arsenic content in rice-based foods for children is high enough to increase the risk of contracting cancer. In addition, dietary exposure to inorganic arsenic in children under three years of age is about 2-3 times higher than in adults because of increased consumption of food relative to their body weight (EFSA 2009).

Pre-cooked, milled rice is a dominant carbohydrate source to weaning babies up to 1 year of age due to a range of its blandness, material properties, low allergen potential and nutritional value (Mennella et al., 2006). Therefore, babies have much higher inorganic arsenic intakes than adults, also due to their much lower body masses. In children under 3 years, exposure to As through foods made from rice is generally estimated at about 2 to 3 times that of adults. This because in the diets of infants rice is the main constituent of products for the weaning and hypo-allergenic (rice milk), but is also used in foods in the diets of children (crackers, biscuits, puffed cereals, and puddings). Andrew A. Meharg and his colleagues, (2008) in a study conducted at the University of Aberdeen, tested the levels of inorganic arsenic in rice-based products for children. The study, funded by the Natural Science Foundation of China and the Royal Society of London and Edinburgh, assumes that only in the last three or four years researchers have shown that rice is one of the most important food sources of inorganic As and, since it is a carcinogenic agent, the precautionary principle should be adopted until safety limits are established.

Moreover, considering the body mass index, infants and children can be exposed through their diet to higher levels of inorganic arsenic than adults.

In the UK the average content of inorganic arsenic in rice used in food for children is 0.11 mg / kg. In terms of the ratio of inorganic arsenic to total arsenic, it was found that the concentrations of inorganic As in a linear increase up to 0.25 mg / kg of total arsenic, then stabilized at 0.16 mg / kg at the highest concentrations of total arsenic. Meharg and colleagues analyzed samples of 3 different brands of rice-based products used for infants (4-12 months), all from supermarkets in Aberdeen. More than a third of the samples contained levels of inorganic As, the most toxic of the two forms of arsenic are normally found in rice, at or above the legal limit of arsenic allowed in food products in China, where there is a more severe legislation than in the 'European Union and the United States (the regulatory limit of i-As 0.15 mg / kg of total arsenic). Investigators have observed that the levels of inorganic As in the test samples ranged from 0.50 to 2.66 mg / kg of body weight per day. In China, 35% of these products would be illegal, while the EU and U.S. regulations do not exist for the levels of inorganic As levels in food, even if the threshold of 1 mg / kg of body weight per day is often cited in literature as a safe level for arsenic in foods, and specifically for rice (The Stationery Office, UK 1959).

The researchers argue that the limits of arsenic in food are outdated and do not take recent scientific studies into account. The current rules in the EU and the United States for the content of As in drinking water and the WHO recommend an intake of no more than 2 g / kg body weight of i-As.

The researchers calculated a one-year-old of average weight eating three servings of rice a day-As has higher levels i-As than the maximum permissible exposure for adults through the intake of drinking water. Currently in the United Kingdom most of the rice used for the preparation of baby foods comes from Europe, but researchers believe that it would be better to use low-As rice from other countries (Indian subcontinent, California, Cadiz and Seville Spain). EFSA in 2009 at the conclusion of its scientific advice about arsenic in food said that the dietary exposure should be reduced to the inorganic As. The manufacturers of baby food rice could for example use the low-As. In this the directions of scientific research undoubtedly play a key role.

3. ARSENIC'S MECHANISM OF ACTION

Inorganic As is quickly absorbed after oral exposure by most mammalian species, while absorption of inorganic As after inhalation is relatively lower than after oral exposure and is even more limited after dermal exposure (NRC 1999; WHO 2001).

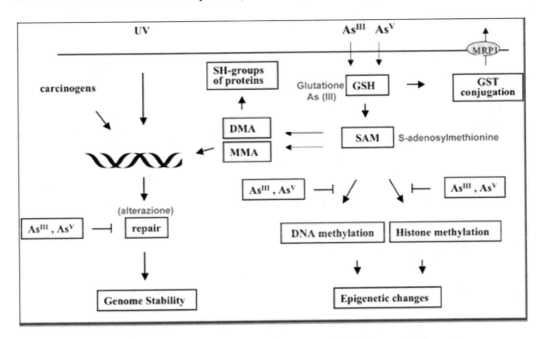

Figure 10. Mechanism of entry and methylation of inorganic As in the liver cell. The main processes, altered after exposure to As, affect the DNA repair and epigenetic changes, (modified from Salnikow, 2007).

After oral absorption, inorganic As is primarily methylated in the liver and excreted in urine by most species (Figure 10). The metabolic pathway of inorganic As involves two sequential reactions: first there is an electron reduction of AsV species with formation of compounds such as monomethylarsonic acid (MMAV) and dimethylarsinic acid (DMAV). Afterwards the oxidative methylation involves AsIII species, monomethylarsonous acid (MMAIII) and dimethylarsinous acid (DMAIII). Furthermore only in rats is there also the formation of oxide trimetilarsina (TMAO) (Wanibuchi, 1996). The AsV species are reduced to AsIII species *in vitro* nonenzymatically by thiols and enzymatically by purine nucleotide phosphorylase (PNP) and MMAV reductase. The oxidative methylation of AsIII species is an enzymatic process. Arsenic-methylating enzymes from rabbit and rat livers have been characterized. For *in vitro* activity, these enzymes require a thiol and the methyl donor S-adenosylmethione (SAM) (Aposhian et al. 2004; Thomas et al. 2001). The rabbit enzyme appears to have two distinct methylating activities, one each for inorganic AsIII and MMAIII. The rat enzyme (AS3MT) has methyltransferase and reductase activities. Mice and humans have orthologues of the rat gene that encodes for AS3MT (Thomas et al. 2004). Recently, a novel metabolic pathway was shown in rodents, which is basically different to the classical pathway but valid also for humans (Hayakawa, 2005). According to this pathway, inorganic AsIII reacts with GSH and becomes arsenic triglutathione (ATG), which is methylated by AS3MT by transfer of the methyl group from SAM, which becomes methylarsenic diglutathione (MADG). MADG, further methylated by AS3MT, leads to the formation of dimethylarsenic glutathione (DMAG) or becomes MMAIII after reacting with GSH. MADG e MMA species are in a chemical equilibrium in the presence of glutathione as well as DMAG species and DMA. Both MADG and DMAG were unstable in solution when the GSH concentration was lower than 1 mM, and were hydrolyzed and oxidized to MMAV and DMAV, respectively (Kobayashi, 2005).

Regardless of the pathway of arsenic metabolism, exposure to inorganic As results in the urinary excretion of predominantly DMAV and smaller amounts of inorganic and organic AsV and AsIII species, including TMAO.

Metabolism of inorganic As varies between species and human populations (Loffredo et al. 2003; Vahter 2000; Vahter et al. 1995). Dogs and mice are rapid methylators of inorganic As and excrete ≥ 80% of the dose as DMAV in urine. Humans excrete relatively more MMAV than other species, suggesting that humans are slower methylators of inorganic As. This may explain in part why humans are more sensitive to inorganic As than other species. Arsenic excreted by humans tends to be 10–20% MMAV, whereas that of dogs, hamsters, mice, rabbits, and rats is 1–5% MMAV.

The distribution of arsenic in human urine is generally 10–30% inorganic As, 10–20% MMAV, and 60–70% DMAV. Some populations excrete varying amounts of MMAV, both considerably less and more, in urine (Vahter 2000; Vahter et al. 1995). This suggests that there are genetic polymorphisms in the regulation of the enzyme(s) that metabolize arsenic, which may lead to differences in toxicity related to arsenic exposure.

It would be useful if studies on metabolism *in vitro* and on the mechanisms of action were conducted using cells representative of human tissue (primary cell cultures), in order to correlate the data obtained with biomarkers of exposure and of effect. The tissue-specific metabolism in target organs should include any pharmacokinetic and pharmacodynamic effects in kinetic models and dynamics, even for an application in the assessment of the risk of Arsenic exposure (Huges, 2007).

Methylation of inorganic As to DMAV facilitates excretion of arsenic. DMAV has been identified as being > 20-fold less acutely toxic than inorganic As, suggesting that methylation of inorganic As is a detoxication reaction. Improvements in analytical techniques have resulted in detection of MMAIII and DMAIII in the urine of individuals exposed to inorganic As (Aposhian et al. 2000; Del Razo et al. 2001b; Mandal et al. 2001). Thus, trivalent organic arsenicals are not the transitory intermediates previously believed, although their stability is an issue. MMAIII and DMAIII are potent in vitro and in vivo toxicants (Petrick et al. 2000, 2001; Styblo et al. 2000). DMAV, albeit at relatively high doses, is a multiorgan tumor promoter in rodents and a complete bladder carcinogen in rats (Wanibuchi et al. 2004). Methylation of inorganic As may not be a mechanism of detoxification but one of activation.

In toxicological studies performed on individuals exposed to arsenic, extensive global changes in gene expression have been highlighted (Andrew et al. 2008; Bailey et al. 2009; Bourdonnay et al. 2009; Xie et al. 2007), in particular, the authors describe a change in epigenetic processes, which are major regulators of gene expression. Epigenetic alteration does not involve changes in DNA sequence, so it does not cause a genotoxic effect, but can lead to heritable phenomena in the regulation of gene expression (Feinberg and Tycko 2004).

Recent studies have proposed the involvement of altered epigenetic regulation in gene expression changes and cancer induced by arsenic exposure, affecting both people exposed to arsenic directly and those of future generations in a heritable manner, without directly altering the genome. (Xuefeng et al. 2011)

Different epigenetic mechanisms, such as altered DNA methylation, histone modification and microRNA (miRNA) expression have been proposed to play a role in arsenic-induced carcinogenesis. In mammals DNA methylation is the basic process on regulation of gene expression and maintenance of genome stability; it occurs predominantly in CpG islands that are cytosine-rich gene regions (Yoder et al. 1997).

The enzyme DNA methyltransferase (DNMTs) catalyzes the transfer of a methyl group from *S*-adenosyl methionine (SAM) cofactor to the cytosine nucleotide, producing 5′-methylcytosine and *S*-adenosyl homocysteine (Razin and Riggs 1980).

Three different families of DNMT genes have been identified so far: DNMT1, DNMT2, and DNMT3 (Robertson and Wolffe 2000).

Altered DNA methylation in particular global genomic DNA hypomethylation has been associated with several human diseases (Robertson 2005), resulting in illegitimate recombination events and causing transcriptional deregulation of affected genes (Robertson 2005).

According to Reichard et al. global DNA hypomethylation is expected to result from arsenic exposure through both SAM insufficiency and reduction of DNMT gene expression (Reichard et al. 2007). In fact, because SAM is the unique methyl group donor in each conversion step of biomethylation of arsenic, long-term exposure to arsenic may lead to SAM insufficiency and global DNA hypomethylation (Coppin et al. 2008; Goering et al. 1999; Zhao et al. 1997); furthermore arsenic can cause a reduction of mRNA levels and DNMTs activities.

Different studies on human skin and bladder cancer (Chanda et al. 2006, Chen et al. 2007; Marsit et al. 2006c) point out how arsenic exposure is also associated with DNA hypomethylation or hypermethylation of promoters of some genes; this mechanism could mediate carcinogenesis respectively through up-regulation of oncogene expression or down-regulation of tumor suppressor genes.

MicroRNAs (miRNAs) consisting of short non-coding RNA sequences have been widely studied in recent years because of their post-transcriptional regulatory activity, that can significantly alter gene expression. The exact mechanisms by which expression is repressed are still under investigation but may include the inhibition of protein synthesis, the degradation of target mRNAs, and the translocation of target mRNAs into cytoplasmic processing bodies (Jackson and Standart 2007).

A reduction or elimination of miRNAs that target oncogenes could result in the inappropriate expression of those oncoproteins; conversely, the amplification or overexpression of miRNAs that have a role in regulating the expression of tumor suppressor genes could reduce the expression of such genes.

Several studies have shown that exposure to exogenous chemicals can alter miRNA expression (Kasashima et al. 2004; Pogribny et al. 2007; Shah et al. 2007).

ROS generation resulting from arsenic exposure is thought to play a large role in arsenic-induced carcinogenesis and toxicity (Flora et al. 2007; Hei and Filipic 2004) and could potentially alter these miRNAs in a similar manner.

In vivo studies performed in *C.elegans* model system in order to evaluate arsenic effect on some miRNA families homologous to humans profiles showed an up-regulation of some microRNAs such as lin-4 (human homologue mir125), let-7 (human homologue let-7, d, e, a, i, b), mir-57 (human homologues: mir-100, mir-99), mir-234 (human homologue mir-137), which were found to be up-regulated in some conditions of stress as reported in a recent study of Simone et al., 2009.

The up-regulation of these microRNAs, caused by the treatment with sodium arsenite, confirms the hypothesis that oxidative stress underlies the mechanism of exposure to this xenobiotic, and that miRNAs play an important role on cellular response to oxidative stress in non-genotoxic substances.

Particularly a deregulation of specific miRNAs involved in Arsenic resistance pathways has been observed: SKN-1, AIP-1 and PMK-1. The SKN-1 pathway consists of a conserved family of proteins that regulate the major oxidative stress response, probably through the direct control of the expression of a g-glutamyl-cysteine synthetase heavy chain gene that encodes the enzyme required for the rate-limiting step in the synthesis of glutathione. Then, SKN-1 activates transcription of the AIP-1 gene encoding a protein carrying the RING finger domain, which is required for protecting cells from arsenite toxicity. At the same time another important pathway activated by arsenite stress that protects worms from arsenite by functioning in the intestine is the PMK-1; it encodes p38 MAPK in *C.elegans*. The evolutionarily conserved p38 MAPK cascade is an integral part of the processes of the response to a variety of environmental stresses. (Boccia et al. 2011)

The induced changes in miRNA expression were not stable and returned to baseline levels upon removal of the stress conditions, suggesting that chronic exposure may be necessary to permanently alter expression of miRNAs (Marsit et al. 2006a).

A recent study evaluated that exposure to cellular factors can lead to developmental abnormalities or cancer, and can also vary the profile of miRNAs (Marsit, 2006).

It was found that folic acid deficiency or exposure to sodium arsenite induced an increased miRNAs expression, while a γ-ray irradiation did not cause any alteration.

It has been shown that chronic exposure to inorganic arsenic and its metabolites through drinking water increased the development of skin cancer in K6/ODC transgenic mice. In order to identify potential biomarkers and possible mechanisms of carcinogenesis, studies of gene expression profiles have been conducted in these animals after administration of 0 - 0.05 to 0.25 and 10 ppm sodium arsenite.

It was found that folic acid deficiency in animals amplifies the response to treatment with As, resulting in interference with normal cell proliferation and differentiation.

This experiment confirms that folic acid deficiency is a nutritional susceptibility factor in the genesis of skin cancer induced by As (Nelson, 2007).

Another study showed that the interaction between miRNA and target, which leads to translational repression of the gene, can easily be suppressed in particular oxidative stress conditions including exposure to 0.1 mM sodium arsenite (Bhattacharyya, 2006).

This observation, if confirmed *in vivo*, further complicates the understanding of the mechanisms of action of this interferent.

It was reported that interindividual variations in arsenic metabolism might be due to the presence of polymorphisms in genes that encode the enzymes involved in the methylation of arsenic (Vahter, 2000). This is important since the organic arsenical species are less toxic than the inorganic ones (detoxification process). In the rats, it is the S-adenosyl-l-methionine-dependent enzyme that catalyzes the methylation of arsenite (AS3MT). In the human AS3MT gene the following single nucleotide polymorphisms (SNPs) are known: R173W (C517T); M287T (T860C) and T306I (C917T) in African-American and Caucasian-American subjects (Wood, 2006). The polymorphism M287T has been studied particularly in Asian, African and Caucasian populations (Fujihara, 2008).

The frequency of the polymorphism results is significantly lower in the Asian population. This suggests that the risk connected to As exposure could be inferior in these populations, despite the concentration of the element in drinking water, which was the most elevated to be found in literature (also > of 1000 ppbs).

Another study has underlined that the individual variability can be connected to the susceptibility to develop skin cancer or exposure lesions to As. Besides the gene AS3MT, the gene that encodes for purine nucleotide phosphorylase (PNP) also works as an arsenato reductase, and has been involved in the metabolism of the As. Three exonic polymorphisms of the PNP gene (His20His, Gly51Ser and Pro57Pro) have been connected to forms of cutaneous arsenicosis (De Chaudhuri, 2008). Based on the extent to which it is underlined it appears evident that the study of polymorphisms is very important to the knowledge of the mechanisms of transformation and metabolism of As in humans and in the evaluation of the risks connected to exposure.

4. EFFECTS OF ARSENIC EXPOSURE

The toxic effects of arsenic are related to its oxidation state (Hughes 2002; WHO 2001); trivalent arsenicals are more toxic than pentavalent arsenicals, even if the mechanism of action for arsenic toxicity is not clearly known. Trivalent arsenicals react directly with sulfhydryls groups of proteins causing the alteration of the quaternary structure. Other arsenite binding sites are selenocysteines, selenium atoms and molybdenum atoms (Kitchin, 2001). Arsenate (inorganic and pentavalent) has properties similar to those of phosphate (Dixon 1997). And it may replace phosphate in critical biochemical processes that could lead to a toxic effect.

The LD50 (median lethal dose) of inorganic As in rodents, depending on its oxidation state, ranges from 10 to 90 mg/kg (WHO 2001). Arsenite is 3–4 times more potent than arsenate. MMAIII administered intraperitoneally is more acutely toxic than inorganic AsIII in hamsters (Petrick et al. 2000). MMAV and DMAV are less potent than inorganic As in rodents, with LD50 values, depending on the route of administration, ≥ 470 mg/kg (WHO 2001). Adverse noncancerous effects of inorganic As include embryo and fetal toxicity, teratogenicity, genotoxicity (by indirect DNA- or chromosome-damaging mechanisms), and cardiovascular toxicity (WHO 2001).

In humans, the oral LD50 is estimated to be 1–2 mg/kg (Ellenhorn 1997). Chronic exposure to inorganic As may result in cutaneous, developmental, hematologic, reproductive, and vascular effects ((ATSDR, 2000; NRC, 1999; WHO, 2001) and questions have been raised about its role in causing specific birth anomalies such as neural tube defects (DeSesso et al. 1998). An altered intellectual function has been reported in children of Bangladesh chronically exposed to inorganic As (Wasserman, 2004). The potential relationship between toxic effects and the concentrations of trivalent arsenic species methylated in biological samples should be confirmed in people exposed to arsenic.

The International Agency for Research on Cancer (1987) and the U.S. EPA (1993) have classified inorganic As, based on human evidence alone, as a group 1 and group A carcinogen, respectively.

The mechanism of action for inorganic As-induced carcinogenicity is not known, but there is evidence that exposure can generate free radicals and other reactive species in biological systems.

Proposed mechanisms include genotoxicity, oxidative stress, inhibition of DNA repair, tumor promotion, cocarcinogenesis, cell proliferation, and altered signal transduction or DNA methylation (Hughes 2002; Kitchin 2001; Rossman 2003). More than one of these mechanisms may occur, and some may work together.

Many cancer bioassays conducted in several species, which administered inorganic As through the diet or drinking water or after oral intubation, have been negative (Hughes 2002; Kitchin 2001; Rossman 2003). The lack of an animal model for the study of carcinogenicity induced by inorganic As has not made it possible to identify the mechanism capable of inducing such effect.

However, recent studies were performed on transgenic mice after administration of As using relatively elevated doses of ultraviolet rays as co-carcinogenic (exposure in uterus) (Burns, 2004; Waalkes, 2004).

Under such conditions exposed animals have developed tumors. Rats administrated with elevated levels of DMAV with the diet or through drinking water have developed bladder tumors (Arnold, 1999; Wei, 1999). The DMAV results also caused multiorgan tumors in rodents (Wanibuchi et al. 2004). In rats administered with TMAO in drinking water, the frequency of the hepatocellular adenoma was increased significantly compared with the value of the controls (Shen, 2003b). In contrast, the administration of MMAV through the diet (Arnold et al. 2003) or drinking water (Shen et al. 2003a) was not carcinogenic for the rodents.

The onset of lung cancer has been observed after occupational exposure to arsenic in smelter workers, in miners and in pesticide manufacturers (NRC, 1999; WHO, 2001). The exposure due to the ingestion of contaminated water with inorganic As can lead to the development of cancer in skin, bladder, lung, kidney and other internal organs (NRC, 1999 and 2001; WHO, 2001). Quantitative estimates of the risk of development of skin cancer from oral exposure to inorganic As corresponds to an Oral Slope Factor of 1.5 mg/Kg/die, while the unit risk for ingested water has been estimated as $5x10^{-5}\mu g/l$ (U.S. EPA 1993). The unit risk for the development of cancer after As inhalation (studies exposed professional workers) is $4.3x\ 10^{-3}\mu g/m^3$ (U.S. EPA 1993).

A National Academy of Sciences committee analyzed the health effects of inorganic As in drinking water and reported the maximum theoretical probability of risk for bladder and lung cancer (NRC, 2001). At the concentration of 10 μg/l of As in water, the incidence of bladder cancer in 105 people corresponds to 12 in females and to 23 in males. In lung cancer, the rate of incidence in 105 people is 18 for females and 14 for males. These risk estimates are greater than those used by the EPA to establish the reduction of the maximum level of As (MCL-maximum contaminant level) in drinking water from 50 to 10 μg As/l.

Arsenic (As) contamination of drinking water is considered a serious health problem because of its association with the worldwide environmental health threat that is related to increased disease risks - including skin, lung, bladder - and to an increased risk of developing different pathologies (skin, lung and bladder cancer, other tumors, type 2 diabetes) vascular and cardiovascular diseases, reproductive, developmental, neurological and cognitive effects. Increased health risks may occur at low doses, corresponding to 10-50μg/l. Nevertheless, it is a relief that in experimental animal and in cell culture systems biological effects have been observed to be at markedly lower levels.

Arsenic can act as a powerful endocrine disruptor, altering gene regulation through close interactions with the receptors of steroid hormones [glucocorticoid (GR), mineralocorticoid (MR), progesterone (PR) and androgen (AR)] (Bodwell, 2006; Bodwell, 2004; Kaltreider, 2001). Alterations of glucocorticoids metabolism can result in negative consequences on development and in adverse effects to the health. Very low doses of As (0.1-1μM) enhance hormone-mediated gene transcription, while slightly higher but still non-cytotoxic doses (1-5μM) can determine suppressive effects. Using GR as a representative model for all steroid receptors (SRs), it has been underlined that Arsenic is capable of altering DNA-dependent but not DNA-independent regulation of transcription, suggesting that the target is the transcriptional machinery (Bodwell, 2004). Mutational analysis on GRs confirmed that the As activity involves activation steps prior to DNA binding the enhancement by As may involve activation steps prior to DNA binding, whereas the suppressive effects may involve steps downstream of activation and DNA binding. These observations, despite the striking similarity in response from the four receptors of steroid hormones, suggest that the target for these As effects is a part of the machinery they use to regulate gene expression rather than the receptors themselves.

An endocrine disruptor can develop its own action according to different formalities: 1) increasing or reducing the quantity of hormone produced and the metabolic activity; 2) interfering in the bond of the hormone to its receptors.

To this end, Kaltreider et al. have not confirmed that the Arsenic is able to interfere with the complex hormone-receptor before the entrance of the complex into the nucleus. The data reported by these authors indicated that the changes that happen inside the nucleus imply a selective inhibition of the DNA transcription that should, under normal conditions, be stimulated by the complex glucocorticoid-GR (Figure 11).

The detailed mechanism of the interference is not yet known (Kaltreider, 2001).

Arsenic interferes with estrogenic receptor (ER), both *in vivo* and in cell culture. At non-cytotoxic doses (1–50 mmol/kg arsenite) As strongly suppressed ER-dependent gene transcription of the 17β-estradiol (E2)–inducible vitellogenin II (study performed in chick embryo liver).

In cell culture, non-cytotoxic levels (0.25–3mM, ~ 20–225 ppb) of As significantly inhibited E2-mediated gene activation of an ER-regulated reporter gene and the native ER-regulated GREB1 gene in human breast cancer MCF-7 cells.

The effects of As on ER-dependent gene regulation are similar to those of the other steroid receptors. A specific difference is, particularly, the lack of meaningful enhancement of gene expression at the lowest doses, even if the mechanism(s) through which the As alters the gene regulation, binding ERs and the other SRs, is not known today.

As is clearly not only an endocrine-disrupting chemical (EDC) but also acts by a unique mechanism quite distinct from those of previously characterized organic EDCs, most of which act as hormone mimetic and as either agonists or competitive antagonists. In fact the studies about As and SRs show that the element is not an agonist in the activation of the SRs and doesn't act as competitive antagonist or not competitive (Bodwell, 2006; Bodwell, 2004).

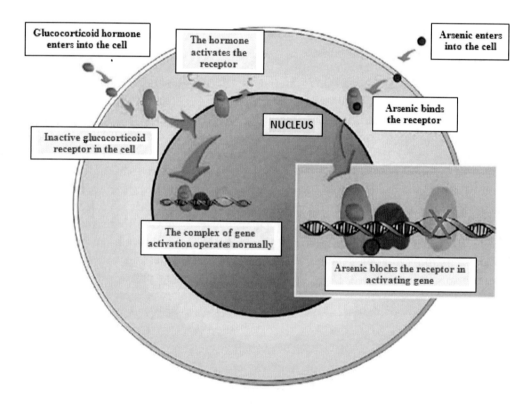

Figure 11. Mechanism of action of Arsenic as an endocrine disruptor. As acts by a different mechanism than organic compounds with characteristics of EDCs, many of which act as mimetic hormone agonists or competitive antagonists. In fact, studies on As and SRs (steroid receptors) indicate that the element is not an agonist of SR activation and does not act as a competitive or non-competitive antagonist. The As altered the ability of the SRs -As complex to bind DNA in the regulation of gene transcription.

Arsenic alters the ability of the complex SRs-As to bind the DNA during the regulation of the gene transcription process. These effects of Arsenic on ER and the other four steroidal receptors (GR, AR, PR and MR) are very similar, and small absolute sequences or structural identity within the DBDs (DNA Binding Domain) could be represent a common As target (Bodwell, 2006).

Recent studies (J. C. Davey, 2008) have underlined similar effects of As on the TR (Thyroid Hormone Receptor) and on the RAR (Retinoic Acid Receptor) whose domains of DNA binding differ significantly from those of GR. This suggests that the receptors themselves are not the target of As; other proteins or pathways of regulation are the real target. In a model animal system low concentrations of Arsenic can affect developmental processes that involved TR. Both RA (Retinoic Acid) and thyroid hormone (TH), which represent critical factors for normal development and for adult functions and alterations in the pathways regulated by these substances, have been associated with different disease processes (Davey, 2008). A serious thyroid hormone deficiency in humans at birth leads to cretinism with characteristic mental retardation, short stature and hearing loss. Severe deficiencies detected at birth can be resolved with the administration of hormones but the possibility of neglecting the least evident effects to the birth is not excluded (Galton, 2005; Oppenheimer and Samuels, 1983; Raz and Kelley, 1997).

Epidemiological studies performed in children exposed to high doses of As through drinking water have also underlined effects on cognitive functions (Wasserman, 2004).

In the past it has been assumed that the consequential effects of arsenic exposure were transient and reversible, considering that the element does not accumulate in the body, as it happens with persistent organic mixtures or metals such as mercury and lead. Also because in the past, the ability of As to act on the DNA, both inducing mutations that epigenetically change the genome, was not yet known.

More recent studies support the conviction that As acts as the hormonal signal during different phases of development, for which long-term effects are not excluded (Liu, 2006a; Shen, 2006; Waalkes, 2004a). Consequently, considering the evidence on the role of As as an endocrine disruptor, it is important to evaluate the general impact on human health, in order to reach to a characterization of the possible risk in exposure, particularly for categories of susceptible subjects. (Smith, 2006; Wasserman, 2004). The ERs play an important role in etiological and therapeutic answers to breast, ovarian and uterus cancer. They directly influence the growth and aggressiveness of breast cancer while the agonists and antagonists of the ERs play a role in the promotion and inhibition of breast cancer and of other ER dependent tumors. These considerations suggest that chemical mixtures that are able to modify the ERs activity can contribute to etiology, progression or regression of the pathology. Considering the actual knowledge, there is not enough evidence to affirm that As exposure through the daily consumption of contaminated water can determine an increase of breast cancer or other ER dependent tumor. At the same time, the role of As in causing alterations in the signalling is not excluded, a condition to which some authors have also associated possible cardiovascular pathologies. For example, a recent study performed by Waalkes et al. has underlined that the exposure to inorganic As in mice increases the incidence of cancer in the offspring. The hypo-methylation of promoters of the ER receptors causes an over expression of P-ER (Chen, 2004; Liu, 2006a; Liu, 2006b; Shen, 2006). The ERs play an important role in normal development and function, but also in carcinogenesis of the liver in experimental systems. In fact, the As exposure from water for human consumption has been associated with the increment of risk to develop cancer of the liver, so it is possible that the As effects on the ERs can affect this pathology in a meaningful way. The study of Waalkes on fetal exposure to As suggests the existence of a correlated imprinting to the ERs, playing an important role in the induction of specific pathologies in the adult subject. As previously underlined, metabolism and molecular toxicity of As are tightly associated with the carcinogenicity and this is also the base of its anticancer activity in therapeutic trials. It has been hypothesized that the epigenetic changes induced by As could constitute it as "key events" of this paradox. It has long been known that arsenic inhibits numerous enzymes (NCR, 1999): DNA repair enzymes, antioxidant-related enzymes (thioredoxin reductase), several methyltransferases, in the urine of arsenic exposed individuals, and lipidic peroxidation. Low concentrations of As inhibit both DNMT1 (Dna methyltransferases 1) and DNMT3A (Dna methyltransferases 3A) in human keratinocytes. Arsenic-induced changes in DNA methylation may have severe consequences for the development of health effects both before and after birth. It has also been shown that arsenic would be able to induce epigenetic changes through the sequestration of the methyl groups, limiting in such way the reactions catalyzed by other cellular methyltransferases, with consequent hypomethylation of the cytosine. Changes observed in cancer occur in DNA methylation where global methylation of the genome is reduced and some gene-specific promoter methylation is increased (tumor suppressor genes). The promoters of several mammal genes are characterized by the presence of islands CpG. The methylation of the cytosine in dinucleotidis CpG induces gene silencing. Pharmacological therapy inhibits the DNA methyltransferases allowing the reactivation of the silenced genes and the inhibition of the tumor growth or the sensitization to other anticancer therapies. More recently the anticancer mechanism related to As2O3 has been underlined and it occurs through the reactivation of silenced tumor suppressor genes through DNA demethylation, over that through the induction of apoptosis, as previously underlined.

Arsenic trioxide (As2O3) has been reintroduced as an anticancer agent after pioneer studies performed in China based on the administration of an As2O3-containing herbal mixture for the treatment of acute promyelocytic leukemia (APL). Such searches have shown the marked activity of monotherapy with As2O3 in the treatment of APL and they have stimulated further examinations by the Food and Drug Administration (September 2000) for the treatment of the recidives and of the APL refractory. As2O3 can also be used in combination with other anticancer agents to improve its potential use. Phase I and II of Clinical Trials are evaluating the feasibility, the safety and potential effects of As in several types of human cancer (Xing, 2008).

REFERENCES

Alais C. 2000 *Scienza del latte*. 3rd ed. Milano: Tecniche Nuove.
Alam MC. Arsenic and heavy metal contamination of vegetables grown in Samta village, Bangladesh. *Sci Total Environ* 2003; 308:83-96.
Andrew AS, Jewell DA, Mason RA, Whitfield ML, Moore JH, Karagas MR. 2008. Drinking-water arsenic exposure modulates gene expression in human lymphocytes from a U.S. population. *Environ Health Perspect* 116:524–531.

APAT 2003 Prima ricognizione sulla presenza di sostanze pericolose nei corpi idrici in Italia – Attuazione Direttiva 76/464/CEE, Rapporti 34/2003

Aposhian HV. Occurrence of monomethylarsonous acid in urine of humans exposed to inorganic arsenic. *Chem Res Toxicol* 2000; 13: 693-697.

Argese E. Distribution of arsenic compounds in Mytilus galloprovincialis of the Venice lagoon (Italy). *Sci Total Environ* 2005; 348:267-77.

Arnold LL. Chronic studies evaluating the carcinogenicity of monomethylarsonic acid in rats and mice. *Toxicology* 2003; 1990: 197-219.

Arnold LL. Effects of dietary dimethylarsinic acid on the urine and urothelium of rats. *Carcinogenesis* 1999; 20: 2175-2179.

Ayar A. The trace metal levels in milk and dairy products consumed in middle Anatolia-Turkey. *Environ Monit Assess* 2008; in stampa.

Bailey K, Xia Y, Ward WO, Knapp G, Mo J, Mumford JL, et al. 2009. Global gene expression profiling of hyperkeratotic skin lesions from Inner Mongolians chronically exposed to arsenic. *Toxicol Pathol* 37(7):849–859.

Beni C., Pennelli B., Ronchi B., Marconi S. 2007 Xenobiotics concentration and mobility in bovine milk from Italian farms. *Prog. Nutr.* 9(1), 39-45.

Bhattacharyya SN. Relief of microRNA-mediated translational repression in human cells subjected to stress. *Cell* 2006; 125: 1111-1124.

Boccia P., Zanellato M., Meconi C., Mercurio G., Sturchio E. 2011. *Evaluation of the arsenic effects on C.elegans model system.* GeoMed 2011- 4th International Conference on Medical Geology.

Bodwell EJ. Arsenic at very low concentrations alters glucocorticoid receptor (GR) mediated gene activation but not GR mediated gene repression: complex dose-response effects are closely correlated with levels of activated GR and require a functional functional GR DNA binding domain. *Chem Res Toxicol* 2004; 17: 1064-1076.

Bodwell EJ. Arsenic Disruption of Steroid Receptor Gene Activation: Complex Dose-Response Effects Are Shared by Several Steroid Receptors. *Chem Res Toxicol* 2006; 19: 1619-1629.

Bourdonnay E, Morzadec C, Sparfel L, Galibert MD, Jouneau S, Martin-Chouly C, et al. 2009. Global effects of inorganic arsenic on gene expression profile in human macrophages. *Mol Immunol* 46(4):649–656.

Buchet JP. Assessment of exposure to inorganic arsenic, a human carcinogen, due to the consumption of seafood. *Arch Toxicol* 1996; 70: 773-778.

Burns FJ. Arsenic-induced enhancement of UV radiation carcinogenesis in mouse skin: a dose-response study. *Environ Health Perspect* 2004; 112: 599-603.

Cerruti G. 1999 Residui addittivi e contaminanti negli alimenti. *Ist. Ed Milano: Tecniche Nuove.*

Chanda S, Dasgupta UB, Guhamazumder D, Gupta M, Chaudhuri U, Lahiri S, et al. 2006. DNA hypermethylation of promoter of gene p53 and p16 in arsenic-exposed people with and with¬out malignancy. *Toxicol Sci* 89(2):431–437.

Chen H. Chronic inorganic arsenic exposure induces hepatic global and individual gene hypomethylation: implications for arsenic hepatocarcinogenesis. *Carcinogenesis* 2004; 25: 1779-1786.

Chen WT, Hung WC, Kang WY, Huang YC, Chai CY. 2007. Urothelial carcinomas arising in arsenic-contaminated areas are associated with hypermethylation of the gene promoter of the death-associated protein kinase. *Histopathology* 51(6):785–792.

Coppin JF, Qu W, Waalkes MP. 2008. Interplay between cellular methyl metabolism and adaptive efflux during oncogenic transformation from chronic arsenic exposure in human cells. *J Biol Chem* 283(28):19342–19350.

Cullen WR. Arsenic speciation in the environment. *Chem Rev* 1989; 89: 713-764.

D'Amato M. Identification and quantification of major species of arsenic in rice. *J AOAC Int* 2004; 87: 238-243.

Dall'Aglio M. 1996 Problemi emergenti di geochimica ambientale e salute in Italia con particolare riferimento all'arsenico. Atti del II° Convegno Nazionale sulla protezione e gestione delle acque sotterranee. Metodologie, tecnologie e obiettivi. Modena, 17/19 maggio1995, volume 4, Quad. Geol. Appl., 1, gennaio-giugno 1996, pp. 85-95, Pitagora Editrice, Bologna.

De Chaudhuri S. Genetic variants associated with arsenic susceptibility: study of purine nucleoside phosphorylase, arsenic (+3) methyltransferase, and glutathione s-transferase omega genes. *Environ Health Perspect* 2008; 116: 501-505.

De Sesso JM. An assessment of the developmental toxicity of inorganic arsenic. *Reprod Toxicol* 1998; 12: 385-433.

Del Razo LM. Determination of trivalent methylated arsenicals in biological matrices. *Toxicol Appl Pharmacol* 2001b; 174: 282-293.

Diaz O. Contents in raw potato, cooked potato and industrialized products as mashed potatoes. *Rev Clin Nutr* 1989; 17: 116-121.

Dixon HBF. The biochemical action of arsonic acids especially as phosphate analogues. *Adv Inorg Chem* 1997; 44: 191-227.

EFSA 2009, Scientific Opinion on Arsenic in Food; EFSA Panel on Contaminants in the Food Chain (CONTAM), *EFSA Journal*; 7 (10): 1351.

Ellenhorn MJ. 1997. *Ellenhorn's Medical Toxicology: Diagnosis and Treatment of Human Poisoning.* 2nd ed. Baltimore: Williams and Wilkins, 1540.

E-PRTR The European Pollutant Release and Transfer Register, 2009 [on line] http://prtr.ec.europa.eu/PollutantReleases.aspx (accessed 24/10/2011)

Fattorini D. Chemical speciation of arsenic in different marine organisms: Importance in monitoring studies. *Mar Environ Res* 2004; 58: 845-50.

Feinberg AP, Tycko B. 2004. The history of cancer epigenetics. *Nat Rev Cancer* 4(2):143–153.

Flora SJ, Bhadauria S, Kannan GM, Singh N. 2007. Arsenic induced oxidative stress and the role of antioxidant supple¬mentation during chelation: a review. *J Environ Biol* 28(2 suppl):333–347.

Francesconi KA. Arsenic and marine organism. *Adv Inorg Chem* 1997; 44: 147-189.

Fujihara J. Asian specific low mutation frequencies of the M287T polymorphism in the human arsenic (+3 oxidation state) methyltransferase (AS3MT) gene. *Mutat Res* 2008;

Galton VA. The roles of the iodothyronine deiodinases in mammalian development. *Thyroid* 2005;15: 823-834.

Ghidini S. Livelli ed evoluzione di cadmio, mercurio ed arsenico nei pesci dell'alto Adriatico. 2000. Università di Parma- Facoltà di Medicina Veterinaria. http://www.unipr.it/ arpa/facvet/annali/ 2000/ghidini/ghidini/ghidini.htm

Goering PL, Aposhian HV, Mass MJ, Cebrian M, Beck BD, Waalkes MP. 1999. The enigma of arsenic carcinogenesis: role of metabolism. *Toxicol Sci* 49(1):5–14.;

Hayakawa T. A new metabolic pathway of arsenite: arsenic-glutathione complexes are substrates for human arsenic methyltransferase Cyt19. *Arch Toxicol* 2005; 79: 183-191.

Hei TK, Filipic M. 2004. Role of oxidative damage in the geno¬toxicity of arsenic. *Free Radic Biol Med* 37(5):574–581.

Heinrich-Ramm R. Arsenic species excretion in a group of persons in northern Germany-contribution to the evaluation of reference values. *Int J Hyg Environ Health* 2001; 203: 475-477.

Heitkemper DT. Determination of total and speciated arsenic in rice by ion chromatography and inductively coupled plasma mass spectrometry. *J Anal At Spectrom* 2001; 16: 299-306.

Hsiung TM e Huang CW. Quantitation of Toxic Arsenic Species and Arsenobetaine in Pacific Oysters Using an Off-line Process with Hydride Generation-Atomic Absorption Spectroscopy. *J Agric Food Chem* 2006; 54: 2470-2478.

Huges MF. Research approaches to address uncertainties in the risk assessment of arsenic in drinking water. *Toxicology and Applied Pharmacology* 2007; 222: 399-404.

Hughes MF. Arsenic toxicity and potential mechanisms of action. *Toxicol Lett* 2002; 133:1-16.

IARC Monographs on the Evaluation of Carcinogenic Risks to Humans. Some Drinking-water Disinfectants and Contaminants, including Arsenic. *The monographs: Arsenic in Drinking-water* 2004; 84: 40-267.

INES 2006 Inventario Nazionale delle Emissioni e loro Sorgenti, Registro 2006 validato ed aggiornato all'8/04/2010, available on http://www.eper.sinanet.apat.it/site/it-IT/ (accessed 24/05/2010)]

International Agency for Research on Cancer. 1987. Arsenic and arsenic compounds. IARC Monogr Eval Carcinog Risks Hum 23(suppl 7): 100-103.

ISPRA 2010a, Banca Dati BRACE - Misure di qualità dell'aria a livello nazionale, available on http://www.sinanet.apat.it/it/aria (accessed 19/04/2010)

ISPRA 2010b Risorsa Inventaria Manuale e linee guida degli inventari locali e database dei fattori di emissione, available on: http://www.sinanet.apat.it/it/inventaria/ disaggregazione_prov2005/ (accessed 11/05/2010)

Jackson RJ, Standart N. 2007. How do microRNAs regulate gene expression? Sci STKE 2007(367):re1; doi:10.1126/stke. 3672007re1 [Online 2 January 2007].

Kaltreider RC. Arsenic alters the function of the glucocorticoid receptor as a transcription factor. *Environ Hlth Perspect* 2001; 109: 245- 251.

Karagas MR. Design of an epidemiologic study of drinking water arsenic exposure and skin and bladder cancer risk in a U.S. population. *Environ Health Perspect* 1998; 106: 1047-1050.

Kasashima K, Nakamura Y, Kozu T. 2004. Altered expression profiles of microRNAs during TPA-induced differentiation of HL-60 cells. *Biochem Biophys Res Commun* 322(2):403–410.

Kitchin KT. Recent advances in arsenic carcinogenesis: modes of action, animal models and methylated arsenic metabolites. *Toxicol Appl Pharmacol* 2001; 172: 249-261.

Kobayashi Y. Stability of arsenic metabolites, arsenic triglutathione [As(GS)3] and methylarsenic diglutathione [CH3As(GS)2], in rat bile. *Toxicology* 2005; 211: 115-123.

Kohlmeyer U. Benefits of high resolution IC-ICP-MS for the routine analysis of inorganic and organic arsenic species in food products of marine and terrestrial origin. *Anal Bioanal Chem* 2003; 377: 6-13.

Liu CW. Bioaccumulation of arsenic compounds in aquacultural clams (Meretrix lusoria) and assessment of potential carcinogenic risks to human health by ingestion. *Chemosphere* 2007; 69:128-134.

Liu CW. Bioaccumulation of arsenic compounds in aquacultural clams (Meretrix lusoria) and assessment of potential carcinogenic risks to human health by ingestion. *Chemosphere* 2007; 69:128-134.

Liu J. Global gene expression associated with hepatocarcinogenesis in adult male mice induced by in utero arsenic exposure. *Environ Hlth Perspect* 2006a; 114: 404-411.

Loffredo CA. Variability in human metabolism of arsenic. Environ Res 2003; 92: 85-91

Mandal BK. Arsenic round the world: A review. *Talanta* 2002; 58: 201-235

Mandal BK. Identification of dimethylarsinous and monomethylarsonous acids in human urine of the arsenic-affected areas in West Bengal, India. *Chem Res Toxicol* 2001; 14: 371-378.

Mandal BK. Impact of safe water for drinking and cooking on five arsenic-affected families for 2 years in West Bengal, India. *Sci Total Environ* 1998; 218: 185-201.

Marsit CJ, Karagas MR, Schned A, Kelsey KT. 2006c. Carcinogen exposure and epigenetic silencing in bladder cancer. *Ann NY Acad Sci* 1076:810–821

Marsit CJ. MicroRNA responses to cellular stress. *Cancer Res* 2006; 66:10843-10848.

Meacher DM. Estimation of Multimedial inorganic arsenic intake in the US population. *Hum Ecol Risk Assess* 2002; 8: 1697-1721.

Meharg AA e Rahman MM. Arsenic contamination of Bangladesh paddy field soils: implications for rice contribution to arsenic consumption. *Environ Sci Technol* 2003; 37: 229-34.

Meharg AA e Rahman MM. Arsenic contamination of Bangladesh paddy field soils: implications for rice contribution to arsenic consumption. *Environ Sci Technol* 2003; 37: 229-34.

Meharg AA. Arsenic in rice-understanding a new disaster for South-East Asia. *Trends Plant Sci* 2004; 9: 415-417. No abstract available.

Meharg AA. Inorganic arsenic levels in rice milk exceed EU and US drinking water standards. *J Environ Monit* 2008a; 10: 428-431.

Meharg AA. Speciation and localization of arsenic in white and brown rice grains. *Environ Sci Technol* 2008c; 42: 1051-1057.

Meliker JR. Major contributors to inorganic arsenic intake in southeastern Michigan. *Int Hyg Environ Health* 2006; 209: 399-411.

Mennella, J.A., Ziegler, P., Briefel, R., Novak, T., 2006. *Feeding infants and toddlers study: the types of foods fed to hispanic infants and toddlers.* American Dietetic Association 106, S96eS106.

Mohri T. Arsenic intake and excretion by Japanese adults: a 7-day duplicate diet study. *Food Chem Toxicol* 1990; 28:521-529.

Morton J. Speciation of arsenic compounds in urine from occupationally unexposed and exposed persons in the U.K. using a routine LC-ICP-Ms method. *J Anal Tox* 2006; 30: 293-301.

Muñoz O. Total and inorganic arsenic in fresh and processed fish products. *J Agric Food Chem* 2000; 48: 4369-4376.

Nelson GM. Folate deficiency enhances arsenic effects on expression of genes involved in epidermal differentiation in transgenic K6/ODC mouse skin. *Toxicology* 2007; 241: 134-145.

Nordstrom DK. Worldwide occurrences of arsenic in groundwater. *Science* 2002; 296: 2144-2145.

NRC. 1999. *Arsenic in Drinking* Water. Washington, DC: National Research Council.

NRC. 2001. *Arsenic in Drinking Water*. Update. Washington, DC: National Research Council

Oppenheimer JH. 1983. Molecular Basis of Thyroid Hormone Action. New York: Academic Press.

Pearson GF. Rapid arsenic excretion speciation using ion pair LC-ICPMS with a monolithic silica column reveals increased urinary DMA excretion after ingestion of rice. *J Anal At Spectrom* 2007; 22: 361-369.

Petrick J.S., Monomethylarsonous acid (MMA(III) and arsenite: LD(50) in hamsters and in vitro inhibition of pyruvate dehydrogenase. *Chem Res Toxicol* 2001; 14: 651-666.

Petrick JS. Monomethylarsonous acid (MMA(III)) is more toxic than arsenite in Chang human hepatocytes. *Toxicol Appl Pharmacol* 2000; 163: 203-207

Petrick JS. Monomethylarsonous acid (MMA(III)) is more toxic than arsenite in Chang human hepatocytes. *Toxicol Appl Pharmacol* 2000; 163: 203-207.

Pogribny IP, Tryndyak VP, Boyko A, Rodriguez-Juarez R, Beland FA, Kovalchuk O. 2007. Induction of microRNAome deregulation in rat liver by long-term tamoxifen exposure. *Mutat Res* 619(1–2):30–37.

Raz Y. Effects of retinoid and thyroid receptors during development of the inner ear. Stem *Cell Dev Biol* 1997; 8: 257-264.

Razin A, Riggs AD. 1980. DNA methylation and gene function. *Science* 210(4470):604–610.

Reichard JF, Schnekenburger M, Puga A. 2007. Long term low-dose arsenic exposure induces loss of DNA methylation. *Biochem Biophys Res Commun* 352(1):188–192.

Retinoic Acid Receptor-and Thyroid Hormone Receptor-Mediated Gene Regulation and Thyroid Hormone-Mediated Amphibian Tail Metamorphosis. *Environmental Health Perspectives* 2008; 116: 165-172.

Robertson KD, Wolffe AP. 2000. DNA methylation in health and disease. *Nat Rev Genet* 1(1):11–19.

Robertson KD. 2005. DNA methylation and human disease. *Nat Rev Genet* 6(8):597–610.

Rossman TG. Mechanism of arsenic carcinogenesis: an integrated approach. *Mutat Res* 2003; 533: 37-65.

Roychowdhury T. Survey of arsenic and other heavy metals in food composites and drinking water and estimation of dietary intake by the villagers from an arsenic affected area of West Bengal, India. *Sci Total Environ* 2003; 308: 15-35.

Schmeisser E. Human metabolism of arsenolipids present in cod liver. Anal Bioanal Chem 2006; 385: 367-76.

Schoof RA. Dietary arsenic intake in Taiwanese districts with elevated arsenic in drinking water. *Hum Ecol Risk Asses* 1998; 4:117-135.

Shah YM, Morimura K, Yang Q, Tanabe T, Takagi M, Gonzalez FJ. 2007. Peroxisome proliferator-activated receptor alpha regulates a microRNA-mediated signaling cascade responsible for hepatocellular proliferation. *Mol Cell Biol* 27(12):4238–4247.

She L. Arsenic contents in some Malaysian vegetables. *Pertanika* 1992; 15: 171-173.

Shen J. Fetal Onset of Aberrant Gene Expression Relevant to Pulmonary Carcinogenesis in Lung Adenocarcinoma Development Induced by In Utero Arsenic Exposure. *Toxicol Sci* 2006; 89:108-119.

Shen J. Induction of glutathione S-transferase placental form positive foci in liver and epithelial hyperplasia in urinary bladder, but not tumor development in male Fischer 344 rats treated with monomethylarsonic acid for 104 weeks. *Toxicol Appl Pharmacol* 2003a; 193: 335-345.

Shen J. Liver tumorigenicity of trimethylarsine oxide in male Fischer 344 rats—association with oxidative DNA damage and enhanced cell proliferation. *Carcinogenesis* 2003b; 24: 1827-1835.

Signes-Pastor AJ. Arsenic Speciation in Food and Estimation of the Dietary Intake of Inorganic Arsenic in a Rural Village of West Bengal, India. *J Agric Food Chem* 2008; in stampa. Devesi, 2001;

Simsek O. 2000 The effects of environmental pollution on the heavy metal content of raw milk. *Nahrung-Food;* 44: 360-371.

Smadley P.L., Kinninburgh D.G. 2002 A review of the source, behaviour and distribution of arsenic in natural waters. Applied Geochemistry 17, p.517-568.

Smith AH. Increased mortality from lung cancer and bronchiectasis in young adults after exposure to arsenic in utero and in early childhood. *Environ Health Perspect* 2006; 114:1293-1296.

Smith NM. Inorganic arsenic in cooked rice and vegetables from Bangladeshi households. *Sci Total Environ* 2006; 370: 294-301.

Soleo L. Significance of urinary arsenic speciation in assessment of seafood ingestion as the main source of organic and inorganic arsenic in a population resident near a coastal area. *Chemosphere* 2008; 73: 291-299.

Storelli MM e Marcotrigiano GO. Total, organic, and inorganic arsenic in some commercial species of crustaceans from the Mediterranean Sea (Italy). *J Food Prot* 2001; 64: 1858-62.

Sturchio E. 2011 Evaluation of arsenic effects on soil-plant system. *Chemistry and Ecology.* Vol. 27, Supplement 1, pp. 67-78(12)

Styblo M. Comparative toxicity of trivalent and pentavalent inorganic and methylated arsenicals in rat and human cells. *Arch Toxicol* 2000; 74: 289-299.

Sun GX. Survey of arsenic and its speciation in rice products such as breakfast cereals, rice crackers and Japanese rice condiments. *Environ Int* 2008; in stampa.

The Stationary Office, 1959. *The arsenic in food regulations* 1559 (S.I. 1959/831) as amended, London.

Thomas DJ. Elucidating the pathway for arsenic methylation. *Toxicol Appl Pharmacol* 2004; 198: 319-326.

Thomas DJ. The cellular metabolism and systemic toxicity of arsenic. *Toxicol Appl Pharmacol* 2001; 176:127-144.

U.S. EPA 1993. Arsenic, Inorganic (CASRN 7440-38-2), Integrated Risk Information System. Washington, DC:U.S. Environmental Protection Agency. Available: http://www.epa.gov/iris/subst/0278.htm

U.S. EPA 2001. National primary drinking water regulations: arsenic and clarifications to compliance and new source contaminants monitoring. *Final rule. Fed Reg* 66:6976-7066.

U.S. EPA 2002, Arsenic Treatment Technologies for Soil, Waste and Water, EPA-542-R-02-004, September 2002 [on line http://www.clu-in.org/download/remed/542r02004/arsenic_report.pdf] accessed 19/10/2011

Vaessen HA e van Ooik A. Speciation of arsenic in Dutch total diets: methodology and results. *Lebensm Unters Forsch* 1989; 189: 232-235.

Vahter M. A unique metabolism of inorganic arsenic in native Andean women. *Eur J Pharmacol* 1995; 293: 455-462.

Vahter M. Genetic polymorphisms in the biotransformation of inorganic arsenic and its role in toxicity. *Toxicol Lett* 2000; 112:209-217.

Vahter M. Genetic polymorphisms in the biotransformation of inorganic arsenic and its role in toxicity. *Toxicol Lett* 2000; 112:209-217.

Vahter M. Role of metabolism in arsenic toxicicity. *Pharmacol Toxicol* 2001; 89: 1-5.

Waalkes MP. Animal models for arsenic carcinogenesis: inorganic arsenic is a transplacental carcinogen in mice. *Toxicol Appl Pharmacol* 2004; 98: 377-384.

Waalkes MP. Induction of tumors of the liver, lung, ovary and adrenal in adult mice after brief maternal gestational exposure to inorganic arsenic: promotional effects of postnatal phorbol ester exposure on hepatic and pulmonary, but not dermal cancers. *Carcinogenesis* 2004a; 25: 133-141.

Wanibuchi H. Promoting effects of dimethylarsinic acid on Nbutyl-N-(4-hydroxybutyl)nitrosamine-induced urinary bladder carcinogenesis in rats. *Carcinogenesis* 1996; 17: 2435-2439.

Wanibuchi H. Understanding arsenic carcinogenicity by the use of animal models. *Toxicol Appl Pharmacol* 2004; 198: 366-376.

Wasserman GA. Water arsenic exposure and children's intellectual function in Araihazar, Bangladesh. *Environ Health Perspect* 2004; 112: 1329-1333.

Wei M. Urinary bladder carcinogenicity of dimethylarsinic acid in male F344 rats. *Carcinogenesis* 1999; 20: 1873-1876.

WHO 2001. Arsenic and arsenic compounds. *Environmental Health Criteria 224*. World Health Organization, Geneva.

Williams PN. Market basket survey shows elevated levels of As in South Central U.S. processed rice compared to California: consequences for human dietary exposure. *Environ Sci Technol* 2007a; 417: 2178-2183

Williams PN. Market basket survey shows elevated levels of As in South Central U.S. processed rice compared to California: consequences for human dietary exposure. *Environ Sci Technol* 2007a; 417: 2178-2183.

Williams PN. Variation in arsenic speciation and concentration in paddy rice related to dietary Exposure. *Environ Sci Technol* 2005; 39: 5531-5540.

Williams PN. Variation in arsenic speciation and concentration in paddy rice related to dietary Exposure. *Environ Sci Technol* 2005; 39: 5531-5540.

Wood TC Human arsenic methyltransferase (AS3MT) pharmacogenetics, *J Biol Chem* 2006; 11:7364-7373.

Xie Y, Liu J, Benbrahim-Tallaa L, Ward JM, Logsdon D, Diwan BA, et al. 2007. Aberrant DNA methylation and gene expression in livers of newborn mice transplacentally exposed to a hepatocarcinogenic dose of inorganic arsenic. *Toxicology* 236(1–2):7–15.

Xing C. Metabolism and the Paradoxical Effects of Arsenic: Carcinogenesis and Anticancer. *Current Medicinal Chemistry* 2008; 15: 2293-2304.

Xuefeng Ren, Cliona M. McHale, Christine F. Skibola, Allan H. Smith, Martyn T. Smith, and Luoping Zhan 2011. An Emerging Role for Epigenetic Dysregulation in Arsenic Toxicity and Carcinogenesis. *Environmental Health Perspectives* 119 (1): 11-19.

Yoder JA, Walsh CP, Bestor TH. 1997. Cytosine methylation and the ecology of intragenomic parasites. *Trends Genet* 13(8):335–340.

Zhao CQ, Young MR, Diwan BA, Coogan TP, Waalkes MP. 1997. Association of arsenic-induced malignant transformation with DNA hypomethylation and aberrant gene expression. *Proc Natl Acad Sci USA* 94(20):10907–10912.

Reviewed by: Dr. Claudio Beni - Senior Researcher, CRA-RPS. Via della Navicella, 2 - 00184 Rome (Italy) +3967008260 fax +3967005711 claudio.beni@entecra.it

In: Arsenic
Editor: Andrea Masotti

ISBN: 978-1-62081-320-1
© 2013 Nova Science Publishers, Inc.

Chapter 2

ARSENIC SPECIATION IN ROCK-FORMING MINERALS DETERMINED BY EPR SPECTROSCOPY

Yuanming Pan[*]

Department of Geological Sciences, University of Saskatchewan,
Saskatoon, Canada

ABSTRACT

The mobility, bioavailability and toxicity of arsenic in aqueous environments depend upon its physical distribution and chemical speciation in the source rocks and soils. In this context, enormous efforts have been devoted to investigate arsenic speciation in arsenic-rich minerals, such as sulfarsenides, arsenites, arsenates and sulfides. Most rock-forming minerals (especially silicates), on the other hand, were thought to contain negligible amounts of arsenic and, consequently, were generally ignored with respect to their roles in arsenic contamination and potential applications for remediation, except that clay minerals and zeolites have been investigated for their surface sorption properties. This contribution reviews the present status of EPR-based understanding of arsenic speciation in rock-forming minerals (i.e., carbonates, silicates, phosphates, sulfates and borates) and demonstrates its advantages (e.g., superior sensitivity and unambiguous distinction between lattice-bound and surface-sorpted arsenic species) over other state-of-the-art structural techniques such as synchrotron X-ray absorption spectroscopy. In particular, identification of four arsenic-centered oxyradicals and discussion about the oxidation states, site assignments, local structural environments and substitution mechanisms of their diamagnetic precursors, in selected rock-forming minerals are presented.

INTRODUCTION

Arsenic, usually a trace element in the natural environment, is known to occur in multiple oxidation states from -3 to -1, 0, $+3$ and $+5$. The mobility, bioavailability and toxicity of

---

[*] Tel: (306) 966-5699, Fax: (306) 966-8593, email: yuanming.pan@usask.ca

arsenic in the environment are known to depend upon its speciation. Therefore, ability to determine arsenic speciation in various hosts at dilute concentrations is important to understanding, management and remediation of arsenic contamination in the environment.

Structural techniques that are commonly used to determine arsenic speciation in minerals and other solids include X-ray and neutron diffraction analyses, synchrotron X-ray absorption spectroscopy, Raman spectroscopy, and nuclear magnetic resonance spectroscopy. However, most of these structural techniques are applicable to arsenic-rich materials only. The arsenic contents in most rock-forming minerals, on the other hand, are usually below 10 ppm [1] and therefore present a major challenge to most structural techniques, including the highly sensitive synchrotron X-ray absorption spectroscopy. In this regard, electron paramagnetic resonance (EPR) spectroscopy that measures interactions between unpaired electrons and electromagnetic radiation is unique, because of its unparallel sensitivity for structural studies of paramagnetic centers at concentrations as low as $\sim 10^{11}$ spins or sub-ppm level [2, 3]. Conventional EPR experiments are done using a constant microwave frequency and scanning the externally applied magnetic field over a suitable range, while maintaining a continuous microwave power incident on the sample (hence known as continuous-wave or CW-EPR). In addition, several related techniques such as electron nuclear double resonance (ENDOR) and electron spin echo envelope modulation (ESEEM) spectroscopy provide structural information about paramagnetic centers containing nuclei with non-zero nuclear spins. Unlike the CW techniques, however, modern ENDOR and ESEEM experiments are commonly made by use of various radiofrequency pulses to excite electron spins [4]. This contribution presents a review of arsenic speciation in rock-forming minerals as determined by those EPR techniques, in comparison with, where available, data from other structural techniques such as synchrotron X-ray absorption spectroscopy.

PARAMAGNETIC ARSENIC-CENTERED RADICALS

There are a large number of organic and inorganic arsenic species with diverse geometric configurations [5], but none of them is intrinsically paramagnetic. Therefore, detection of arsenic by the EPR techniques requires transformation of these diamagnetic species to paramagnetic centers, which can be accomplished by a wide variety of physico-chemical methods such as photolysis and radiolysis. For example, irradiation using X-ray, gamma-ray, electrons or neutrons is commonly applied for converting diamagnetic ions and molecules to paramagnetic centers for EPR studies. On the other hand, arsenic species, when paramagnetic, can often be unambiguously identified by EPR, because of their diagnostic ^{75}As hyperfine structures arising from interaction with the ^{75}As nucleus with the nuclear spin number $I = 3/2$ and natural isotope abundance of 100%. Also, ^{75}As, unlike other $I=3/2$ nuclei such as ^{11}B and ^{23}Na, is known for its considerable nuclear quadrupole effect [3, 6], which is potentially helpful for identifying arsenic-related paramagnetic species as well.

To date, four arsenic-centered oxyradicals: $[AsO_4]^{2-}$, $[AsO_4]^{4-}$, $[AsO_3]^{2-}$ and $[AsO_2]^{2-}$ have been documented in minerals ([6-17]; Table 1), although numerous organic radicals containing arsenic are known [18, 19]. Of these, the first one is the only hole-like center with $g_{avg} > g_e = 2.0023$, whereas the remaining three are all electron-like centers (Table 1).

Table 1. Spin Hamiltonian parameters of four arsenic-centered oxyradicals in crystalline hosts

Host	g_1	g_2	g_3	A_1/h (MHz)	A_2/h (MHz)	A_3/h (MHz)	Unpaired spin (%)			Ref.
							$4s$	$4p_z$	$4p_x$	
$[AsO_4]^{2-}$										
Scheelite (77K)	2.0470	2.0122	2.00070	64.4	53.2	52.4	0.39	1.0		7
(NH$_4$)$_2$HPO$_4$	2.030(5)	2.023(5)	2.016(5)	87(3)	48(3)	20(3)	0.35	3.3		8
Hemimorphite	2.02407(1)	2.00982(1)	2.00326(1)	60.48(3)	52.15(3)	47.02(3)	0.36	0.7		6
$[AsO_4]^{4-}$										
KDA(4.2K)	2.011(5)	1.995(5)	2.000(4)	3199(2)	2895(3)	2833(3)	20.3	36.6	6.2	9
KDA(296K)	2.001(5)	2.000(5)	2.004(5)	3253(2)	2926(3)	2922(3)	20.7	32.9		9
ADA (296K)	2.006(5)	1.993(5)	1.998(4)	3247(3)	2943(5)	2890(5)	20.7	35.7	5.3	9
RDA(296K)	1.998(4)	1.998(4)	1.998(4)	3183(3)	2870(4)	2867(4)	20.3	31.4		9
CDA(296K)	1.996(5)	1.996(5)	1.996(5)	3115(3)	2805(4)	2805(4)	19.8	31.0		9
Hemimorphite	2.00490(4)	2.00130(4)	1.99568(4)	3683.3(3)	3415.7(3)	3362.1(3)	23.8	32.1	5.4	6
$[AsO_3]^{2-}$										
Calcite (77K)	2.00162(5)	2.00195(5)	2.00195(5)	2626.7(1)	2108.8(2)	2108.8(2)	15.6	51.8		10
Na$_2$HAsO$_4$.7H$_2$O	2.004(2)	2.005(2)	2.005(2)	2033(2)	1590(2)	1590(2)	11.9	44.3		11
KH$_2$AsO$_4$ (K)	1.996(5)	1.996(5)	1.996(5)	2573	2102	2079	15.4	49.3	2.1	9
KH$_2$AsO$_4$ (L)	1.999(5)	1.999(5)	1.999(5)	2145	1704	1693	12.6	45.1	0.9	9
KH$_2$AsO$_4$ (M)	2.000(5)	2.000(5)	2.000(5)	1852	1401	1391	10.6	46.0	0.9	9
Struvite (I)	2.0064(2)	2.0085(2)	2.0058(2)	1694(6)	1523(6)	1378(6)	10.4	31.4	14.7	12
Struvite (II)	2.0070(2)	2.0076(2)	2.0058(2)	1697(6)	1525(6)	1366(6)	10.4	33.0	15.9	12
betaine arsenate	2.001(4)	1.998(4)	1.998(4)	1848(10)	1447(10)	1447(10)	10.8	40.1		13
Haidingerite(I)	1.999	1.999	1.999	2388	1885	1885	13.9	53.2		14
Haidingerite(II)	1.998	1.993	1.993	2635	2120	2120	15.6	51.6		14
Gypsum	2.00537(4)	2.00371(5)	2.00134(3)	1952.0(2)	1492.6(2)	1488.7(2)	11.2	46.2		15
$[AsO_2]^{2-}$										
Calcite (4.2K)	2.0150(2)	1.9991(1)	1.9910(2)	614(1)	−152(2)	−173(2)	0.6	78.5	2.1	16
C$_2$H$_6$AsNaO$_2$	2.012	1.995	1.967	513	−210	−238	0.1	75.0	3.0	17
Gypsum	2.01484(2)	1.99962(2)	1.9958(1)	475.5(2)	−211.1(2)	−229.5(5)	0.1	70.5	1.8	15

KDA, ADA, RDA and CDA denote KH$_2$AsO$_4$, NH$_4$H$_2$AsO$_4$, RbH$_2$AsO$_4$, and CsH$_2$AsO$_4$, respectively. The distribution of unpaired spins in the $4s$ and $4p$ orbitals of arsenic has been re-calculated to the a_0 and b_0 values of 14660 and 333 MHz, respectively (Weil and Bolton 2007).

However, several of the electron-like $[AsO_4]^{2-}$, $[AsO_3]^{2-}$, and $[AsO_2]^{2-}$ radicals in Table 1 have one, two or all three principle g values >2.0023. This is attributable to the spin-orbit coupling expected for As, although analytical uncertainty arising from difficulty in the calibration of the magnetic field for these radicals with large hyperfine structures may be a contributing factor as well. Another arsenic-centered oxyradical $[AsO_2]^0$, which was established in the X-ray-irradiated neon matrix [20], has not been documented in minerals and will not be considered further here. Similarly, other inorganic radicals involving As such as AsF_3^-, $AsCl_3^-$, and AsH_3F [21] are unlikely to be important in rock-forming minerals and are not considered either.

The $[AsO_4]^{2-}$ radical, which was first discovered in gamma-ray-irradiated scheelite $CaWO_4$ [7], is isoelectronic with $[PO_4]^{2-}$, $[SO_4]^-$ and $[SeO_4]^-$[22] and is characterized by trapping of the unpaired electron largely on one of the four oxygen atoms, hence a variety of the classic O^- centers [6, 23, 24]. Its ^{75}As hyperfine coupling constants, with a principle $A(^{75}As)$ axis along one As–O bond direction, are expected to be small (Table 1) and arise from spin polarization [8,24]. In scheelite the precursor $[AsO_4]^{3-}$ relative to the host $[WO_4]^{2-}$ contains one excess electron and therefore is an ideal candidate for losing an electron to form the $[AsO_4]^{2-}$ radical during radiolysis [7]:

$$[AsO_4]^{3-} - e \rightarrow [AsO_4]^{2-} \tag{1}$$

The $[AsO_4]^{4-}$ radical, isoelectronic with PF_4, $[PO_4]^{4-}$, $[SeO_4]^{3-}$ and $[SO_4]^{3-}$, was first identified by Hampton et al. [25] in X-ray-irradiated KH_2AsO_4 (KDA) at room temperature. Similar to the $[AsO_4]^{2-}$ radical, the diamagnetic precursor for the $[AsO_4]^{4-}$ radical is undoubtedly the $[AsO_4]^{3-}$ group as well [9, 25-27]:

$$[AsO_4]^{3-} + e \rightarrow [AsO_4]^{4-} \tag{2}$$

Here, the $[AsO_4]^{3-}$ precursor in arsenates and phosphates has all the valence shells of the oxygen atoms fully filled, so that the additional electron trapped during irradiation can be accommodated only by the central arsenic via a steric transformation involving its $4d$ orbital [27]. Therefore, sp^3 hybridization results in significant amounts of the unpaired spin on the $4s$ and $4p$ orbitals, yielding large ^{75}As hyperfine coupling constants (Table 1).

The $[AsO_3]^{2-}$ radical, isoelectronic with $[PO_3]^{2-}$, $[NO_3]^{2-}$, $[SO_3]^-$, $[SeO_3]^-$ and ClO_3[22], was first identified in X-ray-irradiated Na_2HAsO_4[11]. This radical has a pyramidal structure involving $sp3$ hybridization with large amounts of the unpaired spin on the $4s$ and $4p$ orbitals of the arsenic atom, hence its large ^{75}As hyperfine coupling constants. However, the ^{75}As hyperfine coupling constants of the $[AsO_3]^{2-}$ radical are notably smaller than those of the $[AsO_4]^{4-}$ center (Table 1). Moreover, the spin-density ratios between the $4s$ and $4p$ orbitals of the $[AsO_3]^{2-}$ radical are significantly smaller than those of the $[AsO_4]^{4-}$ center (Table 1). In gamma-ray irradiated haidingerite ($CaHAsO_4 \cdot H_2O$), Gilinskaya and Afanas'eva [12] observed two varieties of the $[AsO_3]^{2-}$ radical and suggested the following radio-chemical reaction for their formation:

$$[HAsO_4]^{2-} - e \rightarrow [AsO_3]^{2-} + OH^- \tag{3}$$

The $[AsO_2]^{2-}$ radical is isostructural with the ClO_2, NO_2^{2-} and SeO_2^- centers [16] and has an approximately planar configuration with sp^2 hybridization, resulting in almost all the unpaired electron localized on the $4p$ orbital of arsenic (Table 1). The $[AsO_2]^{2-}$ radical, unlike the $[AsO_4]^{2-}$, $[AsO_4]^{4-}$ and $[AsO_3]^{2-}$ radicals from the $[AsO_4]^{3-}$ precursor, apparently forms from the $[AsO_3]^{3-}$ group via a reaction of the type:

$$[AsO_3]^{3-} - O^{2-} + e \rightarrow [AsO_2]^{2-} \qquad (4)$$

Therefore, four types of arsenic-centered oxyradicals are readily distinguishable on the basis of their characteristic ^{75}As hyperfine coupling constants, although small variations within each type are present (Table 1) and are probably attributable to the crystal field effects. The ^{75}As hyperfine coupling constants of these radicals are all consistent with the expectations of their respective electronic structures. Also, Reactions 1-4 show that the diamagnetic precursors to these arsenic-centered oxyradicals are all known. Therefore, their EPR detection immediately provides detailed information about arsenic such as its oxidation state, coordination number, and the next-nearest nuclei if superhyperfine structure(s) are observed. Spin Hamiltonian parameters from single-crystal EPR studies also provide additional information about site assignments and formation mechanisms of individual arsenic-centered oxyradicals in minerals investigated.

CARBONATES

The presence of elevated arsenic contents in natural calcite ($CaCO_3$) has attracted numerous recent studies using a variety of the state-of-the-art structural techniques and sophisticated first-principles calculations to investigate the roles and applications of this mineral for remediation of arsenic contamination in the environment [28-35]. These studies have provided compelling evidence for the incorporation of both ^{V}As and ^{III}As at the carbon site in calcite. In this regard, it is surprising that these recent studies appeared to have overlooked the fact that CW-EPR studies of Serway and Marshall [10] and Marshall and Serway [16] established lattice-bounded As(V) and As(III) at the carbon site in natural calcite some 40 years ago.

Serway and Marshall [10], in their single-crystal EPR study of gamma-ray-irradiated calcite containing 1 to 40 ppm As, observed an axially symmetrical center characterized by a four-component hyperfine structure. They noted that the symmetry axis of this center coincides with the crystal c-axis direction and reported its spin Hamiltonian parameters at 77 K (Table 1), including the nuclear quadrupole coupling constant $e^2qQ = -2.4$ MHz. These results allowed Serway and Marshall [10] to interpret this center to be the AsO_3^{2-} radical as a substitution for the CO_3^{2-} group in the calcite lattice. Serway and Marshall [10] noted that this radical is not detectable in crystals that were irradiated and measured at 77 K but is observed after the samples have been warmed to room temperature or by performing irradiation at room temperature. Also, the growth of the AsO_3^{2-} radical at room temperature in crystals irradiated at 77 K is accompanied by the simultaneous decay of the CO_3^{3-} center. These results led Serway and Marshall [32] to suggest that the AsO_3^{2-} radical in calcite is not a primary radiation product but probably formed from charge trapping.

Marshall and Serway [6] observed the AsO_2^{2-} radical in single-crystal EPR spectra of the same suite of gamma-ray-irradiated calcite, measured at 4.2 K. The g and $A(^{75}As)$ matrices of this radical are approximately axial in symmetry and have their unique principle axes along the crystal c-axis direction and a C–O bond direction, respectively, confirming its location at the carbonate ion site. All spin Hamiltonian parameters of this radical, including the nuclear quadrupole coupling constant $e^2qQ = 16.2$ MHz and the asymmetric parameter $\eta \cong 0$, are consistent with the AsO_2^{2-} model with a planar configuration involving sp^2 hybridization and the unpaired electron largely localized in the $4p$ orbital of the arsenic atom.

Di Benedetto et al. [30] measured ESEEM spectra for Mn- and As-bearing calcite (i.e., travertine) in both the field-domain and time-domain (two-pulse Hahn echo, 2p) modes, at temperature of 4.2 K and operating frequency of 9.717 GHz. Their two-pulse ESEEM spectra revealed superimposed nuclear modulations, which have been shown by spectral simulations to arise from the presence of ^{75}As at the carbon site with a Mn-As distance comparable to the Ca-C distance in calcite [32]. Di Benedetto et al. [30] suggested a substitution of $[AsO_3]^{3-} \rightarrow [CO_3]^{2-}$ for incorporating arsenic in calcite, and their confirmation of As(V) came from a subsequent synchrotron X-ray absorption spectroscopic study [31].

SILICATES

Rock-forming silicates were thought to contain ≤3 ppm As [1], although rare arsenosilicates such as ardennite, mediate and tiragalloite are known but do not involve any isomorphous substitution between As and Si[36,37]. Direct evidence of isomorphous substitutions between $[^VAsO_4]^{3-}$ and $[SiO_4]^{2-}$ in silicates has been documented only recently. For example, Vergasova et al. [38] and Filatov et al. [39] reported filatovite $K[(Al,Zn)_2(As,Si)_2O_8]$, which is the first arsenate of the feldspar group and involves the coupled substitution $Zn^{2+} + As^{5+} = Al^{3+} + Si^{4+}$, from the Tolbachik volcano, Kamchatka peninsula, Russia. Charnock et al. [40] reported data from synchrotron As K-edge X-ray absorption spectra for the substitution of As^{5+} for Si^{4+} in garnet containing 2,500 ppm As, from the Central Oslo Rift. Interestingly, the synchrotron X-ray absorption spectroscopic studies of Hattori et al. [41] and Pascua et al. [42] documented elevated As contents, including the identification of As^{5+}, in smectite and antigorite but were unable to determine whether arsenic in these cases represents isomorphous substitution for Si, surface absorption or contamination from impurity phases.

Single-crystal EPR measurements by Mao et al. [6] revealed the presence of the $[AsO_4]^{4-}$ and $[AsO_4]^{2-}$ radicals with characteristic ^{75}As hyperfine structures (Fig. 2a) in a suite of gamma-ray-irradiated hemimorphite $(Zn_4(Si_2O_7)(OH)_2 \bullet H_2O)$ from various Zn-Pb-Cu-Ag-Au deposits. Mao et al. [6] noted that the half lifes of the $[AsO_4]^{4-}$ and $[AsO_4]^{2-}$ radicals are 21 and 29 days, respectively. Each hyperfine component of the $[AsO_4]^{4-}$ radical in the single-crystal EPR spectra is resolved into four lines when the magnetic field is rotated away from crystal axes, indicative of a triclinic site symmetry in the orthorhombic hemimorphite. On the other hand, each hyperfine component of the $[AsO_4]^{2-}$ radical is resolved into at most two lines, indicating a monoclinic site symmetry. In addition, spectral simulations confirmed that the weak doublet accompanying each hyperfine component of the $[AsO_4]^{2-}$ radical (Fig. 2b) represents the ^{29}Si superhyperfine structure arising from interaction with a single ^{29}Si nucleus with I = ½ and natural isotope abundance = 4.7%. Spectral simulations showed that additional weak lines of the $[AsO_4]^{2-}$ radical arise from a considerable nuclear quadrupole effect (Fig. 2b), which is characteristic of the ^{75}As nucleus and thus providing further confirmation for the identification of this arsenic-associated radical. Moreover, each main line of the $[AsO_4]^{2-}$ radical in a few spectra measured at specific crystal orientations is resolved into three components with an intensity ratio of 1:2:1, attributable to ^{1}H superhyperfine structure arising from interaction with two equivalent or nearly equivalent hydrogen nuclei (I=1/2 and isotope abundance = ~100%).

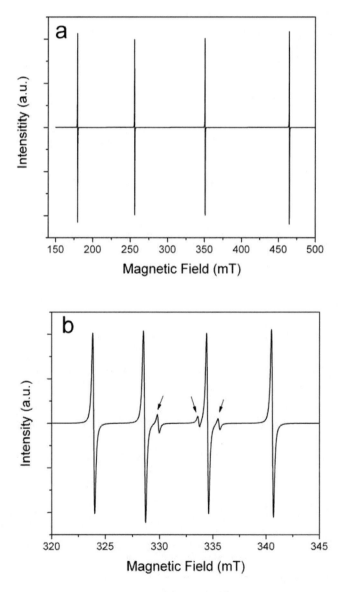

Figure 1. Simulated single-crystal EPR spectra of a) the $[AsO_3]^{2-}$ radical and b) the $[AsO_2]^{2-}$ radical in calcite, with the magnetic field along the crystal c-axis direction and microwave frequency = 9.388 GHz: Note that weak lines marked by arrows in b) arising from the nuclear quadrupole effect (data from [10] and [16]).

The best-fit spin Hamiltonian parameters of the $[AsO_4]^{4-}$ radical have the orientations of the $g_{intermediate}$ and $A_{maximum}$ principle axes both approximately along one of the pseudo-tetrad axes of the SiO_4 tetrahedron in hemimorphite, confirming the location of arsenic at the Si site. Similarly, the best-fit orientations of the $g_{maximum}$ and $A_{maximum}$ principle axes of the $[AsO_4]^{2-}$ radical are close to the Si−O4 bond direction, also confirming the location of this radical at the SiO_4 tetrahedral position and the unpaired electron largely on the non-bonding orbital of the O4 atom. This structural model for the $[AsO_4]^{2-}$ radical is further supported by the observed ^{29}Si and ^1H superhyperfine structures (Fig. 2b). For example, the magnitude of ^{29}Si superhyperfine splitting at ~1 mT (Figure 2b) supports the presence of a nearest-neighbor Si atom. Also, the O4 atom has two equivalent hydrogen atoms at ~3.5 Å, which is consistent with the dipole-dipole-model prediction from the observed ^1H superhyperfine splitting of ~0.3 mT[6].

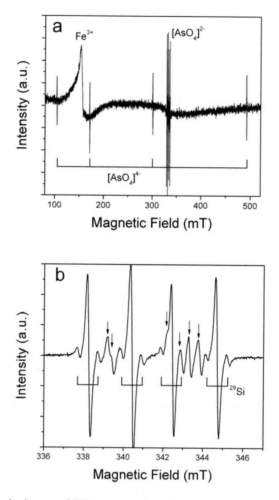

Figure 2. Representative single-crystal EPR spectra of gamma-ray-irradiated hemimorphite with the magnetic field parallel to the crystal c-axis and: a) wide-scan spectrum at a microwave frequency of 9.36 GHz showing the Fe^{3+} center and the $[AsO_4]^{2-}$ and $[AsO_4]^{4-}$ radicals, and b) narrow-scan spectrum at a microwave frequency of 9.61 GHz showing the $[AsO_4]^{2-}$ radical with the ^{29}Si superhyperfine structure and additional weak lines (marked by arrows) arising from the nuclear quadrupole effect (modified from [6]).

Mao et al. [6] interpreted the $[AsO_4]^{2-}$ radical in hemimorphite to originate from a substitutional $[AsO_4]^{3-}$ group at the SiO_4 tetrahedral position without an immediate charge compensator. Such an $[AsO_4]^{3-}$ group in comparison with the $[SiO_4]^{2-}$ group that it replaces for contains one extra electron and therefore is an ideal candidate for trapping an hole to form the $[AsO_4]^{2-}$ radical during gamma-ray irradiation. Similarly, Mao et al. [6] interpreted the $[AsO_4]^{4-}$ radical in hemimorphite to form from a substitutional $[AsO_4]^{3-}$ group at the SiO_4 tetrahedral position as well. However, the $[AsO_4]^{3-}$ precursor for the formation of the $[AsO_4]^{4-}$ radical most likely has a neighboring charge compensator (e.g., Cu^+ at the Zn^{2+} site), yielding a neutral configuration that is capable for trapping an electron during gamma-ray irradiation.

Interestingly, inductively coupled plasma mass spectrometric (ICPMS) analyses of the hemimorphite samples yielded 6 to 338 ppm As [6]. Powder EPR spectrum of the sample containing only 6 ppm As shows that the $[AsO_4]^{4-}$ radical is readily detectable, whereas the $[AsO_4]^{2-}$ radical is superimposed by other radiation-induced centers in the central magnetic field (Figure 3). Spectral simulations with spin Hamiltonian parameters obtained from the single-crystal EPR study [6] reproduce the powder EPR spectrum of the $[AsO_4]^{4-}$ radical very well (Figure 3).

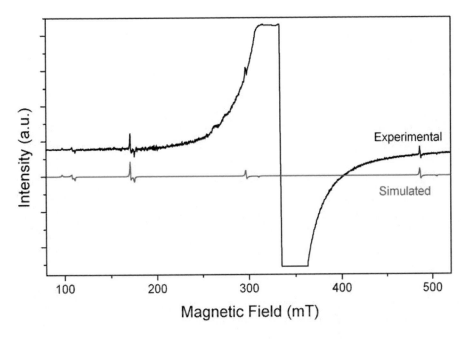

Figure 3. Powder EPR spectrum of a hemimorphite sample containing 6 ppm As measured at $v=9.387$ GHz, in comparison with a simulated spectrum using the spin Hamilton parameters of the $[AsO_4]^{4-}$ radical from [6]. Note that the $[AsO_4]^{2-}$ radical in the central region of the magnetic field is not visible owing to the presence of other paramagnetic centers.

PHOSPHATES

Chemical similarities between the arsenate and phosphate groups are well known, and extensive substitutions of $[^VAsO_4]^{3-}$ for $[PO_4]^{3-}$ in apatites ($Ca_{10}(PO_4)_6(F,Cl,OH)_2$) and other phosphates are well documented[43,44]. Table 1 shows that arsenic-centered oxyradicals are well established in phosphates, and EPR studies have identified a variety of paramagnetic radicals such as CO_3^{3-}, CO_3^-, SO_3^-, SiO_3^- and CrO_4^{3-} in the PO_4^{3-} position in apatites [43, 45, 46]. Therefore, it is surprising to note that arsenic-centered oxyradicals have not yet been detected in apatites.

Struvite $(NH_4)Mg(PO_4) \cdot 6H_2O$, a major constituent of kidney, bladder and urinary stones in animals and human and an increasingly important fertilizer recovered from sewer and wastewater systems [47,48], is a common mineral in guano and dung deposits, peat beds and organic-rich sediments. Struvite having an arsenate analog $(NH_4)Mg(AsO_4) \cdot 6H_2O$ also has been proposed to be useful for the removal of As and other metalloids [49, 50]. Xu [12], on the basis of single-crystal EPR measurements, reported two $[AsO_3]^{2-}$ radicals in gamma-ray-irradiated, arsenate-doped struvite (Table 1). However, careful examination of the descriptions in [12] reveals that they are two magnetically non-equivalent sites of the same $[AsO_3]^{2-}$ radical, rather than two distinct species. Moreover, the EPR spectra reported in [12] appear to include an uncharacterized $[AsO_2]^{2-}$ radical. Therefore, a further EPR study of struvite is warranted to investigate whether this important biomineral is capable of accommodating both As(V) and As(III) in its lattice.

SULFATES

Similarly, the arsenate and sulfate groups share chemical similarities, and extensive substitutions of $[^V AsO_4]^{3-}$ for $[SO_4]^{2-}$ have been documented in a number of sulfates such as jarosite [51, 52]. Gypsum $(CaSO_4 \cdot 2H_2O)$, which is the most common sulfate mineral, is a major by-product containing elevated arsenic contents, from the mining and milling processes of borate, phosphate and uranium deposits. For example, production of boric acid from Turkish borate deposits yields an annual output of 2.5 billion tonnes of arsenic-boron-bearing gypsum sludge (also known as "arsenical borogypsum wastes" [53, 54]). Several studies [28, 55-58] investigated incorporation of arsenic in gypsum but reported contradictory results. For example, Rodriguez-Blanco et al. [55], in a SEM-EDS study of co-crystallized pharmacolite $(CaHAsO_4 \cdot 2H_2O)$ and gypsum, did not detect any substitution between As^{5+} and S^{6+}. On the other hand, Fernandez-Martinez et al. [53, 54], in their studies of synthetic gypsum by combined neutron and X-ray diffraction analyses, synchrotron X-ray absorption spectroscopy (XAS) and density functional theory (DFT) modeling, reported substitution of As^{5+} for S^{6+} (see also [28]). This discrepancy most likely stems from different sensitivities of the analytical techniques employed in previous studies [55-58].

Single-crystal EPR studies of gamma-ray-irradiated arsenic-doped gypsum have identified both the $[AsO_3]^{2-}$ and $[AsO_2]^{2-}$ radicals [15]. The $[AsO_3]^{2-}$ center is characterized by its unique $A(^{75}As)$ axis along the S-O1 bond direction, and contains an ^1H superhyperfine structure arising from interaction with at least four neighboring protons, which has been confirmed by pulse electron nuclear double resonance (ENDOR) spectra measured at 20 K. These results show that the $[AsO_3]^{2-}$ center formed from electron trapping on the central As^{5+} ion of a substitutional $(AsO_4)^{3-}$ group after removal of an O1 atom. The $[AsO_2]^{2-}$ center is characterized by its unique $A(^{75}As)$ axis approximately perpendicular to the O1-S-O2 plane and the A_2 axis along the S-O2 bond direction, consistent with electron trapping on the central

As^{3+} ion of a substitutional $(AsO_3)^{3-}$ group after removal of an O2 atom [15]. It is also interesting to note that the $[AsO_3]^{2-}$ center predominates over its $[AsO_2]^{2-}$ counterparts in the crystal synthesized with an arsenate in the starting material, but reverse is true in another crystal from an arsenite-bearing starting material. This result supports Reactions 3 and 4 that the $[AsO_3]^{2-}$ and $[AsO_2]^{2-}$ radicals originate from the arsenate and arsenite precursors, respectively. Therefore, this EPR study not only confirms previous suggestions for lattice-bound As^{5+} in gypsum [28, 55, 56], but demonstrates incorporation of As(III) in gypsum as well.

BORATES

The few reported analyses of arsenic in borates were based on bulk-chemical techniques [59] and may not be reliable, because these minerals commonly occur in intimate association with sulfarsenides [60, 61]. Synchrotron micro-X-ray fluorescence mapping and analyses [61] revealed up to 125 ppm As in colemanite $(CaB_3O_4(OH)_3 \cdot H_2O)$, a major ore mineral in world-class borate deposits. Modeling of As K-edge X-ray absorption spectra suggests that both As(III) and As(V) are present and preferentially occupy the triangular B1 site and the tetrahedral B2 and site, respectively [61].

Single-crystal EPR spectra of gamma-ray-irradiated colemanite, measured at 40 K, show the presence of an $[AsO_3]^{2-}$ radical with an angular dependence indicative of its accommodation in the crystal lattice [61]. However, quantitative spin Hamiltonian analysis for this radical in colemanite was not possible owing to its spectra having exceedingly low signal-to-noise ratios, even after high doses of gamma-ray irradiation. This example illustrates the limitation of the EPR method in cases where conversion from diamagnetic precursors to paramagnetic centers is inefficient. In such cases, the EPR method would be best combined with synchrotron X-ray absorption spectroscopy, provided that the arsenic contents in the minerals investigated are sufficiently high for the latter technique.

CONCLUSION

In summary, EPR spectroscopy, including related techniques such as pulse ENDOR and ESEEM, is known for its superior sensitivity over other structural techniques and provides a wealth of information about arsenic speciation in rock-forming minerals. In particular, single-crystal EPR studies allow not only unambiguous identification of As^{3+} and As^{5+} but determination of their site assignments and local structural environments in rock-forming minerals.

ACKNOWLEDGMENTS

I would like to thank Mao Mao for provision of the experimental EPR spectrum in Figure 3, and Rong Li, Jinru Lin, Mao Mao and Mark J. Nilges for collaborations on the EPR studies of arsenic in rock-forming minerals.

REFERENCES

[1] P. L. Smedley and D.G. Kinninburg, *Appl. Geochem.* 17, 517 (2002).
[2] Y. Pan, N. Chen, J. A.Weil and M.J. Nilges, *Amer. Mineral.* 87, 1333, (2002).
[3] J. A. Weil and J. R. Bolton *Electron Paramagnetic Resonance: Elementary Theory and Practical Applications.* John Wiley and Sons, New York (2007).
[4] A. Schweiger and G. Jeschke *Principles of Pulse Electron Paramagnetic Resonance.* Oxford University Press, New York (2001).
[5] P. A. O'Day, *Elements* 2, 77 (2006).
[6] M. Mao, J. Lin and Y. Pan, *Geochim. Cosmochim. Acta* 74, 2943 (2010).
[7] P. R. Edwards, S. Subramanian and M. C. R. Symons, *Chem. Commun.* 799 (1968).
[8] S. Subramanian, P. N. Murty, and C. P. K. Murty, *J. Phys. Chem. Solids* 38, 825 (1977).
[9] N. S. Dalal, J. R. Dickson and C. A. McDowell, *J. Chem. Phys.* 57, 4254 (1972).
[10] R. A. Serway and S. A. Marshall, *J. Chem. Phys.* 45, 2309 (1966).
[11] W. C. Lin and C. A. McDowell, *Mol. Phys.* 7, 223 (1964).
[12] R. Xu, *Phys. Stat. Sol. (b)* 172, K15 (1992).
[13] A. Pöppl, O. Tober and G. Völkel, *Phys. Stat. Sol. (b)* 183, K63 (1994).
[14] L. G. Gilinskaya and V. I. Afanas'eva, *J. Struct. Chem.* 22, 187 (1981).
[15] J. Lin, N. Chen, M. J. Nilges and Y. Pan, (ms)
[16] S. A. Marshall and R. A. Serway, *J. Chem. Phys.* 50, 435 (1969).
[17] M. Geoffroy and A. Llinares, *Helvet. Chim. Acta* 62, 1605 (1979).
[18] E. Furimsky, J. A. Howard and J. R. Morton, *J. Am. Chem. Soc.* 95, 6574 (1973).
[19] M. Geoffroy, A. Llinares and J. R. Morton, *Helv. Chim. Acta* 66, 673 (1983).
[20] L. B. Jr. Knight, G. C. Jones, G. M. King, R. M. Babb and A. J. McKinley, *J. Chem. Phys.* 103, 497 (1995).
[21] A. J. Colussi, J. R. Morton and K. F. Preston, *J. Phys. Chem.* 79, 1855 (1975).
[22] P. W. Atkins and M. C. R. Symmons,*The Structure of Inorganic Radicals: An Application of Electron Spin Resonance to the Study of Molecular Structure.* Elsevier, New York (1967).
[23] O. F. Schirmer, *J. Phys. Conden. Matt.* 18, 667 (2006).
[24] Z. Li and Y. Pan, *Phys. Rev. B* 84, 115112 (2011).
[25] M. Hampton, F. G. Herring, W. C. Lin and C. A. McDowell, *Mol. Phys.* 10, 565 (1966).
[26] N. S. Dalal, J. A. Hebden, D. E. Kennedy and C. A. McDowell, *J. Chem. Phys.* 6, 4425 (1977).
[27] P. N. Murty, C. P. K. Murty and S. Subramanian, *Phys. Stat. Sol. (a)* 39, 675 (1977).
[28] G. Roman-Ross, L. Clarlet, G. J. Cuello and D. Tisserand, *J. Phys. IV France* 107, 1153(2003).
[29] G. Roman-Ross, G. J. Cuello, X. Turrillas, A. Fernandez-Martinez and L. Charlet, *Chem. Geol.* 233, 328 (2006).
[30] F. Di Benedetto, P. Costagliola, M. Benvenuti, P. Lattanzi, M. Romanelli and G. Tanelli, *Earth Planet. Sci. Lett.* 246, 458 (2006).
[31] V. G. Alexandratos, E. J. Elzinga and R. J. Reeder, *Geochim. Cosmochim. Acta* 71, 4172 (2007).

[32] M. Romanelli, M. Benvenuti, P. Costagliola, F. Di Benedetto and P. Lattanzi, *Atti del 1 Meeting SIMP-AIC, Sestri Levante, 7–12 Settembre* (2008).

[33] H. U. Sø, D. Postma, R. Jakobsen and F. Larsen, *Geochim. Cosmochim. Acta* 72, 5871 (2008).

[34] Y. Yokoyama, S. Mitsunobu, K. Tanaka, T. Itai and Y. Takahashi, *Chem. Lett.* 38, 910 (2009).

[35] F. Bardelli, M. Benvenuti M, P. Costagliola, F. Di Benedetto, P. Lattanzi, C. Meneghini, M. Romanelli and L. Valenzano, *Geochim. Cosmochim. Acta* 75, 3011 (2011).

[36] C. M. Gramaccioli, G. Liborio and T. Pilati, *Acta Crystal.* 37, 1972 (1981).

[37] M. Nagashima and T. Armbruster, *Mineral. Mag.* 74, 55 (2010).

[38] L. P. Vergasova, S. V. Krivovichev, S. N. Britvin, P. C. Burns and V. V. Ananiev, *Eur. J. Mineral.* 16, 533 (2004).

[39] S. K. Filatov, S. V. Krivovichev, P. C. Burns and L. P. Vergasova, *Eur. J. Mineral.* 16, 537 (2004).

[40] J. M. Charnock, D. A. Polya, A. G. Gault and R. A. Wogelius, *Amer. Mineral.* 92, 1856 (2007).

[41] K. Hattori, Y. Takahashi, S. Guillot and B. Johanson, *Geochim. Cosmochim. Acta* 69, 5585 (2005).

[42] C. Pascua, J. Charnock, D. A. Polya, T. Sato, S. Yokoyama and M. Minato, *Mineral. Mag.* 69, 897 (2005).

[43] Y. Pan and M. E. Fleet, *Rev. Mineral. Geochem.* 48, 13 (2002).

[44] Y. J. Lee, P. W. Stephens, Y. Tang, W. Li, B. L. Phillips, J. B. Parise and R. J. Reeder, *Amer. Mineral.* 94, 666 (2009).

[45] S. Nokhrin, Y. Pan and M. J. Nilges, *Amer. Mineral.* 91, 1425 (2006).

[46] L. G. Gilinskaya, *Phosphorus, Sulfur, and Silicon and the Related Elements* 51, 438 (2009).

[47] L. Shu, P. Schneider, V. Jegatheesan and J. Johnson, *Biores. Tech.* 97, 2211 (2006).

[48] K. P. Fattah, Y. Zhang, D. S. Mavinic and F. A. Koch, *Can. J. Civil Eng.* 37, 1271 (2010).

[49] M. Weil, *Crystal. Res. Tech.* 43, 1286 (2008).

[50] A. A. Rouff and D. Maza, *Geochim. Cosmochim. Acta* 74, A887(2010).

[51] D. Paktunc and J. E. Dutrizac, *Can. Mineral.* 41, 905 (2003).

[52] K.S. Savage, D. K. Bird and P. A. O'Day, *Chem. Geol.* 215, 473 (2005).

[53] M. Delfini, M. Ferrini, A. Manni, P. Massacci and L. Piga, *Miner. Eng.* 16, 45 (2003).

[54] I. Alp, H. Deveci, Y. H. Süngün, E. Y. Yazici, M. Savaş and S. Demirci, *Environ. Sci. Tech.* 43, 6939 (2009).

[55] A. Fernández-Martínez, G. Roman-Ross, G. J. Cuello, X. Turrillas, L. Charlet, M. R. Johnson and F. Bardelli, *Physica B* 385-386, 935 (2006).

[56] A. Fernández-Martínez, L. Charlet, G. J. Cuello, M. R. Johnson, G. Roman-Ross, F. Bardelli and X. Turrillas, *J. Phys. Chem.* A112, 5159 (2008).

[57] J. D. Rodríguez-Blanco, A. Jiménez and M. Prieto, *Crystal Growth Design* 7, 2756 (2007).

[58] J. D. Rodríguez, A. Jiménez, M. Prieto, L. Torre and S. García-Granda, *Amer. Mineral.* 93, 928 (2008).

[59] D. E. Garrett, *Borates: Handbook of Deposits, Processing, Properties, and Use.* Academic Press (1998).

[60] C. Helvaci and R.N. Alonso, *Turkish J. Earth Sci.* 9, 1 (2000).

[61] J. Lin, Y. Pan, N. Chen, M. Mao, R. Li and R. Feng, *Can. Mineral.* 49, 809 (2011).

In: Arsenic
Editor: Andrea Masotti

ISBN: 978-1-62081-320-1
© 2013 Nova Science Publishers, Inc.

Chapter 3

ENVIRONMENTAL SPECIATION OF ARSENIC

Virender K. Sharma[1], Mary Sohn[1], Maurizio Pettine[2] and Barbara Casentini[2]*

[1]Department of Chemistry, Florida Institute of Technology,
Melbourne, Florida, US
[2]Instituto di Ricerca sulle Acque (IRSA) / Water Research Institute Consiglio
Nazionale delle Ricerche (CNR) / National Research Council,
Monterotondo (RM), Italy

ABSTRACT

This chapter reviews the current state of knowledge on the speciation of arsenic in aquatic and terrestrial environments. Due to the enormous differences in the toxicity of different arsenic species, knowledge of arsenic speciation is of utmost importance in terms of environmental management and remediation. The effects of natural organic matter, pH, Eh, adsorption, desorption, biological transformations and key inorganic interactions are considered.

INTRODUCTION

Arsenic (As) is a ubiquitous element, which is naturally present in the earth's crust at low abundance (0.0001 %). Of the four oxidation states in the environment, +V (arsenate), +III (arsenite), 0 (arsenic), and –III (arsine), inorganic arsenite and arsenate are considered the most toxic forms. In addition to inorganic forms, organic compounds of As have also been of environmental concern, and include methylated derivatives of arsenite and arsenate, "fish arsenic" (arsenobetaine, AB and arsenocholine, AC) and arsenosugars [1]. Arsenite and arsenate have been determined in water, sediments, soil, and plants [2, 3]. The concentration of arsenic is typically < 2 μg L^{-1}, and 0.15 – 0.45 μg L^{-1} in seawater and freshwater respectively while unpolluted surface water and groundwater range 1-10 μg L^{-1} and thermal

---

* Email: vsharma@fit.edu

waters have had concentrations reported to be as high as 8.5 mg L^{-1} [2]. Elevated levels of As in drinking water sources around the world are usually related to natural geological sources through biogeochemical and hydrological processes [4]. The United States Environmental Protection Agency (U.S.E.P.A) and the World Health Organization (WHO) have set a guideline of 10 μg L^{-1} As as the drinking water standard.

Arsenic is toxic to humans and is believed to be a contributing cause to kidney, liver, lung, skin, and bladder cancer [5]. Interestingly, the concentration of arsenic was found to be inversely related to the fecal contamination of shallow tube wells in Bangladesh [6]. Arsenic may induce oxidative stress resulting in the generation of reactive oxygen species (ROS) such as superoxide anion ($O_2^{\bullet-}$), singlet oxygen ($^1O_2$), hydrogen peroxide (H_2O_2), and hydroxyl radical ($^{\bullet}OH$) [7]. Basically, the cycling between different oxidation states of arsenic and other metals (e.g. iron, copper, and chromium), interaction of metals with antioxidants, and increasing inflammation may produce ROS. An example is the role of arsenic in the production of ROS in mitochondria (Figure 1).

The reaction of molecular oxygen with semiubiquinone in the mitochondrial respiratory chain produces events which can generate ROS. The oxidation of arsenite to arsenate can also generate ROS. Another possible pathway involves the conversion of $O_2^{\bullet-}$ to H_2O_2, which can then possibly react with iron to produce highly reactive $^{\bullet}OH$. The presence of redox active iron is associated with the reaction of ferritin with methylated arsenic species [7]. A recent study postulates the formation of ROS by the interactive effect of arsenic and fluoride on cardio-respiratory disorders in male rats [8].

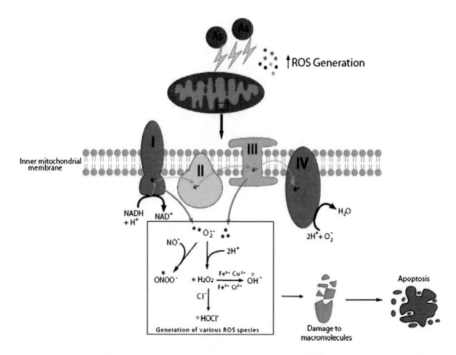

Figure 1. Arsenic-induced mitochondrial generation of various reactive oxygen species and their effects. Arsenic induces significant ROS generation mainly through complex I and complex II of the electron transport chain (ETC). Superoxide radical generated through the ETC reacts with various other radicals present in the cell to form stable and long-lived reactive species, damaging macromolecules and inducing apoptosis via various pathways (Reproduced from [7] with the permission of Elsevier Inc.).

Knowledge of the speciation of arsenic is critical to understanding the toxicity as well as its transport in the environment. Arsenite is generally considered more toxic than arsenate, however arsenite adsorbs more strongly to solid surfaces in comparison with arsenate and can thus be more effectively removed from aqueous solution by adsorption. Arsenic speciation depends on many different environmental parameters, which include, pH, E_h, concentrations of inorganic and organic constituents of water, and the nature of solid surfaces. This chapter summarizes the effects of these parameters on the speciation of arsenic, followed by case studies of arsenic in natural waters.

EFFECT OF PH

The acid-base equilibria of As(III), As(V), monomethylarsenic acid (MMA^V), and dimethylarsenic acid (DMA^V) are presented by Eqs. 1-9.

$$As(OH)_3 \Leftrightarrow As(OH)_2O^- + H^+ \quad pK_{a1} = 9.2 \tag{1}$$

$$As(OH)_2O^- \Leftrightarrow As(OH)O_2^{2-} + H^+ \quad pK_{a2} = 12.1 \tag{2}$$

$$As(OH)O_2^{2-} \Leftrightarrow AsO_3^{3-} + H^+ \quad pK_{a3} = 12.7 \tag{3}$$

$$AsO(OH)_3 \Leftrightarrow H^+ + AsO_2(OH)_2^- \quad pK_{a1} = 2.3 \tag{4}$$

$$AsO_2(OH)_2^- \Leftrightarrow H^+ + AsO_3(OH)^{2-} \quad pK_{a2} = 6.8 \tag{5}$$

$$AsO_3(OH)^{2-} \Leftrightarrow H^+ + AsO_4^{3-} \quad pK_{a3} = 11.6 \tag{6}$$

$$CH_3AsO(OH)_2 \Leftrightarrow H^+ + CH_3AsO_2(OH)^- \quad pK_{a1} = 4.1 \tag{7}$$

$$CH_3AsO_2(OH)^- \Leftrightarrow H^+ + CH_3AsO_3^{2-} \quad pK_{a2} = 8.7 \tag{8}$$

$$(CH_3)_2AsO(OH) \Leftrightarrow H^+ + (CH_3)_2AsO_2^- \quad pK_{a1} = 6.2 \tag{9}$$

The dissociation of As(OH)$_3$ in water occurs according to Eqs 1-3 [9] and the speciation of As(III) is shown in Figure 2.

At neutral pH, As(OH)$_3$ is the major species while As(OH)$_2$O$^-$ is a minor species (< 1.0%). Fractions of As(OH)O$_2^-$ and AsO$_3^-$ are insignificant. The equilibrium reactions of the triprotic acid AsO(OH)$_3$ are described by Eqs 4-6. As shown in Figure 2, nearly equal concentrations of AsO$_2$(OH)$_2^-$ and AsO$_3$(OH)$^{2-}$ are present at neutral pH. In contrast to inorganic arsenic species, organic arsenic species, MMA^V and DMA^V are diprotic and monoprotic acids, respectively (Eqs 7-9) [10]. At neutral pH, CH$_3$AsO$_2$(OH)$^-$ and CH$_3$AsO$_3^{2-}$ are the major and minor species of MMA^V, respectively. Comparatively, both (CH$_3$)$_2$AsO(OH) and (CH$_3$)$_2$AsO$_2^-$ species for the case of DMA^V are present at pH 7 (Figure 2).

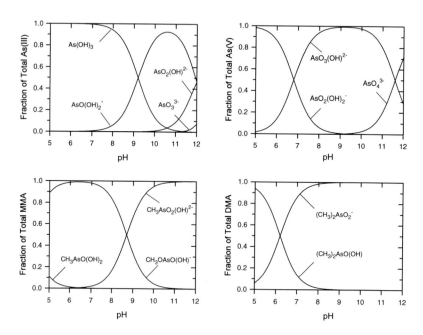

Figure 2. Distribution of As(III), As(V), MMA, and DMA hydroxides species as a function of pH at 25°C (Reproduced from [2] with the permission of Elsevier Inc.).

EFFECT OF E_H

Besides pH, redox potential (E_h) also affects arsenic speciation in the natural environment [11]. This is demonstrated in the E_h-pH diagram for inorganic As species (Figure 3).

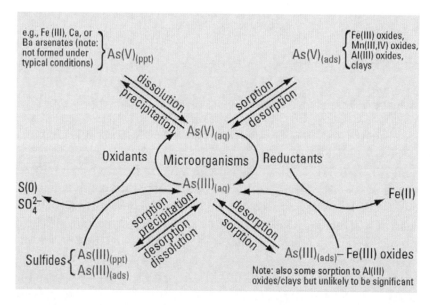

Figure 3. Possible processes in biogeochemical cycling of arsenic (reproduced from [23] with permission of the American Chemical Society).

In natural waters inorganic arsenic occurs primarily as H_3AsO_4 (iAs^V) at pH < 2, and both $H_2AsO_4^-$ and $HAsO_4^{2-}$ species exist in the pH range from 2-11 at high E_h values (oxidizing conditions). However, H_3AsO_3 is the predominant inorganic arsenic species (iAs^{III}) at low E_h values (reducing conditions). Other inorganic species, As_2S_3, arsine and elemental arsenic may also exist under strong reducing conditions and in the presence of sulfur or hydrogen sulfide. Different arsenosulfur species have been reported in arsenite- and sulfide-containing solutions [12, 13]. Examples of arsenosulfur species include $HAsS_2O^-$, $HAsS_3^-$, $As_3S_6^{3-}$, and monomeric and trimeric thioarsenite [14-16]. Thioarsenates are formed in sulfidic groundwaters and artificial arsenite /hydrogen sulfide solutions [17, 18]. Mono-, di-, tri-, and tetrathioarsenates have also been observed in the geothermal waters of Yellowstone National Park [19]. A detained examination of (oxy)thioarsenates in sulfidic waters has been assembled [20], which suggests that complex nonequilibrium distributions of arsenic oxyanions, thioarsenites, and thioarsenates, as a result largely of biochemical processes, are involved in arsenic speciation in sulfidic waters [20, 21]. A very recent study suggests that zero-valent sulfur may play an important role in the speciation of arsenic in reducing environments [22].

BIOGEOCHEMICAL PROCESSES

In addition to oxidizing versus reducing conditions, processes such as sorption, adsorption, precipitation, and biological mediation are also participative in determining distributions of inorganic arsenic in natural waters. This is illustrated in Figure 4 [23]. Immobilization of dissolved aqueous arsenic may occur through sorption onto solid aquifer components and/or by precipitation of solids involving the incorporation of arsenic. Redox conditions strongly influence the sorption process due to the presence of different arsenic species as a function of pH and E_h. For example, sorption is likely at mildly acidic to neutral pH and is strongly influenced by electrostatic interactions with positively charged surfaces which would exist at pH values below the point of zero charge of the surface. Comparatively, the electrostatic effects on sorption are more pronounced for As(V) than for As(III). Redox conditions can also have a pronounced effect on the stability of solid surfaces. For example, iron(III) oxides and sulfide minerals are unstable under reducing and oxidizing conditions, respectively. The levels of microbial activity also control the reactions rates of transformation between different oxidation states of Fe, S, and As (Figure 4), and thus greatly affect the extent of mobilization of arsenic.

One recent study reveals how competitive microbially and Mn oxide mediated redox processes control the speciation and partitioning of arsenic in surface and subsurface environments [3]. The redox dynamics in this study were examined in a diffusively controlled system using a Donnan reactor in which *Shewanella sp.* ANA-3 and birnessite (an oxide mineral of manganese that also contains calcium, potassium and sodium and a strong oxidizing agent) were isolated by an arsenic semipermeable membrane. This study could provide information on biogeochemical cycling under variable redox and microbial respiration processes. Initially, injected As(III) in the birnessite chamber was rapidly oxidized to As(V), which then diffused into the chamber of *Shewanella*. Reduction of As(V) occurred to As(III) in this chamber, which then diffused back to the birnessite chamber, thus completing the redox cycle of arsenic. A rapid decrease in the rate of oxidation of As(III) was observed which was caused by the passivity of the birnessite surface. Results provided evidence that microbial respiration increased the concentrations of Mn(II) and $CO_3^{2,}$, which resulted in precipitation of rhodochrosite to passivate the surfaces to Mn oxide surfaces. Overall, microorganisms and solid surfaces might produce the biogeochemical conditions conducive to cycling of arsenic in the natural environment.

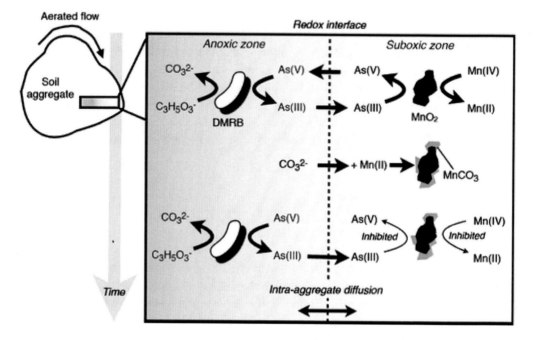

Figure 4. Schematic illustration of diffusion limited transport and formation of redox gradients in soil aggregates (A) and schematic overview of arsenic cycling between a suboxic zone (containing birnessite) and an anoxic zone (containing dissimilatory metal reducing bacteria, *Shewanella*) as determined by Donnan cell experiment (B). (Reproduced from [3] with the permission of the American Chemical Society).

EFFECTS OF ORGANIC MATTER

The omnipresence of organic matter in terrestrial and aquatic environments underscores the importance of considering its effects on arsenic geochemistry in the environment. The pH affected functional groups associated with both low and high molecular weight organic matter can influence arsenic speciation through complexation, adsorption and redox reactions [24]. The study on interaction between Suwanee River NOM (Natural Organic Matter) and As(III) found evidence for the formation of As(III)-NOM soluble complexes which would enhance the environmental mobility of arsenic. On the other hand, organic coatings on suspended solid surfaces may enhance arsenic adsorption and thereby reduce arsenic mobility and due to solid surface charge effects, arsenic adsorption is pH dependent. The use of humic acid in arsenic and heavy metal remediation has been proposed by Wang and Mulligan [25] in a study which suggested arsenic mobilization by the formation of metal-bridging mechanisms with humic acid. Arsenic mobilization through metal ion bridging structures involving Fe(III) has also been described by Wang and Mulligan (2006) [25] in accounting for complexation of soluble anionic arsenic by carboxylate and phenolate groups associated with NOM. In a recent study by Bauer and Blodau [26], the distribution of arsenic in the dissolved, colloidal and particulate size fractions of solutions rich in dissolved organic matter and Fe(III) were investigated. It was found that in the presence of dissolved organic matter the precipitation and sedimentation of As containing colloidal Fe was impeded and was associated with the formation of aqueous Fe complexes. The enhancement of As in the different dissolved/colloidal/particulate pools was dependent on the Fe/C ratio, pH and the size of the dissolved organic molecules.

Observations of redox interactions between NOM and arsenic most commonly involve oxidation from As(III) to As(IV) or reduction from As(V) to As(III). While Redman et al. [24] found NOM oxidation of As(III) to As(VI). Tongesay and Smart [27] reports the reduction of As(V) to As(III) by Suwanee River fulvic acid. The redox effects of organic matter on arsenic speciation is of especial concern due to the greater toxicity of As(III) relative to As(V), which may be due to an interaction of As(III) with sulfhydryl groups on proteins. The presence of methylated forms of arsenic such as monomethyl and dimethyl forms of arsenic (MMA and DMA respectively) as well as arsenosugars are known to occur in natural aquatic environments [28].

CASE STUDIES OF AS-RICH GROUNDWATER AND RISKS ASSOCIATED WITH ITS USE

The presence of arsenic in groundwater is often reported as one of the major problems for drinking and cooking purposes. Arsenic affected areas are very unevenly distributed between the continents. In terms of the exposed population, by far the worst pollution is found in Asia, especially in a band running from Pakistan, along the southern margins of the Himalayan and Indo-Burman ranges, to Taiwan. The World Health Organization (WHO) described the situation in Bangladesh as 'the largest poisoning of a population in history' [29]. It is estimated that in 1998–99 around 27 million people were drinking water containing more than the national standard of 50 μgL^{-1} As. In the Americas, the USA is the most affected country, but the level of population exposure is lower when compared to Asia. In Europe the most exposed population resides in the Pannonian basin (Hungary, Romania, Slovakia, Croatia and Serbia) where 25% exposure to concentrations over 10 $\mu g/L$ was calculated for Hungary, Romania and Slovakia [30]. In South America two areas of severe arsenic pollution (the Pampean Plains of Argentina and the Pacific Plains of Chile) have resulted in extensive arsenicosis incidence. In Africa and Australasia some cases of groundwater contamination have been reported but related health effect were not visible or detected [31].

Few geogenic materials are significant contributors to high As concentrations in groundwater: organic-rich or black shales, Holocene alluvial sediments with slow flushing rates, mineralized and mined areas (most often gold deposits), volcanogenic sources and thermal springs [32]. The main mechanisms controlling arsenic release into aquifers are reductive dissolution, alkali desorption, sulfide oxidation and induced mobilization from parent rocks by geothermal waters [31]. A detailed overview of some of those mechanisms is given in Table 1. Background concentrations of As in groundwater are in most countries less than 10 µg/L and sometimes substantially lower. However, values quoted in the literature show a very large range from <0.5 to 5000 µg/L (i.e. four orders of magnitude). Arsenic is mostly found as inorganic forms (oxyanions) of trivalent arsenite As(III) or pentavalent arsenate As(V).

Table 1. Summary of biotic and abiotic As mobilization processes

As mobilization triggers in aquifers	Reference
Reductive dissolution of As-bearing Fe(III) oxohydroxides promoted by dissimilatory Fe(III)-reducing bacteria under anoxic conditions. This process is typical of reducing alluvial and deltaic environments such as in Bangladesh and India, and it may also be driven by young carbon sources brought to the depth by recent irrigation pumping	[57–62]
Onset of oxidizing conditions in reducing environments where As-sulfide minerals (arsenopyrite, realgar and orpiment) are present, following infiltration of oxygenated ground waters as well as an increased diffusion of atmospheric oxygen into the aquifer driven by the lowering of the groundwater table and the drilling of wells	[60, 63, 64]
Arsenate reduction driven by Sulfate-reducing bacteria under prolonged anoxic conditions. This process may drive the release of adsorbed As(V) in groundwater since As(III) has a lower sorption capacity than As(V) on Fe(III) hydroxides at circumneutral pH values. Production of sulfide may also induce sulfide-driven arsenic mobilization from arsenopyrite and black shale pyrite	[57, 65–67]
Natural organic matter (NOM) may induce arsenic release by competing for adsorption sites, forming aqueous complexes, changing the redox potential of surface adsorption sites	[68]
Anion exchange with competitive ions such as phosphate, carbonate and possibly some anionic organic compounds	[69–73]
pH increase in semiarid or arid environments to values >8.5 as a result of the combined effects of mineral weathering and high evaporation rates leads to desorption of adsorbed As [mainly As(V)] and other anions (V, B, F, Mo, Se and U) from mineral oxides; need for irrigation in semiarid regions may also drive As mobilization	[33, 74–76]
As mobilization by the formation of arsenocarbonate complexes, suggested by strong correlation between As and HCO_3^-; the latter correlation may also be an indirect result of the microbial mediated reduction of Fe(III) hydroxides, which produces both the As and HCO_3^- species	[77–79]
Groundwater age and aging of Fe oxides leading to more ordered forms such as goethite and hematite that have larger crystal sizes and reduced surface areas may lead to a release of As adsorbed on young Fe oxides	[80–83]

Organic As forms may be produced by biological activity, mostly in surface waters, but are rarely quantitatively important in groundwater. The presence of either As(III) or As(V) is not as straightforward as one might think, indicated by thermodynamic predictions, and perhaps the most that can be said at present is that the existence of As(III) implies reducing conditions somewhere in the system [33]. The kinetics of abiotic and biotic redox processes determine As speciation and usually they are driven by a redox potential imposed by the major elements (O, C, N, S and Fe). In natural environments often As(III)/As(V) ratios are biotically controlled. For instance in most geothermal systems, As is discharged mainly in the reduced form. Microbiological activity catalyzes the oxidation of As(III) to As(V) during the ascension of geothermal hot waters and their mixing with shallow oxidized groundwater. A half-life of about 20 mins for As(III) oxidation to As(V) was calculated for As in Hot Creek geothermal springs [34, 35]. This explains why As found in aquifers directly connected to geothermal reservoirs is mainly present as As(III) and after mixing aquifer water with a geothermal contribution, As(V) dominates [33, 36, 37].

In arid and semiarid regions where no alternative water sources are available for drinking or irrigation, the presence of contaminated aquifers represents a management challenge to local administrations. Two illustrative case studies in Mediterranean geothermal areas are presented below.

a) Arsenic occurrence in Central Italy Volcanic Province and management challenges for drinking water supply

In Central Italy, the Roman Magmatic Province is highly affected by As presence in groundwater, causing population exposure through drinking water supply, especially in the municipalities located in the provinces of Rome and Viterbo. In this volcanic region widespread hydrothermal circulation has been reported as underlined by the occurrence of geothermal fields [38]. Angelone et al. (2009) [39] reported As concentrations in the range 176-371 μgL^{-1} for thermal waters and 1.6 and 195 $\mu g/L$ in cold waters around Vico Lake (Northern Latium, Central Italy). The authors' evidence supports a relationship between the As content of groundwater and the hydrothermal processes of the volcanic region. Where the shallow volcanic aquifer is open to deep-rising thermal fluids, relatively high arsenic concentrations (20–100 μgL^{-1}) are found. This occurs close to areas of more recent volcano-tectonic structures. In a nearby area, Vivona et al. (2007) [40] studied As trends along a section cutting from volcanic aquifers to sedimentary aquifers. They reported a positive correlation among As, V and F and an inverse correlation with Ca. Based purely on thermodynamic predictions they hypothesized the precipitation of apatite, scavenging F and As within the sedimentary aquifer. Mechanisms of As groundwater enrichment in this area are not clear. Assuming possible release from volcanic host rocks, weathering was studied by batch leaching experiments [41, 42]. Surface volcanic rock samples, including lava, peperino, red tuff with black scoria (known also as "Ignimbrite C") and an unspecified fallout deposit (U.F.D.), were collected from geological outcrops of Vico Lake. The As total contents were low, with 13, 14, 10 and 17 $mgkg^{-1}$ in lava, peperino, tuff and U.F.D., respectively. Mobilized arsenic in simple MilliQ solution was negligible for observation times up to 2 months, except minimal hydroxyl promoted exchange at alkaline pH. The additional effects of a number of inorganic anions (F^-, HCO_3^-, $B(OH)_4^-$, Cl^-, I^-) and of the siderophore DFOB on As release from volcanic rocks has been investigated. Siderophores are small-molecule compounds generally <1 kDa produced by many organisms to strongly bind Fe, inducing mineral dissolution and shuttle this element to cell surfaces [43, 44]. Nodules of Fe minerals are not uncommon in volcanic rocks, hence siderophore promoted mineral dissolution could trigger indirect abiotic induced release of adsorbed As into groundwater and fluoride, bicarbonate and DFOB have shown a significant influence on As release from those rocks. The fluoride effect was more pronounced at pH 6 and in the case of lava, the net As release increases from 3.7 to about 10% by increasing the concentration of F^- in solution from 0.01 M to 0.05 M.

The bicarbonate showed a significant induced As release only at pH 8.5. Increased temperature exerts higher influence on As release at pH 8.5 rather than at pH 6.1, reflecting a corresponding higher temperature dependence of dissolution rates of silicate minerals at that pH. Siderophore promoted dissolution favors As release from volcanic rocks within one week. The most favorable pH for As release was pH 6 while concentrations above 250 μM DFOB considerably enhanced As and Fe concentrations in solution. At pH 6 and 1000 μM DFOB, the released As fractions were 10.3%, 2.8%, 5.6% and 7.8% from lava, tuff, U.F.D. and peperino, respectively, while the corresponding release at pH 8 were 4.7%, 3.4%, 7.0% and 6.7% [42].

The European Directive 98/83/EC, implemented in Italy as D. Lgs. 31/2001, allows maximum 10 μgL^{-1} As in water supplied for drinking purposes. In the province of Rome and Viterbo, the widespread geogenic presence of As concentrations exceeding this limit (see Figure 5), is causing serious management problems for drinking water distribution at the municipality level. The calculated natural background concentration in the volcanic aquifer was 33.53 μgL^{-1} and this emphasizes the difficulties in finding alternative drinking water sources [45]. Since 2001 the European Union twice excepted some EU As affected countries, Italy included, to comply with the 10 μg/L limit by enhancing the limit up to a provisional value of 50 μgL^{-1}. With the recent Decision of 28/10/2010, the EU granted exceptions up to 20 μg/L to a reduced number of towns in Italy, none of which are located in the provinces of Rome or Viterbo, where 91 municipalities that asked for a derogation allowing 50 μg/L cannot further supply drinking water. The last derogation up to 20 μgL^{-1} will be operative up to the end of 2012. Local administrations of drinking water companies are thus urgently requested to install appropriate treatment units to comply with the 98/83/EC limit.

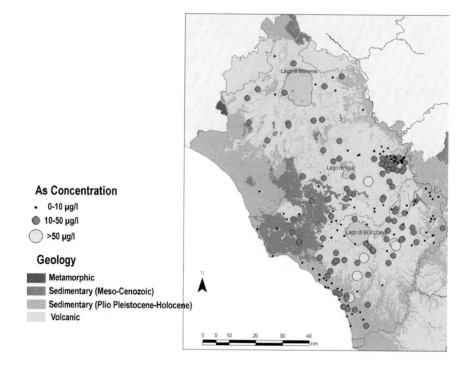

Figure 5. Arsenic distribution in Northern Roman Magmatic Province (modified from [45]).

b) Arsenic occurrence in the geothermal Chalkidiki area (Northern Greece) and As accumulation on irrigated topsoils

There is growing worldwide concern on the risks derived from the use of contaminated groundwater in areas where irrigation relies on As-rich groundwater [46–48]. In Asia and Southeast Asia, the dynamics of As accumulation in flooded rice fields has been extensively studied due to widespread high As-concentrations in groundwater and due to the large number of affected people [47, 49, 50]. The use of As-rich soil in agriculture has been shown to be associated with the accumulation of As in crops and reduced plant growth [51]. An average rice yield loss of 16% has been estimated in Bengal basin rice fields containing 10 to 70 $mgkg^{-1}$ As in topsoil [52]. A reduction of wheat root growth in Chinese soils was observed at 50 $mgkg^{-1}$ As [53]. No observable adverse effect was calculated to occur at <30 $mgkg^{-1}$ As in 16 European barley fields [54].

Casentini et al. (2011) [55] studied As accumulation in agricultural topsoil due to the use of As-rich groundwater for irrigation in the selected Nea Triglia region (Chalkidiki Prefecture). This western part of Ckalkidiki Prefecture is known for the presence of geothermal activities and high hydraulic pressures which force the flow of geothermal waters from Katsika massive limestone formations along the coastal alluvial deposits towards the sea, impacting mainly the municipalities of Nea Triglia, Petralona, Nea Kallikratia, Sozopolis, Nea Silata, Eleachoria, Nea Tenedos and Nea Plagia. This region is highly exploited for agricultural purposes for crops such as wheat, olives, pistachios, tomatoes, cabbages, eggplants, lettuce, onions and cotton. In this area the average groundwater concentration in 2008-2009 was 1000 μgL^{-1} As with a peak of 3960 μgL^{-1}, with As(V) prevalence in almost all the wells. While this groundwater is no longer used for drinking, it represents the sole source for irrigation. The As content in sampled soils ranged from 20-513 mg/kg inside to 5-66 $mgkg^{-1}$ outside the geothermal area with good correspondence of higher As concentrations in well waters and related irrigated soils (see Figure 6) [55].

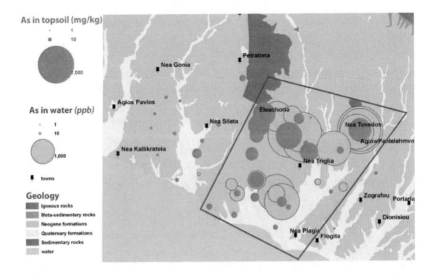

Figure 6. Arsenic distribution in soil and groundwater in the Nea Triglia geothermal fields, Northern Greece. The geothermal area is characterized by high As concentrations and delimited by a red rectangle (modified from [55]).

Around irrigation sprinklers, high As concentrations extended horizontally to distances of at least 1.5 m, and to 50 cm in depth, evidencing a serious impact of irrigation practices on topsoil As accumulation. Simulated rain-irrigation cycles were performed on column experiments of field collected soils (low, medium and high As) to study their ability to mobilize and retain As. During simulated rain events in soil columns (pH=5, 0 µgL^{-1} As), accumulated As was quite mobile, resulting in porewater As concentrations of 500-1500 µg/L with consequent plant root exposure to high As(V) concentrations. In experiments with irrigation water (pH=7.5, 1500 µgL^{-1} As), As was strongly retained (50.5-99.5%) by the majority of the soils. Uncontaminated soils (<30 mgkg^{-1} As) kept soil porewater As concentrations to below 50 µg/L. The column leaching experiments suggest a quite mobile As fraction during rainfall and a good ability of the less contaminated soils for retaining As during irrigation periods, posing serious risk for As buildup in these soils with possible consequences for As crop uptake. During irrigation simulation, As retention can be explained by both adsorption on Fe/Mn phases and by coprecipitation with Ca, as the waters of this area are known to be rich in this element. In this work As contents measured in olives (0.3-25 µgkg^{-1}) do not represent a threat to human health. However, soil arsenic concentrations are frequently elevated to far above EU recommended intervention levels (5 and 110 mgkg^{-1} [56], and As uptake in faster growing plants has to be assessed with high urgency if these soils are to be used for crops other than olives.

REFERENCES

[1]   J.C. Ng, Environmental contamination of arsenic and its toxicological impact on humans, *Environ. Chem.* 2, 146-160 (2005).

[2]   V.K. Sharma, M. Sohn, Aquatic arsenic: Toxicity, speciation, transformations, and remediation, *Environ. Int.* 35, 743-759 (2009).

[3]   S.C. Ying, B.D. Kocar, S.D. Griffis, S. Fendorf, Competitive microbially and Mn oxide mediated redox processes controlling arsenic speciation and partitioning, *Environ. Sci. Technol.* 45, 5572-5579 (2011).

[4]   S. Fendorf, H.A. Michael, A. Van Geen, Spatial and temporal variations of groundwater arsenic in South and Southeast Asia, *Science* 328, 1123-1127 (2010).

[5]   K.T. Kitchin, R. Conolly, Arsenic-induced carcinogenesissoxidative stress as a possible mode of action and future research needs for more biologically based risk assessment, *Chem. Res. Toxicol.* 23, 327-335 (2010).

[6]   A. Van Geen, K.M. Ahmed, Y. Akita, M.J. Alam, P.J. Culligan, M. Emch, V. Escamilla, J. Feighery, A.S. Ferguson, P. Knappett, A.C. Layton, B.J. Mailloux, L.D. McKay, J.L. Mey, M.L. Serre, P.K. Streatfield, J. Wu, M. Yunus, Fecal contamination of shallow tubewells in Bangladesh inversely related to arsenic, *Environ. Sci. Technol.* 45, 1199-1205 (2011).

[7]   S.J.S. Flora, Arsenic-induced oxidative stress and its reversibility, *Free Radic. Biol. Med.* 51, 257-281 (2011).

[8]   S.J.S. Flora, V. Pachauri, M. Mittal, D. Kumar, Interactive effect of arsenic and fluoride on cardio-respiratory disorders in male rats: Possible role of reactive oxygen species, *Biometals* 24, 615-628 (2011).

[9] M.L. Pierce, C.B. Moore, Adsorption of arsenite and arsenate on amorphous iron hydroxide, *Water Res.* 16, 1247-1253 (1982).

[10] C.D. Cox, M.M. Ghosh, Surface complexation of methylated arsenates by hydrous oxides, *Water Res.* 28, 1181-1188 (1994).

[11] J.F. Ferguson, J. Gavis, A review of the arsenic cycle in natural waters, *Water Res.* 6, 1259-1274 (1972).

[12] J.T. Hollibaugh, C. Budinoff, R.A. Hollibaugh, B. Ransom, N. Bano, Sulfide oxidation coupled to arsenate reduction by a diverse microbial community in a soda lake, *Appl. Environ. Microbiol.* 72, 2043-2049 (2006).

[13] B. Planer-Friedrich, E. Suess, A.C. Scheinost, D. Wallschläger, Arsenic speciation in sulfidic waters: Reconciling contradictory spectroscopic and chromatographic evidence, *Anal. Chem.* 82, 10228-10235 (2010).

[14] G.R. Helz, J.A. Tossell, Thermodynamic model for arsenic speciation in sulfidic waters: A novel use of ab initio computations, *Geochim. Cosmochim. Acta.* 72, 4457-4468 (2008).

[15] D.K. Nordstrom, Worldwide occurrences of arsenic in ground water, *Science* 296,2143-2144 (2002).

[16] R.T. Wilkin, D. Wallschla□ger, R.G. Ford, Speciation of arsenic in sulfidic waters, *Geochem. Trans.* 4, 1-7 (2003).

[17] S. Stauder, B. Raue, F. Sacher, Thioarsenates in sulfidic waters, *Environ. Sci. Technol.* 39, 5933-5939 (2005).

[18] D.G. Beak, R.T. Wilkin, R.G. Ford, S.D. Kelly, Examination of arsenic speciation in sulfidic solutions using X-ray absorption spectroscopy, *Environ. Sci. Technol.* 42, 1643-1650 (2008).

[19] B. Planer-Friedrich, J. London, R.B. Mccleskey, D.K. Nordstrom, D. Wallschläger, Thioarsenates in geothermal waters of yellowstone National Park: Determination, preservation, and geochemical importance, *Environ. Sci. Technol.* 41, 5245-5251 (2007).

[20] D. Wallschläger, C.J. Stadey, Determination of (Oxy)thioarsenates in sulfidic waters, *Anal. Chem.* 79, 3873-3880 (2007).

[21] J.C. Fisher, D. Wallschläger, B. Planer-Friedrich, J.T. Hollibaugh, A new role for sulfur in arsenic cycling, *Environ. Sci. Technol.* 42, 81-85 (2008).

[22] R. Couture, P. Van Cappellen, Reassessing the role of sulfur geochemistry on arsenic speciation in reducing environments, *J. Hazard. Mater.* 189, 647-652 (2011).

[23] H.J. Reisinger, D.R. Burris, J.G. Hering, Remediating subsurface arsenic contamination with monitored natural attenuation, *Environ. Sci. Technol.* 39, 458A-464A (2005).

[24] A.D. Redman, D.L. Macalady, D. Ahmann, Natural organic matter affects Arsenic speciation and sorption onto hematite, *Environ. Sci. Technol.* 36, 2889-2896 (2002).

[25] S. Wang, C.N. Mulligan, Effect of natural organic matter on arsenic mobilization from mine tailings, *J. Hazard. Mater.* 168, 721-726 (2009).

[26] M. Bauer, C. Blodau, Arsenic distribution in the dissolved, colloidal and particulate size fraction of experimental solutions rich in dissolved organic matter and ferric iron, *Geochim. Cosmochim. Acta.* 73, 529-542 (2009).

[27] T. Tongesayi, R.B. Smart, Abiotic reduction mechanism of As(V) by fulvic acid in the absence of light and the effect of Fe(III), *Water SA.* 33, 615-618 (2007).

[28] L. Meunier, I. Koch, K.J. Reimer, Effects of organic matter and ageing on the bioaccessibility of arsenic, Environ. Pollut. 159, 2530-2536 (2011).

[29] A. H. Smith, E. O. Lingas, M. Rahman, *Bull. World Health Organ.* 78, 1093-1103 (2000).

[30] R. L. Hough, T. Fletcher, G. S. Leonardi, W. Goessler, P. Gnagnarella, F. Clemens, E. Gurzau, K. Koppova, P. Rudnai, R. Kumar and M. Vahter, *International Archives of Occupational and Environmental Health* 83, 471-481 (2010).

[31] P. Ravenscroft, H. Brammer, K. Reichards, *Arsenic Pollution: A Global Synthesis* Wiley-Blackwell (2009).

[32] D. K. Nordstrom, *Science* 296, 2143-2145 (2002).

[33] P. L. Smedley and D. G. Kinniburgh, *Appl. Geochem.* 17, 517-568 (2002).

[34] J. A. Wilkie and J. G. Hering, *Environ. Sci. Technol.* 32, 657-662 (1998).

[35] H. W. Langner, C. R. Jackson, T. R. McDermott, W. P. Inskeep, *Environ. Sci. Technol.* 35, 3302-3309 (2001).

[36] A. Aiuppa, W. D'Alessandro, C. Federico, B. Palumbo, M. Valenza, *Appl. Geochem.* 18, 1283-1296 (2003).

[37] I. A. Katsoyiannis, S. J. Hug, A. Ammann, A. Zikoudi, C. Hatziliontos, *Sci.Total Enviro.* 383, 128-140 (2007).

[38] M. Dall'Aglio, G. Giuliano, D. Amicizia, M. C. Andrenelli, G. B. Cicioni, D. Mastroianni , L. Sepicacchi and S. Tersigni, "Assessing drinking water quality in Northern Latium by trace elements." the *10th International Symposium on Water-Rock Interaction*, 1063-1066 (2001).

[39] M. Angelone, C. Cremisini, V. Piscopo, M. Proposito, F. Spaziani, *Hydrogeol J.* 17, 901-914 (2009).

[40] R. Vivona, E. Preziosi, B. Madé, G. Giuliano, *Hydrogeol. J.* 15 1183-1196 (2007).

[41] B. Casentini, M. Pettine, F. J. Millero, *Aquat Geochem* 16, 373-393 (2010).

[42] B. Casentini and M. Pettine, *Appl. Geochem.* 25, 1688-1698 (2010).

[43] H. Boukhalfa and A. L. Crumbliss, *BioMetals* 15, 325-339 (2002).

[44] S. M. Kraemer, *Aquat. Sci.* 66, 3-18 (2004).

[45] E. Preziosi, G. Giuliano, R. Vivona, *Environ. Earth Sci.* 61, 885-897 (2010).

[46] H. Brammer and P. Ravenscroft, *Environ. Int.* 35, 647-654 (2009).

[47] L. C. Roberts, S. J. Hug, J. Dittmar, A. Voegelin, G. C. Saha, M. A. Ali, A. B. M. Badruzzaman and R. Kretzschmar, *Environ. Sci. Technol.* 41, 5960-5966 (2007).

[48] L. C. Roberts, S. J. Hug, J. Dittmar, A. Voegelin, R. Kretzschmar, B. Wehrli, O. A. Cirpka, G. C. Saha, M. A. Ali and A. B. M. Badruzzaman, *Nature Geosci* 3, 53-59 (2010).

[49] J. Dittmar, A. Voegelin, L.C. Roberts, S.J. Hug, G.C. Saha, M.A. Ali, A.B.M. Badruzzaman and R. Kretzschmar, *Environ. Sci. Technol.* 41, 5967-5972 (2007).

[50] G. C. Saha and M. A. Ali, *Sci. Total Environ.* 379, 180-189 (2007).

[51] S. C. Sheppard, *Water Air Soil Pollut.* 64, 539-550 (1992).

[52] G. M. Panaullah, T. Alam, M. B. Hossain, R. H. Loeppert, J. G. Lauren, C. A. Meisner, Z. U. Ahmed and J. M. Duxbury, *Plant Soil* 317, 31-39 (2008).

[53] Q. Cao, Q. H. Hu, C. Baisch, S. Khan, Y. G. Zhu, *Environ. Toxicol. Chem.* 28, 1946-1950 (2009).

[54] J. Song, F. J. Zhao, S. P. McGrath, Y. M. Luo, *Environ. Toxicol. Chem.* 25, 1663 (2006).

[55] B. Casentini, S. J. Hug, N. P. Nikolaidis, *Sci. Total Environ.* 409, 4802-4810 (2011).

[56] C. Carlon, M. D'Alessandro, F. Swartjes, *Derivation methods of soil screening values in Europe. A review and evaluation of national procedures towards harmonisation opportunities* European Commission Joint Research Centre, Ispra (Italy), (2007).

[57] D. E. Cummings, Caccavo, S. Fendorf, R. F. Rosenzweig, *Environ. Sci. Technol.* 33, 723-729 (1999).

[58] H. McCreadie, D. W. Blowes, C. J. Ptacek, J. L. Jambor, *Environ. Sci. Technol.* 34, 3159-3166 (2000).

[59] R. T. Nickson, J. McArthur, W. Burgess, K. M. Ahmed, P. Ravenscroft and M. Rahmann, *Nature* 395, 338 (1998).

[60] R. T. Nickson, J. M. McArthur, P. Ravenscroft, W. G. Burgess, K. M. Ahmed, *Appl. Geochem.* 15, 403-413 (2000).

[61] C. F. Harvey, C. H. Swartz, A. B. M. Badruzzaman, N. Keon-Blute, W. Yu, M. A. Ali, J. Jay, R. Beckie, V. Niedan, D. Brabander, P. M. Oates, K. N. Ashfaque, S. Islam, H. F. Hemond and M. F. Ahmed, *Science* 298, 1602-1606 (2002).

[62] F. S. Islam, A. G. Gault, C. Boothman, D. A. Polya, J. M. Charnock, D. Chatterjee and J. R. Lloyd, *Nature* 430, 68-71 (2004).

[63] T. Roychowdhury, G. K. Basu, B. K. Mandal, B. K. Biswas, G. Samanta, U. K. Chowdhury, C.R. Chanda, D. Lodh, S. L. Roy, K. C. Saha, S. Roy, S. Kabir, Q. Quamruzzaman and D. Chakraborti, *Nature* 401, 545-546 (1999).

[64] N. K. Foley and R. A. Ayuso, *Geochemistry: Exploration, Environment, Analysis* 8, 59-75 (2008).

[65] H. W. Langner and W. P. Inskeep, *Environ. Sci. Technol.* 34, 3131-3136 (2000).

[66] R. S. Oremland and J. F. Stolz, *Science* 300, 939-944 (2003).

[67] W. Zhu, L. Y. Young, N. Yee, M. Serfes, E. D. Rhine and J. R. Reinfelder, *Geochim. Cosmochim. Ac.* 72, 5243-5250 (2008).

[68] S. Wang and C. Mulligan, *Environ. Geochem. Health* 28, 197-214 (2006).

[69] X. Meng, G. P. Korfiatis, S. Bang, K. W. Bang, *Toxicol. Lett.* 133, 103-111 (2002).

[70] L. E. Williams, M. O. Barnett, T. A. Kramer, J. G. Melville, *J. Environ. Qual.* 32, 841-850 (2003).

[71] S. Zhang, W. Li, X. Q. Shan, A. Lu, P. Zhou, *Water, Air, Soil Pollut.* 167, 111-122 (2005).

[72] Y. Tao, S. Zhang, W. Jian, C. Yuan, X. Q. Shan, *Chemosphere* 65, 1281-1287 (2006).

[73] K. Saeki, *Bulletin of Environmental Contamination and Toxicology* 81, 508-512 (2008).

[74] R. T. Nickson, J. M. McArthur, B. Shrestha, T. O. Kyaw-Myint, D. Lowry, *Appl. Geochem.* 20, 55-68 (2005).

[75] N. Madhavan, V. Subramanian, in *Groundwater: resource evaluation, augmentation, contamination, restoration, modeling and management*, Ch.6, Springer, (2007).

[76] M. W. Busbee, B. D. Kocar, S. G. Benner, *Appl. Geochem.* 24, 843-859 (2009).

[77] M.-J. Kim, J. Nriagu, S. Haack, *Environ. Sci. Technol.* 34, 3094-3100 (2000).

[78] H. M. Anawar, J. Akai, H. Sakugawa, *Chemosphere* 54, 753-762 (2004).

[79] T. Radu, J. L. Subacz, J. M. Phillippi, M. O. Barnett, *Environ. Sci. Technol.* 39, 7875-7882 (2005).

[80] R. M. Cornell and U. Schwertmann, *The Iron Oxides: Structure, Properties, Reactions, Occurrence and Uses* VCH, Weinheim, (1996).

[81] S. Dixit and J. G. Hering, *Environ. Sci. Technol.* 37, 4182-4189 (2003).

[82] H. D. Pedersen, D. Postma, R. Jakobsen, *Geochim. Cosmochim. Ac.* 70, 4116-4129 (2006).

[83] K. A. Radloff, Z. Cheng, M. W. Rahman, K. M. Ahmed, B. J. Mailloux, A. R. Juhl, P. Schlosser and A. van Geen., *Environ. Sci. Technol.* 41, 3639-3645 (2007).

In: Arsenic
Editor: Andrea Masotti

ISBN: 978-1-62081-320-1
© 2013 Nova Science Publishers, Inc.

Chapter 4

SCALE ISSUES OF ARSENIC MOBILITY AND STABILIZATION IN A PARIS GREEN CONTAMINATED SOIL

Harald Weigand[1] and Clemens Marb[2]*

[1]Technische Hochschule Mittelhessen - University of Applied Sciences, ZEuUS, Giessen, Germany
[2]Bavarian Environment Agency, Josef-Vogl-Technology Center, Augsburg, Germany

ABSTRACT

Feasibility studies on the stabilization of contaminated soils are typically based on laboratory experiments. Yet, due to sample and site heterogeneity it is unclear whether the results obtained under such conditions are predictive for full-scale applications. This study focuses on the scale dependence of the mobility and stabilization of arsenic in a soil contaminated by production remnants of copper(II)acetoarsenite (Paris Green). Column experiments were performed with packed bed volumes of 1.5 L and 1.5 m^3 employing the grain size fraction < 2 mm and the bulk soil, respectively. Parallel leaching tests were conducted with and without addition of deferrization sludge as a sorbent for arsenic. Pore water samples collected during a five day equilibration period indicated fast arsenic release (control soil) and stabilization kinetics (treated soil). Upon irrigation, identical elution curves with arsenic maxima around 8 mg per liter were obtained in the lab and pilot scale experiments with the control soil. Low effluent concentrations from the treatments pointed to strongly enhanced contaminant retention by deferrization sludge with stabilization efficiencies of 99.5 %. Overall, the results demonstrate successful up-scaling suggesting applicability of a deferrization sludge-based stabilization of arsenic in Paris Green contaminated soils. However, further research is required to optimize amendment levels in view of arsenic groundwater standards.

---

* E-mail address: harald.weigand@kmub.thm.de

INTRODUCTION

The Arsenical Pigment Paris Green

After development of an manufacturing technology by Wilhelm Sattler and Friedrich Ruß in 1814, copper(II)acetoarsenite (Paris Green) was produced at several sites in the region of Lower Franconia, Bavaria, Germany. For its light proofness and thermal stability it was originally used as a pigment and became a fashion color for wallpapers. After cases of workplace contamination and the attribution of chronic diseases to the exposition to Paris Green in the living space its use as a pigment was banned in 1887. However, production continued and the compound was further used in a series of different applications (anti-fouling agent, pesticide in viniculture and potato cultivation etc.) [1].

The pigment was typically obtained as a batch precipitate by reacting copper acetate or copper oxide with arsenic trioxide or arsenous acid in the presence of acetic acid [2]. The water-solubility of fresh Paris-Green is low but soluble decomposition products may be formed in the presence of hydrogen carbonate.

In 1989 a case of intoxication by emissions related to the production of Paris Green was observed in the area of Calcutta, India [3]. At least 1,000 people suffered from acute arsenic toxicity and 10,000 had developed arsenical keratosis due to chronic exposition (e.g. use of drinking water extracted from contaminated wells). One identified source of contamination was the untreated effluent of the production process, another being the dumping of a fine Paris Green precipitate. By analyzing sediment profiles, the authors concluded that As-contaminated seepage water was the primary cause of groundwater contamination. This agrees with the exceptionally high As concentrations of > 50 mg/L observed in the leachate of column experiments with a Paris Green contaminated soil [4].

Stabilization as a Treatment Option for Contaminated Soils

The afore mentioned shows that vadose zone transport of metals and metalloids at contaminated sites may pose a significant risk to groundwater water quality and human health. Since inorganic soil contaminations are not accessible to biodegradation, remediation concepts mostly rely on soil excavation/exchange and disposal of the polluted material. Against this background, stabilization techniques have received considerable attention as a low-cost alternative. Stabilization of contaminated soils aims at minimizing contaminant mobility in soils by enhancing natural retention processes through addition of appropriate amendments. While solidification relies on the application of binders to form soil monoliths of minimized hydraulic conductivity [5] stabilization targets the transfer of contaminants into more leaching-resistant binding forms [6].

Advantages of contaminant stabilization rest with moderate technical requirements and low operating expense. Possible applications encompass in-situ and on-site treatment of contaminated soils with subsequent re-use. This may be particularly useful in cases of extensive contamination inefficiently handled by conventional methods. Alternatively, off-site treatment of excavated material may be conducted prior to landfill. If leaching values are reduced such that the material is no longer a hazardous waste, disposal costs may be cut significantly.

Prospects of stabilization may be assessed by analyzing contaminant binding form patterns, employing, e. g., sequential extraction schemes [7, 8]. If results indicate the presence of labile contaminant pools (i.e. contaminant fractions accessible to relatively mild extractants), carefully selected additives may promote the formation of more recalcitrant binding forms by precipitation/sorption processes.

Contaminant stabilization schemes in soils are mostly derived from waste water treatment technologies and may be divided into precipitation-based and sorption-based approaches [9]. In either case, amendment addition aims at shifting solid-solution equilibria towards lowered aqueous phase concentrations. When contaminant release is controlled by desorption, equilibrium concentrations will establish according to the pertinent isotherm. When dissolution processes control contaminant release, equilibrium concentrations are limited by the solubility of host minerals.

Sorption-based stabilization relies on the addition of amendments with large specific surface area and high charge density. Depending on sorption affinity and uptake capacity aqueous phase contaminant concentrations may be substantially lowered by electrostatic interaction and/or surface complexation. For cationic contaminants clay minerals with permanent negative charge are important sorbents applicable in stabilization schemes [10]. Conversely, the retention capacity of soils for anionic contaminants may be enhanced by adding natural anion exchangers, e.g. iron (hydr)oxides [11].

Stabilization of contaminated soils by precipitation processes relies on the addition of soluble salts to provide counter-ions for the formation of sparingly soluble host minerals. One of the best studied precipitation-based stabilization schemes is the sequestration of lead as pyromorphite by addition of phosphate salts [12]. However, the formation of these minerals has also been questioned [13] and adverse side effects regarding the mobilization of co-contaminants upon addition of phosphate salts have been documented [14].

Asides the solid-solution equilibrium stabilization schemes need to consider process kinetics. Irrespective of the mechanism (precipitation/sorption) contaminant stabilization requires that the rate of immobilization is faster than the rate of release from the contaminant source [9].

Sample Size and Spatial Variability

The selection of additives is typically based on experimental feasibility studies of varying degree of sophistication. Efficiency may be inferred, among others, from comparative batch tests or miscible displacement experiments performed in the laboratory with treated vs. control soil samples [4, 15]. However, due to sample variability promising results at this level do not guarantee success during the implementation of a stabilization scheme at the full scale.

The up-/down-scaling of processes, or the question how information at one scale can be transferred to another, are subjects of ongoing research [16]. Scaling of contaminant mobilization and stabilization needs to consider the heterogeneous distribution of physical, chemical, and biological properties in soils. Heterogeneity has been defined as the degree to which a property or constituent is uniformly/non-uniformly distributed throughout a quantity of the material [17] and transport of fluids and solutes is affected by heterogeneity throughout a hierarchy of scales [18].

The heterogeneous distribution of hydraulic conductivity is responsible for strong deviations of solute transport between lab and larger scales since preferential flow paths may allow for a much faster water and solute movement than the aggregated soil particles. In miscible displacement experiments with monolithic soil samples De Rooij et al. [19] have shown that relatively small regions of the soil volume may control large proportions of solute leaching.

Asides the spatial distribution of hydraulic properties, geochemical heterogeneity, i.e. the uneven distribution of physico-chemical properties of the soil such as surface charge, mineral assemblage, organic matter content etc. may control solute transport [20]. Thus, experiment-based treatability studies for contaminated soils are strongly dependent on the use of representative samples which in turn requires the choice of an appropriate sample support, i.e., mass, volume, or orientation of samples [21].

If sample support is inadequate, conclusions on contaminant mobility in both the untreated and amended soil may be misleading and the efficiency of a stabilization scheme may be misjudged. This situation is depicted in Figure 1. Consider in the left panel a volume of soil with an uneven distribution of a material property relevant to contaminant leaching (e.g. distribution of contaminant hot spots) exposed to atmospheric conditions. If the seepage water draining from the whole volume were collected, this would have an overall concentration C^* that integrates over the heterogeneous distribution of soil properties. Now consider that samples are taken from this volume (right panel) to perform a leaching experiment. Obviously, the sample volumes that contain less or more than the average contaminant level tend to underestimate or overestimate the overall concentration. Only the sample block to the right, containing the average pollutant level may be expected to provide leachate concentrations approximately in the range of C^*.

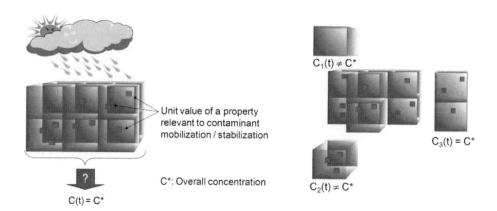

Figure 1. Uncertainties in contaminant transport and stabilization induced by heterogeneous distribution of soil properties that influence leachate concentrations in a contaminated soil.

From the geochemical viewpoint the choice of an appropriate sample support takes account of the fact that relevant constituents often reside in isolated mineral grains. The chance of such particles appearing in the study sample increases with sample size [22]. When sample size is limited by the dimensions of the experimental set-up, heterogeneity reduction may be achieved by sieving the samples and discarding the oversized grain. However, experimental findings obtained under such conditions may not be valid for larger scales when sieving causes a discrimination of relevant soil constituents.

Although much research has been dedicated to the stabilization of contaminated soils, knowledge gaps persist with respect to scaling. Reliable efficiency estimates should consider among others the effects of particle size fractionation and mixing technology as well as rate limitations to solute transport. This study focuses on the scaling of arsenic mobility and stabilization in a contaminated soil by a systematic comparison of leaching experiments with treated and untreated soil carried out on the lab and the pilot scales. Equilibration and elution curves obtained with packed bed volumes of 1.5 L and 1.5 m^3 were compared regarding the validity of lab scale tests for the design of a stabilization scheme based on the addition of iron (hydr)oxides from a drinking water purification plant.

MATERIALS AND METHODS

Site and Soil Characterization

The soil material used in this study originated from a former Paris Green production site in Bavaria, Germany.

Table 1. Selected physico-chemical soil properties

Solid phase	Unit	Value
As[1]	mg/kg	412
Cu[1]	mg/kg	279
ANC pH 4[2]	mmol/kg	349
Surface area	m^2/g	10
C_{org}[3]	g/kg	14
C_{inorg}[4]	g/kg	2
Fe_{DCB}[5]	mg/kg	7,390
$Fe_{oxalate}$[6]	mg/kg	4,860
Aqueous batch extract		
As	μg/L	4,260
Cu	μg/L	86.5
pH	–	8.3
EC[7]	μS/cm	90
E_H[8]	mV	380

[1] Aqua regia digestates
[2] Acid neutralizing capacity determined by 24 h of automated titration to pH 4
[3] Organic carbon according to dry combustion
[4] Inorganic carbon acc. to mass balance of carbon analysis with and without acid addition
[5] Iron in pedogenic (hydr)oxides extracted with dithionite/citrate/bicarbonate
[6] Iron in short-range ordered pedogenic (hydr)oxides extracted with ammonium oxalate
[7] Electrical conductivity
[8] Redox potential normalized to the standard hydrogen electrode

The contamination affected an area of 5,000 m², approximately, with a vertical extension of 4 to 6 m below soil surface. Reported average arsenic contents of the solid phase were in the range of 500 mg/kg with maximum groundwater concentrations of 43 mg/L. The soil developed from Quaternary sediments overlying Middle Triassic limestone [2]. The site was remediated by soil exchange and approximately 5 Mg of the material were provided by the contractor. The material was stored under ambient conditions in a lid-covered container prior to use in the experiments. Soil aliquots for basic characterization and lab scale experiments were obtained as a composed sample of 12 percussion drillings. The material from the drillings was air dried and sieved to < 2 mm. The oversized grain was discarded.

Selected solid phase properties of the soil are presented in Table 1 along with the contents (dry matter basis) and leachate concentrations of the priority contaminants as determined in aqua-regia extracts and batch tests (liquid-to-solid ratio: 10), respectively. The soil is characterized by moderate carbonate contents and acid buffering capacity along with a slightly alkaline pH. In agreement with the source composition arsenic and copper concentrations are increased. While arsenic concentrations in the batch extract indicate high mobility of this contaminant, copper mobilization potential is relatively low.

Ferric Amendment Preparation and Addition

Based on preliminary batch screening tests and in agreement with the well-documented affinity of As towards the surfaces of iron(hydr)oxides [23] deferrization sludge (DFS) was chosen as a soil amendment. The material was obtained at the Donauwörth waterworks, Germany. Groundwater of the region requires treatment in aerated trickle bed reactors to precipitate excess ferrous iron as hydrous ferric oxide. To avoid clogging, the reactors are intermittently back-flushed. The effluent of this procedure consists of a hydrous ferric oxide slurry.

From this slurry, 1 m³ was sampled in a PE container and left to settle for five days. The supernatant was withdrawn with a peristaltic pump and the thickened sludge was drained into a shallow basin (cf. Figure 2). After air-drying for four weeks the material was ground in a jaw breaker and stored at ambient temperature until used in the experiments.

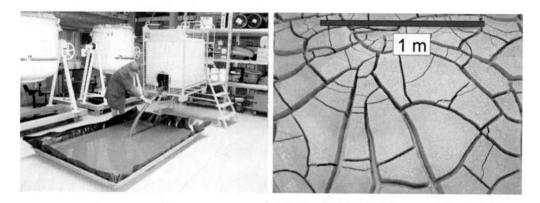

Figure 2. Preparation of deferrization sludge; draining of thickened sludge (left), air-dried product (right).

The specific surface area of the material is 250 m^2/g according to nitrogen adsorption (BET surface). Pedogenic iron content amounts to 218 g/kg as observed by dithionite/citrate/bicarbonate extraction [24]. The largely amorphous character of the oxides evidenced by X-ray powder diffraction suggests dominance of ferrihydrite as the iron-bearing phase.

A DFS level of 50 g/kg and of 15 g/kg was used in the lab and pilot scale study, respectively. In the lab scale study the treated soil was prepared by adding the additive and mixing it with a spatula. Homogeneity was visually controlled. An aliquot of the untreated soil was used for the control experiment. On the pilot scale the amendment addition was performed with a concrete mixer truck. The soil was weighted with a crane balance, delivered to the feed and homogenized in the revolving drum. After 15 min the revolving direction was reversed until the soil came to lie below the feed hopper and the required mass of amendment was added (cf. Figure 3). After another 15 min of mixing the product was delivered into flexible intermediate bulk containers (FIBC) for storage. The control soil was treated likewise but without amendment addition. In the pilot scale treatment eight subsamples from each of the treated soils were taken for quality control. The samples were analyzed for iron as a proxy for the DFS added. Variation coefficients were around 5 % indicating homogeneous mixing.

Figure 3. Feeding the soil from the FIBC into a concrete mixer truck for homogenization (left) and amendment addition (right) for the pilot scale experiments.

Column Experiments

The experiments were conducted as parallel leaching tests using computer controlled column devices (EMC GmbH, Germany; Kaiser and König GmbH, Germany). Lab scale experiments were performed with the fines fraction (< 2 mm) while pilot scale experiments were carried out with the bulk soil including course material (stones, construction debris etc. with diameters of up to 250 mm).

The soils were poured layer-wise into the reactors and compacted manually. A porous PE plate (ecotech, Bonn, Germany) and a strainer covered with a geotextile served as the base for the packed beds in the lab and pilot scale tests, respectively. Nylon suction cups (Eijkelkamp, The Netherlands; ecoTech, Germany) were horizontally inserted in three depths of the packed beds to allow for a sampling of pore water. In the lab scale experiments, 5 mL were taken from each depth and combined to a composite sample. In the pilot scale experiments the samples from each depth (500 mL) were analyzed separately.

The experiments involved three consecutive steps: (1) Water saturation and equilibration, (2) unsaturated percolation with tap water and (3) tracer application with breakthrough monitoring. The experimental conditions are summarized in Table 2.

Saturation was performed from bottom to top and maintained for five days to arrive at defined initial conditions. Intermittent pore water samples were taken to follow the evolution of aqueous phase composition. After five days the columns were drained and irrigation was started. On the lab scale the influent was delivered at a flow rate of 90 mL/h and distributed across the soil surface by a sprinkler equipped with 28 hypodermic needles. On the pilot scale a full cone nozzle (VKL 0.6 90-G1/8-PP, 90° spraying angle, Spraying Systems AG, Germany) was used to apply the influent at a flow rate of 20 L/h. Irrigation rates were chosen regarding homogeneous distribution of the influent and comparable Darcy velocities (ratio of irrigation rate and cross-sectional area).

The effluents were collected by autosamplers and analyzed for metals and metalloids, anions, DOC, and basic hydrochemical parameters. Separation of trivalent and pentavalent arsenic {As(III), As(V)} in suction cup samples was carried out by solid phase extraction with quaternary amine as described earlier [25]. Quantification limits of the method were 2 µg/L for As(III) and 1 µg/L for As(V).

Table 2. Conditions of the lab and pilot scale experiments

Parameter	Unit	Lab scale		Pilot scale	
		Control	Treatment	Control	Treatment
Packed bed mass	kg	2.50	2.50	2,824	2,822
Additive level	g/kg	–	50	–	15
Packed bed height	m	0.25	0.26	1.76	1.81
Cross-sectional area	m^2	$6.4 \cdot 10^{-3}$	$6.4 \cdot 10^{-3}$	0.85	0.85
Bulk density	Mg/m^3	1.56	1.50	1.89	1.80
Water content v/v	–	0.32	0.39	0.23	0.27
Water volume	m^3	$5.1 \cdot 10^{-4}$	$6.2 \cdot 10^{-4}$	0.351	0.414
Irrigation rate	m^3/s	$2.5 \cdot 10^{-8}$	$2.5 \cdot 10^{-8}$	$5.5 \cdot 10^{-6}$	$5.5 \cdot 10^{-6}$
Darcy velocity	m/s	$3.9 \cdot 10^{-6}$	$3.9 \cdot 10^{-6}$	$6.5 \cdot 10^{-6}$	$6.5 \cdot 10^{-6}$
Pore water velocity	m/s	$1.2 \cdot 10^{-5}$	$1.0 \cdot 10^{-5}$	$2.8 \cdot 10^{-5}$	$2.4 \cdot 10^{-5}$
Residence time	s	$2.1 \cdot 10^{4}$	$2.5 \cdot 10^{4}$	$6.3 \cdot 10^{4}$	$7.5 \cdot 10^{4}$

To compare the elution curves of the lab and pilot scale experiments, effluent concentrations were plotted against the number of pore volumes exchanged. This "dimensionless time" expresses the ratio of cumulative irrigation volume and resident water volume of the columns.

The number of pore volumes PV was calculated from the irrigation rate Q [m³/s], the time t_i [s] at which each of the n fractions was sampled, and the resident water volume V_W [m³] of the columns determined by weighting at the end of the experiments, see Equation 1.

$$PV = \frac{Q \cdot \sum_{i=1}^{n} t_i}{V_W} \tag{1}$$

RESULTS AND DISCUSSION

The evolution of arsenic concentrations over equilibration time since completed saturation is shown in Figure 4. The average arsenic pore water concentration of the control soils suggested that quasi-equilibrium conditions were reached within a few hours at a pore water pH of around 8.

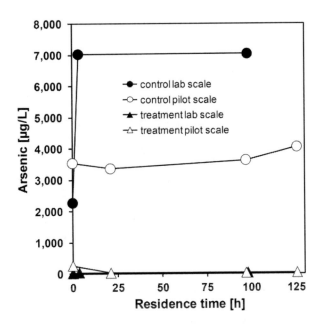

Figure 4. Evolution of average pore water arsenic concentration in the lab and pilot scale experiments during equilibration.

The arsenic levels observed under lab conditions approximately doubled the findings from the pilot scale test. Arsenic release was expected to be primarily dissolution-driven since Paris Green remnants were visually identified in the soil. Grain size reduction by sieving may have increased the availability of arsenic thereby resulting in higher pore water concentrations in the lab experiment.

In the DFS treatment strongly reduced arsenic pore water concentrations were observed. Average pore water concentrations of the treated soil (17 µg/L) correspond to a stabilization efficiency of 99.5 %, irrespective of scale. These findings corroborate the high affinity of arsenic to the surface of deferrization sludge.

In the pilot scale experiments larger pore water volumes could be sampled without affecting soil moisture conditions. Therefore, it was possible to separately analyze the suction cup samples taken from the upper, middle, and lower column sections. The results of the depth-resolved behavior of arsenic during the equilibration phase are shown in Figure 5. The arsenic concentrations of the control soil decreased over time in the lower and increased in the middle and upper column sections. A similar pattern was observed for calcium (not shown) with maximum concentrations of 120 mg/L and 70 mg/L in the upper and lower column sections at the end of the equilibration period.

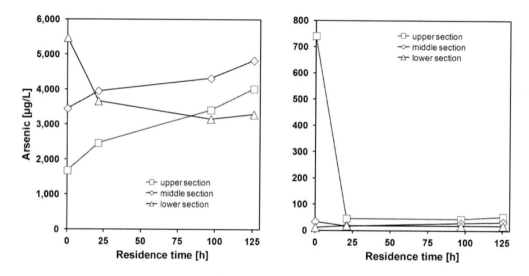

Figure 5. Depth-resolved evolution of pore water arsenic concentration in the pilot scale experiment during equilibration of the control (left) and treated soil (right).

In the treated soil stabilization was accomplished immediately after saturation in the lower and middle sections of the column. In the upper column section arsenic pore water concentrations approached equilibrium within 24 hours of saturation. This may be related to the fact effective residence time increased in the order lower > middle > upper section since saturation proceeded from bottom to top and time to saturation was 30 h, approximately. Thereby, the kinetics of the mixed arsenic release/stabilization process can only be recognized in the upper column section. These findings are in accordance with results from a systematic investigation of the As(III) and As(V) sorption kinetics on ferrihydrite. Depending on pH and arsenic level sorption was completed in the range of 2 h [26].

Pore water samples collected during the equilibration period of the pilot scale experiment were analyzed for arsenic species distribution by the method outlined above. The recovery {sum of As(III) and As(V) divided by arsenic in the original sample} was between 95 and 98 %. The As(III) pore water concentrations in the untreated soil were around 10 µg/L. In the DFS treated soil As(III) was below the limit of quantification (< 2 µg/L) and arsenic levels were exclusively due to As(V). Dominance of As(V) was also found in batch experiments with pure Paris Green aggregates isolated from the soil matrix (not shown). These results point to oxidative alteration of the production remnants since Paris Green contains arsenic in its trivalent form.

The leaching curves obtained after draining the columns and starting the irrigation are shown in Figure 6. The cumulative irrigation volumes of 22.6 L (lab scale) and 6.1 m³ (pilot scale) were equivalent to the exchange of about 45 and 16 pore volumes, depending on the resident water volume of the columns. To improve readability the leaching curves of the lab scale experiments were truncated at 16 pore volumes.

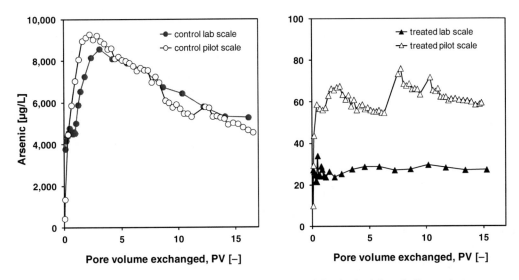

Figure 6. As leaching from the control (left) and treated soil (right) in the lab and pilot scale tests.

Arsenic leaching from the control columns peaked at about 2 to 3 pore volumes (Figure 6, left). In the lab experiment a maximum concentration of 8.5 mg/L was observed compared to 9.2 for the pilot test. These values are higher than those observed during equilibration and indicate that arsenic release from the solid phase was enhanced by the exchange of soil solution. This is supported by pore water samples intermittently collected during the irrigation period of the pilot scale test since after starting the irrigation pore water concentrations increased to a maximum of 7.7 mg/L (not shown).

In the effluent of both the lab and the pilot scale experiments a series constituents peaked in advance of arsenic after exchange of about 0.9 pore volumes (e.g. Ca: ~150 mg/L, Mg: ~20 mg/L, Na: ~30 mg/L, SO_4^{2-}: ~170 mg/L). Possibly, the depletion of these macro-constituents favored arsenic release in the control columns of the lab and pilot scale tests. If this hypothesis holds, arsenic pore water concentrations during the equilibration period may have been limited by the solubility of secondary minerals such as iron arsenates and/or calcium arsenates [27, 28].

Regarding the ratio of calcium to arsenic and the absolute concentrations observed during the equilibration periods it seems possible that in addition to the Paris Green degradation products the calcium arsenate minerals $Ca_5(AsO_4)_3(OH)$ or $Ca_4(OH)_2(AsO_4)_2·4H_2O$ controlled the solubility in the untreated soil [30]. This is consistent with (i) the concomitant release of arsenic and calcium in the depth-resolved pore water samples and (ii) the increase of arsenic effluent concentrations upon leaching of calcium.

Regarding the shape and level of the arsenic leaching curves of the control soils (Figure 6, left) results were almost identical for lab and pilot scale experiments. Such behavior requires two conditions to be met. First, the difference in pore water residence times (ratio of resident water volume and irrigation rate) between the lab and pilot scale must not influence arsenic release. Second, the soil constituents that control the macroscopic contaminant mobility need to be uniformly distributed across the different scales of soil aggregates. Pore water concentrations observed immediately after saturation indicate fast release kinetics. Thereby, the roughly three-fold pore water residence time of the pilot scale tests compared to the lab experiments (cf. Table 2) had no effect on contaminant mobilization. Thus the first condition was met suggesting that the second condition holds and sieving the soil for the lab scale experiments did not discriminated arsenic sources.

Contaminant leaching from the treated soil was strongly reduced confirming the results of the equilibration period (Figure 6, right). Effluent concentrations between 20 and 35 µg/L (lab scale) and between 40 and 60 µg/L (pilot scale) matched the pore water concentrations observed under saturated conditions. In the pilot scale experiment the treatment effect was inferior and yielded a two-fold to three-fold effluent concentration. This is consistent with the corresponding additive levels, in the pilot test only 15 g DFS per kg soil were used compared to 50 g per kg in the lab scale test.

To gain insight into the soil hydraulic properties of the control and treated soil the influent solution was switched to 10^{-2} M potassium bromide towards the end of the leaching experiments and the breakthrough of bromide was monitored in the effluent. Assuming bromide behaves as a conservative tracer, the breakthrough curves reflect the movement of water through the porous media. This may thus be used to infer whether preferential flow paths influenced arsenic leaching from the control and treated soils. Results of the tracer experiments are shown in Figure 7.

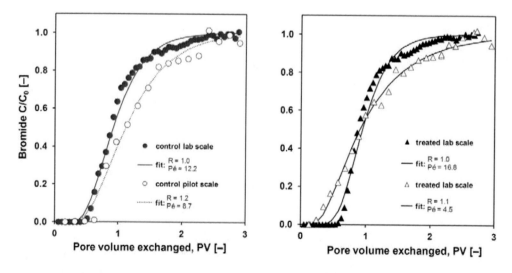

Figure 7. Tracer breakthrough curves (Br⁻, normalized concentration) in column experiments with the control and treated Paris Green soil depending on the study scale; R (retardation coefficient) and Pé (column Péclet number) were obtained by fitting using the numerical code CXTFIT [29].

The measured and fitted bromide breakthrough curves show good agreement. The Péclet numbers (Pé = $l_0 \cdot v / D$, with l: length, v: velocity, D: Dispersion coefficient, subscript 0: characteristic property) characterize a moderately convection-controlled flow regime. Smaller Péclet numbers (higher dispersivity) were obtained on the pilot scale both for the treated and the untreated soil due to the relation between mean particle diameter and the magnitude of dispersion. Overall, the shape of the breakthrough curves and successful fits of the convection-dispersion equation indicate that solute transport was not affected by preferential flow paths. Thus, the arsenic leaching curves give an overall information regarding both the contaminant mobilization and stabilization from the control and treated soils.

CONCLUSION

Scale dependence is not generally a limiting factor for the interpretation of soil column experiments as was demonstrated by the consistent arsenic leaching curves of the Emerald Green soil. When mobilization and immobilization is controlled by chemical equilibrium and contaminant sources and sinks are evenly distributed across aggregate fractions stabilization schemes may be adequately evaluated under lab scale conditions. High efficiency and successful up-scaling suggest applicability of a deferrization sludge-based stabilization of arsenic in Paris Green contaminated soils. Further research is required to optimize amendment levels in view of arsenic groundwater standards.

ACKNOWLEDGMENTS

The study was supported by the Bavarian State Ministry of the Environment and Public Health.

NOTATION

C	concentration, $[M/L^3]$
C_0	concentration at the inlet cross-section, $[M/L^3]$
D	dispersion coefficient, $[L^2/T]$
DOC	dissolved organic carbon, $[M/L^3]$
DFS	deferrization sludge
FIBC	flexible intermediate bulk container
i	index, [–]
l	length, [L]
l_0	characteristic length = height of column, [L]
n	number of fractions, [–]
PE	polyethylene
PV	pore volume exchanged, [–]
Q	irrigation rate, $[L^3/T]$
R	retardation coefficient, [–]

t   time, [T]

v   velocity, [L/T]

V_W water volume, $[L^3]$

Pé   Péclet number, $Pé = l_0 \cdot v / D$, [–]

REFERENCES

[1]   H. Andreas, *Chemie in unserer Zeit* 30, 23 (1996).

[2]   T. Hanauer, *"Untersuchungen zu natürlichen Schadstoffrückhalteprozessen in einem arsenbelasteten Grundwasserleiter"*, PhD thesis, Ruhr Universität Bonn, 177 pp (2008).

[3]   A. Chatterjee and A. Mukherjee, *Sci. Tot. Environ.* 225, 249 (1999).

[4]   C. Gemeinhardt, H. Weigand, S. Müller and C. Marb, *Water Air Soil Pollut.: Focus* 6, 281 (2006).

[5]   A. Al-Tabbaa and A. S. R. Perreira, *Land Contam. Reclamation* 11, 71 (2003).

[6]   J.-Y. Kim, A. P. Davis and K.-W. Kim, *Environ. Sci. Technol.* 37, 189 (2003).

[7]   A. Tessier, P. G. C. Campbell and M. Bisson, *Anal. Chem.* 51, 844 (1979).

[8]   H. Zeien, "Chemische Extraktionen zur Bestimmung der Bindungsformen von Schwermetallen in Böden", *Bonner Bodenkundliche Abhandlungen* 17, Institut für Bodenkunde, Bonn, Germany, 284 pp (1995).

[9]   H. Weigand, C. Gemeinhardt and C. Marb, *"Stabilising inorganic contaminants in soils: considerations for the use of smart additives"*, Proceedings of the International Conference on Stabilisation/Solidification Treatment and Remediation, University of Cambridge (UK), 12-13 April 2005, Balkema Publishers, Leiden, pp357-363 (2005).

[10] B. Lothenbach, G. Furrer, H. Schärli and R. Schulin, *Environ. Sci. Technol.* 33, 2945 (1999).

[11] X. Sun and H. E. Doner, *Soil Sci.* 163, 278 (1998).

[12] S. L. McGowen, N. T. Basta and G. O. Brown, *J. Environ. Qual.* 30, 493 (2001).

[13] M. Chrysochoou, D. Dermatas and D. G. Grubb, *J. Haz. Mat.* 144, 1 (2007).

[14] C. Spuller, H. Weigand and C. Marb, *J. Haz. Mat.* 141, 378 (2007).

[15] H. Weigand, T. Gretzbach, R. Bäumler and C. Marb, *"Stabilization of inorganic contaminants in lead crystal polishing sludge"*, Geotechnics of Waste Management and Remediation, Geotechnical Special Publication 177, American Society of Civil Engineers, Reston, USA, pp708-715 (2008).

[16] M. van der Perk, *"Soil and water contamination. From molecular to catchment scale"*, Taylor and Francis, London, 389 pp (2006).

[17] W. Horwitz, *Pure Appl. Chem.*, 62, 1193 (1990).

[18] C. E. Koltermann and S. M. Gorelick, *Water Resour.Res.*, 32, 2617 (1996).

[19] G. H. de Rooij, O. A. Cirpka, F. Stagnitti, S. H. Vuurens and J. Boll, *Vadose Zone J.*, 5, 1086 (2006).

[20] J. Y. Chen, C.-H. Ko, S. Bhattacharjee and M. Elimelech, *Colloids and Surfaces A: Physicochem. Engin. Aspects* 191, 3 (2001).

[21] T. H. Starks, *Math. Geol.* 18, 529 (1986).

[22] C. O. Ingamells and P. Switzer, P. *Talanta* 20, 547 (1973).

[23] S. Goldberg, *Soil Sci. Soc. Am. J.*, 66, 413 (2002).

[24] O. P. Mehra and M. L. Jackson, *"Iron oxide removal from soils and clays by a dithionite-citrate system buffered with sodium bicarbonate"*, Proc. 7th National Conference on Clays and Clay Minerals, Washington D. C., pp317-327 (1960).

[25] H. Weigand, I. Argut, C. Marb, C. Koch and J. Diemer, *J. Plant Nutr. Soil Sci.*, 170, 250 (2007).

[26] K. P. Raven, A. Jain and R. H. Loeppert, *Environ. Sci. Technol.*, 32, 344 (1998).

[27] P. M. Dove and J. D. Rimstidt, *Am. Mineralogist*, 70, 838 (1985).

[28] J. V. Bothe Jr. and P. W. Brown, *Environ. Sci. Technol.* 33, 3806 (1999).

[29] N. Toride, F. J. Leij and M. Th. van Genuchten, *"The CXTFIT code for estimating transport parameters from laboratory or field tracer experiments, version 2.1."*, Research Print 137, U. S. Salinity Laboratory Riverside, California, USA, 121 pp (1995).

[30] Y. N. Zhu, X. H. Zhang, Q. L. Xie, D. Q. Wang and G. W. Cheng, *Water, Air, Soil Pollution,* 169, 221 (2006).

In: Arsenic
Editor: Andrea Masotti

ISBN: 978-1-62081-320-1
© 2013 Nova Science Publishers, Inc.

Chapter 5

ECOTOXICOLOGY OF ARSENIC IN THE FRESHWATER ENVIRONMENT: CONSEQUENCES AND RISK ASSESSMENT

M. Azizur Rahman[1], Christel Hassler[2], Hiroshi Hasegawa[3] and Richard Lim[1]

[1]Centre for Environmental Sustainability, School of the Environment, Faculty of Science, University of Technology Sydney, Australia
[2]Plant Functional Biology and Climate Change Cluster, School of the Environment, Faculty of Science, University of Technology Sydney, Australia
[3]Institute of Science and Technology, Kanazawa University, Kakuma, Kanazawa, Japan

ABSTRACT

Arsenic is a known environmental toxicant and it occurs in the environment from natural and anthropogenic sources. Arsenic is one of the important environmental issues because of its occurrence, bioaccumulation, toxicity, and trophic transfer in the freshwater food chain. Aquatic organisms accumulate, retain, and transform arsenic when exposed to it through water, their diet, and other routes. Since arsenic toxicity mostly depends on its chemical forms, measurement of arsenic speciation in aquatic organisms is particularly important in assessing the ecological risks of the element. Arsenate (As(V)) comprises the major part of total arsenic in oxic waters. Phytoplankton take up As(V) and subsequently convert it to arsenite (As(III)) and then to less toxic dimethylarsinic acid (DMAA), monomethylarsonic acid (MMAA), and higher order organoarsenicals. Phytoplankton are thought to convert inorganic arsenic species to methylarsenicals and to other organoarsenic compounds (lipids and arsenosugars) to reduce the toxic effects of inorganic arsenicals. Since phytoplankton are a major food source for the organisms of higher trophic levels in the aquatic systems, arsenic is biotransferred from lower to higher trophic levels; while biomagnification of the element in aquatic food chain is not consistent. Other important arsenic forms found in aquatic organisms include arsenocholine (AsC), arsenobetaine (AsB) and arsenosugars (AsS). This review discusses the bioaccumulation, biotransformation, and trophic transfer (biomagnification or diminution) of arsenic in the aquatic food chains in relation to its ecotoxicological risks in the freshwater environment.

1. INTRODUCTION

Ecological risk assessment (ERA) is a process that evaluates the potential for adverse ecological effects occurring as a result of exposure to contaminants or other stressors [1]. The key information required for ERA are the occurrences, exposure pathways, rates of movement of contaminants in the environment, and information on the relationship between contaminant concentrations and the incidence and/or severity of its adverse effects on organisms. Because of the different specific properties and characteristics of environmental contaminants/ stressors, a generalized ERA is not appropriate for all contaminants [1]. Arsenic is one of the well-studied environmental toxicants, which has been found in a wide range of aquatic and terrestrial organisms. Certain metals (*e.g.* copper, iron, zinc etc.) are essential for the biotic health of organisms, and there is an effect threshold for both deficiency and excess of essential metals. Although certain bacteria appear to gain metabolic energy from arsenate under aerobic growth conditions [2, 3], arsenic is not an essential element for the biotic health of most organisms; rather, it produces toxicity by disrupting ATP production through several mechanisms (the details have been discussed in a number of reviews e.g. Jomova et al. [4], Valko et al. [5]). Therefore, it does not have an effect threshold for deficiency, but its excess may produce high toxicity to organisms.

Arsenic can occur in the environment in a variety of forms (inorganic and organic, Fig. 1), and the bioavailability and toxicity of these forms differ [1]. The bioavailability of metals is generally controlled by the external environmental conditions and the biological activities of organisms (mainly for essential metals, e.g. Worms et al., 2006 [6]). Freshwater organisms accumulate, retain, and transform arsenic species inside their bodies when exposed to arsenic through water, their diet, and other routes [7-10]. Some organisms such as phytoplankton and bacteria may have special mechanisms by which they transform inorganic arsenic species to less toxic methylarsenicals (MMAA and DMAA) and/or high-order organoarsenic species such as arsenosugars [11, 12] (Figure 1). Bacteria and yeast detoxify arsenic by reducing As(V) to As(III) by arsenate reductase enzymes (the bacterial ArsC or yeast Acr2p enzymes), and then extruding the toxic oxyanion As(III) from the cell by inducible and selective transporters (ArsB alone or by the ArsAB ATPase) [13, 14]. Although methylation has been considered to be a primary detoxification process of inorganic arsenic [3, 15], a growing body of literature suggests that not all methylated products are less toxic than the inorganic forms of arsenic [3]. For example, MMAA(III) has been reported to be more toxic than inorganic arsenic in Chang human hepatocytes [15].

The extent of arsenic toxicity depends on various factors such as the speciation, dose, individual susceptibility to arsenic and the type of organisms exposed [4]. The toxicity of arsenic species to lake phytoplankton generally decreases in the order of: As(V) > As(III) > DMAA > MMAA [16]. In some cases, organometallic compounds are less toxic than inorganic forms, and therefore, conversion of As(V) to As(III) followed by excretion and biomethylation of inorganic arsenicals are assumed to be the main detoxification/defence mechanisms in phytoplankton [16, 17]. Other organisms can excrete arsenic to avoid the occurrence of toxic effects (e.g. clams, [18]). However, detoxification mechanisms are not common to all aquatic organisms and hence, they are important for the description of arsenic toxicity and the subsequent effects on the functioning of aquatic systems.

Trophic transfer of arsenic has also been reported by a number of researchers [9, 19-22]. Biomagnification (a process whereby contaminant concentrations increase in aquatic organisms for each successive trophic level due to increasing dietary exposures, e.g. increasing concentrations from algae, to zooplankton, to forage fish, to predator fish), of arsenic has been demonstrated in some aquatic food chains [19, 23]. However, other studies have not shown evidence of arsenic biomagnification in [24]. Therefore, distribution, speciation, bioaccumulation and metabolism, and trophic transfer of arsenic in aquatic food chain are important for the determination of its ERA.

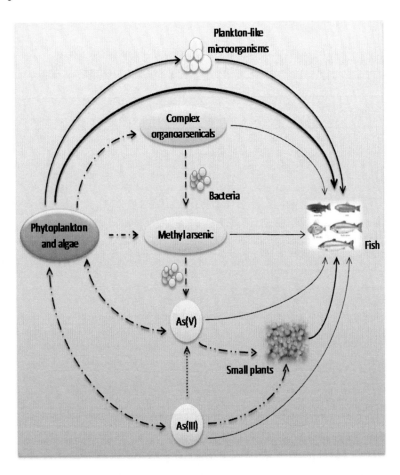

Figure 1. The role of biological activities in the occurrence and biotransformation of arsenic species in aquatic systems. The dotted arrow represents chemically driven process (e.g. oxidation of As(III) into As(V) in oxygenated waters), dashed arrows represent the bacterial transformation of arsenic (e.g. demethylation and remineralisation). The dashed pointed arrows represent biological uptake or release of arsenic forms in water. Arsenic release in water could occur from phytoplankton lysis mediated by viruses, bacteria and grazing by other planktonic microorganisms. Bold arrows schematically represent trophic interactions between phytoplankton, plankton-like microorganisms and small plants to fish (and other aquatic animals) that are relevant for arsenic biomagnification and expression of toxicity. As illustrated here arsenic cycling affects its toxicity and is mainly dominated by biological transformations and interactions.

2. SOURCES AND CONCENTRATIONS OF ARSENIC IN AQUATIC SYSTEMS

Land masses are the main reservoir of arsenic (10^{10} kt), representing 10^4- and 10^{10}-fold more than the arsenic present in the ocean and the atmosphere, respectively [25]. Not surprisingly, the most important fluxes of arsenic are from mining (> 200 kt yr^{-1}), volcanic activities (22 kt yr^{-1}), and human emissions (25 kt yr^{-1}) [25]. Therefore, the main contamination sources of aquatic arsenic are from geogenic and anthropogenic sources, and the occurrence, distribution, and speciation of arsenic in aquatic systems are related to its sources. Arsenic concentrations in aquatic systems are affected by industrial and mining effluents, geothermal water [26], and biological (especially phytoplankton and bacteria) activities [27].

**Table 1. Arsenic concentrations in aquatic systems
(rivers, lakes, and estuaries and marine)**

Aquatic systems and Location	Arsenicconcentrations (average/range(μg L^{-1}))	References
Rivers		
Dordogne, France	0.7	[90]
Po River, Italy	1.3	[91, 92]
Cordoba, Argentina	7-114	[93]
Madison and Missouri rivers, USA	44 (19-67), 10-370	[94, 95]
Waikato, New Zealand	32 (28-36)	[95, 96]
Ron Phibun, Thailand	218 (4.8-583)	[97]
Owens river, CA, USA	85-153	[98]
Lakes		
Lake Echols, Tampa	3.58	[99]
Moira Lake, Ontario, Canada	20.4 (22.0-47.0)	[33]
Lake Biwa, Japan	2.2 (0.6-1.7)	[39]
Estuaries		
Vestfjord, Norway	0.7-1.0	[100]
Bunnefjord, Norway	0.5-1.9	[100]
Saanich Inlet, B.C., Canada	1.2-2.5	[101]
Uranouchi Inlet, Japan	22.0-32.0	[27]
Rhone Estuary, France	2.2 (1.1-3.8)	[90]
Krka Estuary, Yugoslavia	0.1-1.8	[102]
Tamar Estuary, UK	2.7-8.8	[103]
Schelde Estuary, Belgium	1.8-4.9	[104]
Marine		
Deep Pacific and Atlantic	1.0-1.8	[28]
McKay Bay	1.77	[99]
Coastal Malaysia	1.0 (0.7-1.8)	[105]
Southeast coast, Spain	1.5 (0.5-3.7)	[30]
Coastal Nakaminato, Japan	3.1	[106]
Southern coast, Australia	1.3 (1.1-1.6) (inorganic)	[31]

Although arsenic concentrations in aquatic systems (rivers, lakes, estuarine and marine) vary greatly depending on source of occurrence, geochemistry and other environmental conditions [26], the typical levels are mostly within the µg L^{-1} range [28]. The arsenic concentration in open marine systems is around 1.5 µg L^{-1}. Ranges of arsenic concentrations have been reported to be between 1.0 - 1.8 µg L^{-1} in deep Pacific and Atlantic waters [28, 29], 1.5 µg L^{-1} in the southeast coast of Spain [30], and 1.1 - 1.6 µg L^{-1} in coastal waters of southern Australia [31]. Arsenic concentrations in estuarine waters are lower than that in freshwater systems, but higher than those in marine systems (Table 1). Baseline concentrations of arsenic in freshwater systems (lakes and rivers) range between 0.1 and 2.1 µg L^{-1}, though very high concentrations of arsenic have been reported in some contaminated freshwater systems (Table 1). The occurrence of arsenic in marine systems is natural, and is not a toxicological problem to marine organisms. On the other hand, high levels of arsenic in freshwater systems are derived mainly from geothermal inputs, evaporation, groundwater contamination (e.g., Lao River of northern Chile [32]) and mining activity (e.g., Clubs Lake, BC, Canada [33] and Mole River, NSW, Australia [34]), and therefore represent a toxicological problem for freshwater organisms. For this reason, only arsenic in freshwater systems (from lake to groundwater) is discussed further.

3. ARSENIC SPECIATION AND CONCENTRATION

Arsenic has a complex chemistry in freshwater systems influenced by pH, redox potential, and ligands, which affect the metal speciation and coexisting cations (*e.g.* H$^+$, Ca^{2+}) which compete with the metal ions. In the environment, arsenic exists in four oxidation states- arsenate (As(V)), arsenite (As(III)), arsenic (As(0)), and arsine (As(-III)) [35]. The two major inorganic As forms are As(V), which is the thermodynamically stable state in oxic water, and As(III) which is predominant in reduced redox conditions [36]. In freshwater, inorganic As(V) usually dominates with a ratio As(V) to As(III) from 5 to 100 depending on input, redox potential, pH and biological activity [37]. The As(III) form dominates in critically-contaminated groundwater such as in the Bengal Basin [38].

Microorganisms also influence arsenic chemistry in aquatic systems (Fig. 1). Due to differences between the biological transformations occurring in freshwater and marine systems, data on marine systems are included to gain further insight into the impact of biological activity on arsenic chemistry. Freshwater phytoplankton produce As(III), some of which is then converted to often less toxic methylarsenicals and higher-order organoarsenicals. As(III), methylarsenicals, and more complex organoarsenicals are then released to the water following the death of the organisms (e.g. cell lysis promoted by viruses, bacteria and grazers). Therefore, organic forms of arsenic present in freshwaters are mainly derived from the biological activity of phytoplankton [39, 40]. In summer, when the biological activity of phytoplankton is high, organoarsenicals (mainly methylated and ultraviolet-labile fractions) comprise 30-60% of the total arsenic in lake waters. In contrast, inorganic arsenic species (As(III+V)) dominate (about 60-85%) during winter, when the bioactivity of phytoplankton is less [39]. Although the concentrations of MMAA and DMAA in marine systems are generally low, these are the significant forms of arsenic in highly productive freshwaters [29]. The occurrence of hidden arsenic (possibly complex organoarsenicals) in freshwaters has also been reported by a number of researchers [41-44].

Table 2. Chemical forms of arsenic found in aquatic systems (water and organisms)

Name	Abbreviation	Formula/Structure	Reference
Inorganic arsenicals			
Arsenious acid or arsenite	As(III)	$As^{3+}(OH)_3$	[19, 20, 60, 107-109]
Arsenic acid or arsenate	As(V)	$H_3As^{5+}O_4$	[19, 20, 60, 107-109]
Methylated arsenicals			
Monomethylarsonous acid	MMAA(III)	$CH_3As(OH)_2$	[10, 20, 108, 110]
Dimethylarsinous acid	DMAA(III)	$(CH_3)_2AsOH$	[10, 20, 66, 108-110]
Monomethylarsonic acid	MMAA(V)	$AsO(OH)_2CH_3$	[10, 20, 108-110]
Dimethylarsinic acid	DMAA(V)	$AsO(OH)(CH_3)_2$	[10, 20, 60, 66, 108, 110]
Trimethylarsine acid	TMAA	$(CH_3)_3As$	[11, 111, 112]
Organoarsenic Compounds			
Arsenocholine	AsC	$(CH_3)_3As^+CH_2CH_2O$	[113-115]
Arsenobetaine	AsB	$(CH_3)_3As^+CH_2COO^-$	[20, 60, 108, 113, 116, 117]
Arsenosugars	AsS		
Arsenoribosides			[60]
Sulfate arsenoribose			[60]
Sulfonate arsenoribose			[60]
Phosphate arsenoribose			[60]
Glycerol arsenoribose			[60]
Dimethylarsinoylribosides			[117-119]
Trimethylarsonioribosides			[11]
Triaklylarsonioribosides			[117, 119]

Freshwater angiosperms contain mostly inorganic arsenic (As(V) and As(III)) [22, 45-49] and little methylated species [45]. Arsenobetaine is mostly found in marine fishes, crustaceans, molluscs, and polychaetes, while inorganic arsenic predominates in freshwater fishes [50]. The major arsenic species found in aquatic systems are summarised in Table 2.

4. BIOAVAILABILITY AND BIOACCUMULATION

The toxicity of arsenic and its health risks to organisms do not always correlate with the total external exposure dose since virtually all risk estimates ignore the bioavailability of the element in the assessment process. Bioavailability is usually determined by dosing experimental organisms with various concentrations of arsenic and measuring their sub-lethal or lethal responses. Bioaccumulation of arsenic refers to its net accumulation by an aquatic organism as a result of uptake from its environment, and is often use as a surrogate for bioavailability, as arsenic needs first to be accumulated to exert deleterious effects by reacting with sensitive biological sites (or biotic ligands, Worms et al., 2006 [6]). Key determinants of arsenic bioavailability are (i) its release from the environmental matrices, (ii) the chemical forms of the arsenic, with their associated reactivity and solubility, and (iii) the complex interactions between biological and chemical factors [51].

In natural waters, the dominant inorganic arsenic species are incorporated into microorganisms such as phytoplankton by a phosphate transporter, and some of the arsenic species are converted to organoarsenicals [12]. Bacteria also accumulate dissolved As(V) via the phosphate transport system, and excrete it rapidly by another active transport system to prevent net accumulation of As(V) inside the cell [52, 53]. In fact, depletion of phosphate concentrations in surface water can be associated with decreased As(V) concentrations in biologically productive oxic aquatic systems [54-56].

Arsenic bioaccumulation in aquatic organisms/biota is related to the types of organisms, their habitat and many other environmental conditions such as the environmental compartment (water column, sediment pore water), sediment particle type and size, pH, and the presence of other metals [51]. Bioaccumulation of arsenic in aquatic organisms, such as algae and lower invertebrates that are consumed by predator fishes, has been reported in the literature [21, 23, 57, 58]. Rooted aquatic angiosperms are presumed to bioaccumulate comparatively higher amounts of arsenic than unrooted submerged plants since rooted plants are directly associated with sediments, however macrophytes tended to bioaccumulate more arsenic compared to other aquatic flora [45, 56, 59].

Epiphytic algae and fungi associated with mangrove fine roots had higher arsenic concentrations than those on the main roots of mangrove plants [60]. The concentrations of arsenic in detritivores were significantly higher (8.5-55 $\mu g\ g^{-1}$) than those in the major primary producers (0.3-1.5 $\mu g\ g^{-1}$), herbivores (8.0-14.0 $\mu g\ g^{-1}$) and omnivores (2-16.6 $\mu g\ g^{-1}$) [60]. Other studies [49] also reveal that the range of arsenic concentrations in gastropods, crabs, and amphipods of salt marsh ecosystems were similar to those reported in marine herbivorous gastropods and crabs [60], but lower than those normally found in carnivorous gastropods [61]. Thus, arsenic accumulation in aquatic organisms cannot be attributed to their position in the food web or their feeding mode, but is likely to be related to their dietary intake and ability to assimilate, metabolize, and retain arsenic species inside their body [60].

5. BIOACCUMULATION FACTOR

The bioaccumulation factor (BAF) is the ratio of the concentration of a chemical in aquatic organisms in a specified trophic level where both the organism and its food are exposed, to the concentration of the chemical in water [50, 62]. The United States Environmental Protection Agency (USEPA) presented a methodology and guidelines for the estimation of BAFs for various contaminants to reflect the uptake of contaminants by aquatic organisms such as fishes, shellfish, etc. from all sources (e.g., foods, sediment, etc.) rather than just from the water column [50]. The BAF for arsenic can be calculated as:

$$BAF = \frac{C_{tAs}}{C_{wAs}} \tag{i}$$

where; C_{tAs} is the total arsenic concentration in wet tissue (whole organism or specific tissue) and C_{wAs} is the total arsenic concentration in water. The BAF for arsenic in aquatic organisms can be derived from available field data or can be predicted from an acceptable laboratory-measured bioconcentration factor (BCF) using equation (i).

A number of field and laboratory studies have been conducted to estimate BAFs for arsenic in aquatic systems [50, 57, 63-66]. For example, Spehar et al. [63] estimated BCFs for arsenic in freshwater invertebrate species and for rainbow trout exposed to As(III), As(V), DMAA, or MMAA for 28 days. The results showed that stoneflies, snails, and daphnids accumulated greater amounts of arsenic than fish. Henry [50] calculated arsenic BAFs for freshwater lotic organisms from field data of Mason et al. [58], and found that BAFs for herbivorous aquatic insects from western Maryland were higher (393-5619) than for other feeding groups. Chen et al. [65] studied the bioaccumulation and fate of arsenic in large and small zooplankton from numerous lakes in the north-eastern USA, and found that BAFs for arsenic in small zooplankton and large phytoplankton were significantly higher (369-19,487) than those of larger zooplankton (154-2,748).

Wagemann et al. [64] studied arsenic concentrations in several aquatic invertebrate species of lakes in the vicinity of the Yellowknife area (Canada), and the BAFs for arsenic in various invertebrates were found to be consistently higher (28-378) in Grace Lake (reference lake) than in the same species (3-64) in Kam Lake (contaminated lake). This result shows that the BAF for arsenic in an aquatic system might not be related to the extent of contamination; rather, it is more closely related to the bioaccumulation ability of the element by organisms. Due to the importance of environmental factors (physical, chemical, biological (see section 4, Fig. 1)), BAFs for a given species are likely to be dependent on the collection site, as was observed in coastal marine environments in several studies [50, 67, 68].

6. Ecotoxicological Risks of Arsenic Species for Aquatic Organisms

The total arsenic concentration in aquatic environments is not the only determinant required when assessing its ecological risks, since the toxicity is highly dependent on its chemical speciation.

Various physicochemical and biochemical factors influence the bioaccumulation and toxicity of different arsenic species in aquatic organisms [69]. Phosphate concentration, pH, redox potential, temperature and light intensity are important factors which affect arsenic biouptake and toxicity [70, 71]. The ability of aquatic organisms to biotransform and detoxify arsenic also influence its toxicity [20, 71, 72]. For example, in the green microalga, *Stichococcus bacillaris,* the accumulation and toxicity of As(V) and As(III) has been found to depend on environmental variables such as pH, humic acid, chloride and orthophosphate ions [70].

6.1. Algae

Although inorganic arsenic tends to be far more toxic than organic arsenic for organisms [4, 5, 73], Petrick et al. [15] reported that MMAA(III) was the most toxic species for Chang human hepatocytes. The toxicity of arsenic species may not be consistent for all aquatic organisms, since it is related to the individual susceptibility to arsenic and the defence mechanisms of the organism [4]. In general, the toxicity of arsenic species to lake phytoplankton decreases in the order of As(V) > As(III) > DMAA > MMAA [16].

In general, marine microalgae are more sensitive to As(III), while freshwater algae are highly sensitive to As(V) [16, 71, 74]. However, the marine algae *Dunaliella* sp. and *Polyphysa peniculus* are more sensitive to As(V) than to As(III) [75, 76]. On the other hand, As(V) and As(III) have been reported to show equal toxicity to the freshwater alga *Stichococcus bacillaris* at pH 8.2 with phosphate levels between 0.03 and 0.3 mg L^{-1} [70]. Equal toxicity of As(III) and As(V) was also reported for the freshwater alga *Chlorella* sp. at pH 7.6 in a 72-h growth inhibition test, while As(V) was found to be more toxic than As(III) for *Monoraphidium arcuatum* at the same pH [74]. Lower pH has been reported to increase As(V) toxicity to *S. bacillaris* [70], but decrease As(III) accumulation in *Chlorella vulgaris* [77]. Thus pH is also an important factor influencing arsenic toxicity. Mature cystocarp formation was inhibited in the macroalga *Champia parvula* when exposed to 0.095 mg L^{-1} As(III), and the alga was unable to survive when exposed to 0.30 mg L^{-1} As(III) [78]. Phosphate concentration also influences the toxicity of arsenic species to freshwater organisms. Low levels of orthophosphate (0.03 – 0.3 mg L^{-1}) were not found to influence As(V) toxicity to *S. bacillaris* [70]; however a higher concentration of 0.5 mg L^{-1} phosphate decreased the toxicity for the freshwater algae *Chlorella* sp. and *M. arcuatum* [74]. In contrast, a high phosphate concentration (9.1 mg L^{-1}) was not found to be protective against arsenate stress for *Chlorella* sp. isolated from arsenic-contaminated sites [79].

6.2. Animals

Limited data have published for the toxicity of arsenic species to freshwater animals, but some data are available for estuarine and marine species. The acute toxicities (LC_{50}) of As(III) to the estuarine amphipods *Corophium insidiosum* and *Elasmopus bampo* were 1.1 and 2.75 mg L^{-1}, respectively [80]. The median lethal concentration (96-h LC_{50}) of As(III) to juvenile bay scallops *Argopecten irradians* was 3.49 mg L^{-1} [81], while it was 17 mg L^{-1} to the crab *Scylla serrata* [82].

In contrast, the acute lethal concentration of As(V) to a mysid *Mysidopsis bahia* was 2.319 mg L^{-1} [72]. In a study by Madsen [83], As(V) concentrations exceeding 25 mg L^{-1} were found to affect survival of adult shrimp *Crangon crangon*. The respiration rates of the shrimp decreased during short-term exposures to 10 to 50 mg L^{-1} As(V). Growth of *Dunaliella* sp. was inhibited when exposed to 0.1 mg L^{-1} or higher concentrations of As(V) for 3 days, but during continued exposure, growth rate recovered and became normal after 12 days at concentrations up to 2 mg L^{-1}. However, small individuals were more sensitive than are large individuals [72].

7. ECOTOXICOLOGICAL MODELS

Several models have been developed to predict the bioavailability and toxicity of pollutants and trace elements. Firstly, the Free Ion Activity Model (FIAM, Morel, 1983 [84]) assumes a pseudo-equilibrium between the surrounding water and the surface of the organisms, and the toxicity can be predicted from any chemical forms present in the water, including the free ion (e.g., As(V)). The Biotic Ligand Model (BLM, Pagenkopf, 1983 [85]) is an extension of the FIAM that was developed to especially consider competitive interactions such as the one reported for As(V) and phosphate bioaccumulation for many organisms (see references [85]). Implementation of these models in the evaluation of water toxicity (e.g., Environmental Protection Agency for the BLM) represents a real advance in defining the ecotoxicological status of our water. Both models rely on similar assumptions that limit their applicability (see [86, 87] for formulation and critical reviews). For several reasons, the applicability of these models will be limited for arsenic:

1) Organic arsenic is lipophilic and will, therefore, rapidly move into organisms by passive diffusion through the biological membrane, a transport pathway ignored in these models [87].
2) Several organisms (e.g. bacteria, clams) are able to excrete the toxic As(V) form. Active excretion and biological homeostasis is simply ignored in these models [86].
3) Given that the organic forms of arsenic are generally much less toxic than inorganic arsenic (e.g., Tamaki and Frankenberger, 1992 [88]; Qin et al., 2009 [89]), the rapid diffusion of organic forms is likely not important in predicting arsenic toxicity. In fact, a recent study demonstrated that coupling the BLM with a damage assessment model (DAM), which considers arsenic excretion and recovery from damage, was successful in predicting arsenic toxicity in a freshwater clam [18]. This suggests a decoupling between arsenic bioaccumulation and its toxicity.
4) These models, especially the FIAM, are valid only if arsenic is accumulated by or taken up from water. Other pathways of arsenic accumulation (e.g. diet, particle ingestion/uptake) are not considered.

CONCLUSION

Despite model developments and the recent success in describing arsenic toxicity for a single species, we still lack understanding of the complexity of natural systems. As shown here, there is no universal trend in biomagnification or BAFs. Currently there is no research being undertaken on ecotoxicological models to predict arsenic toxicity in freshwater systems. However, considering the widespread contamination of arsenic in freshwater systems, especially in arsenic-contaminated South and South-East Asian countries, determination of arsenic toxicity and its ecological risks are important for aquatic organisms and humans as well.

Arsenic chemistry and biological transformations in freshwater systems are complex and critical in defining its toxicity. These depend on the arsenic source, cycling, and numerous interactions between organisms (e.g., Figure 1), rendering ERA of arsenic difficult. This review shows that the following aspects need to be further investigated to provide a mechanistic understanding of arsenic toxicity in freshwater systems:

➤ The biological cycling of arsenic (Figure 1) with an emphasis on bacterial demethylation/remineralisation, biotransformation inside phytoplankton, and turn-over rates;
➤ The link between arsenic chemical species and their toxicity; and
➤ The impact of trophic transfer and biomagnification (if it occurs) on arsenic toxicity.

Only when the dynamics of arsenic biotransformation in freshwater systems and their control of arsenic toxicity are better understood, can we improve the existing ecotoxicological models to predict the toxicity of arsenic species and their real threat to organisms. Therefore-

1) Studies under well-controlled environmental conditions in the laboratory are required to address clearly-defined working hypotheses to shed light on the mechanisms of arsenic toxicity through the freshwater food chain. Representative organisms of different trophic levels of the food chain should be carefully selected for these studies.
2) Multidisciplinary holistic approaches are required to understand arsenic toxicity and ecological risks in complex natural systems.

ACKNOWLEDGMENTS

The authors wish to thank the University of Technology, Sydney (UTS) for financial support in preparing this review. We are also pleased to acknowledge Dr. Anne Colville (School of Environment, UTS, Australia) and Prof. William Maher (Institute for Applied Ecology, University of Canberra, Australia) for reviewing the manuscript.

REFERENCES

[1] Chapman, P. M.; Wang, F., Issues in ecological risk assessment of inorganic metals and metalloids. *Human and Ecological Risk Assessment* 2000, *6*, 965-988.

[2] Anderson, C. R.; Cook, G. M., Isolation and characterization of arsenate-reducing bacteria from arsenic-contaminated sites in New Zealand. *Current microbiology* 2004, *48*, 341-347.

[3] Stolz, J. F.; Basu, P.; Santini, J. M.; Oremland, R. S., Arsenic and selenium in microbial metabolism. *Annual Review of Microbiology* 2006, *60*, 107-130.

[4] Jomova, K.; Jenisova, Z.; Feszterova, M.; Baros, S.; Liska, J.; Hudecova, D.; Rhodes, C. J.; Valko, M., Arsenic: toxicity, oxidative stress and human disease. *Journal of Applied Toxicology* 2011, *31*, 95-107.

[5] Valko, M.; Morris, H.; Cronin, M. T. D., Metals, toxicity and oxidative stress. *Current Medicinal Chemistry* 2005, *12*, 1161-1208.

[6]   Worms, I.; Simon, D. F.; Hassle, C. S.; Wilkinson, K. J., Bioavailability of trace metals
       to aquatic microorganisms: Importance of chemical, biological and physical processes
       on biouptake. *Biochimie* 2006, *88*, 1721-1731.
[7]   Kuroiwa, T.; Ohki, A.; Naka, K.; Maeda, S., Biomethylation and biotransformation of
       arsenic in a freshwater food chain: Green alga (*Chlorella vulgaris*) shrimp (*Neocaridina
       denticulata*) killifish (*Oryzias iatipes*). *Applied Organometallic Chemistry* 1994, *8*, 325-
       333.
[8]   Maeda, S.; Ohki, A.; Tokuda, T.; Ohmine, M., Transformation of arsenic compounds in
       a freshwater food chain. *Applied Organometallic Chemistry* 1990, *4*, 251-254.
[9]   Suhendrayatna, A. O.; Maeda, S., Biotransformation of arsenite in freshwater food
       chain models. *Applied Organometallic Chemistry* 2001, *15*, 277-284.
[10]   Hasegawa, H.; Sohrin, Y.; Seki, K.; Sato, M.; Norisuye, K.; Naito, K.; Matsui, M.,
       Biosynthesis and release of methylarsenic compounds during the growth of freshwater
       algae. *Chemosphere* 2001, *43*, 265-272.
[11]   Francesconi, K. A.; Edmonds, J. S., Arsenic and marine organisms. In *Advances in
       Inorganic Chemistry*, Sykes, A. G., Ed. Academic Press: 1996; Vol. 44, pp 147-189.
[12]   Maher, W. A.; Clarke, S. M., The occurrence of arsenic in selected marine macroalgae
       from two coastal areas of South Australia. *Marine Pollution Bulletin* 1984, *15*, 111-112.
[13]   Mukhopadhyay, R.; Rosen, B. P.; Phung, L. T.; Silver, S., Microbial arsenic: From
       geocycles to genes and enzymes. *FEMS Microbiology Reviews* 2002, *26*, 311-325.
[14]   Rosen, B. P., Biochemistry of arsenic detoxification. *FEBS Letters* 2002, *529*, 86-92.
[15]   Petrick, J. S.; Ayala-Fierro, F.; Cullen, W. R.; Carter, D. E.; Vasken Aposhian, H.,
       Monomethylarsonous acid (MMA^{III}) is more toxic than arsenite in Chang human
       hepatocytes. *Toxicology and applied pharmacology* 2000, *163*, 203-207.
[16]   Knauer, K.; Behra, R.; Hemond, H., Toxicity of inorganic and methylated arsenic to
       algal communities from lakes along an arsenic contamination gradient. *Aquatic
       Toxicology* 1999, *46*, 221-230.
[17]   Maeda, S.; Kusadome, K.; Arima, H.; Ohki, A.; Naka, K., Biomethylation of arsenic
       and its excretion by the alga *Chlorella vulgaris*. *Applied Organometallic Chemistry*
       1992, *6*, 407-413.
[18]   Chen, W. Y.; Liao, C. M.; Jou, L. J.; Jau, S. F., Predicting bioavailability and
       bioaccumulation of arsenic by freshwater clam *Corbicula fluminea* using valve daily
       activity. *Environmental monitoring and assessment* 2010, *169*, 647-659.
[19]   Foster, S.; Maher, W.; Schmeisser, E.; Taylor, A.; Krikowa, F.; Apte, S., Arsenic
       species in a rocky intertidal marine food chain in NSW, Australia, revisited.
       Environmental Chemistry 2006, *3*, 304-315.
[20]   Goessler, W.; Maher, W.; Irgolic, K. J.; Kuehnelt, D.; Schlagenhaufen, C.; Kaise, T.,
       Arsenic compounds in a marine food chain. *Fresenius' Journal of Analytical Chemistry*
       1997, *359*, 434-437.
[21]   Maeda, S.; Inoue, R.; Kozono, T.; Tokuda, T.; Ohki, A.; Takeshita, T., Arsenic
       metabolism in a freshwater food chain. *Chemosphere* 1990, *20*, 101-108.
[22]   Milton, A.; Johnson, M., Arsenic in the food chains of a revegetated metalliferous mine
       tailings pond. *Chemosphere* 1999, *39*, 765-779.
[23]   Barwick, M.; Maher, W., Biotransference and biomagnification of selenium copper,
       cadmium, zinc, arsenic and lead in a temperate seagrass ecosystem from Lake

Macquarie Estuary, NSW, Australia. *Marine Environmental Research* 2003, *56*, 471-502.

[24]   Maher, W. A.; Foster, S. D.; Taylor, A. M.; Krikowa, F.; Duncan, E. G.; Chariton, A. A., Arsenic distribution and species in two *Zostera capricorni* seagrass ecosystems, New South Wales, Australia. *Environmental Chemistry* 2011, *8*, 9-18.

[25]   Harvey, C. F.; Beckie, R. D., Arsenic: Its biogeochemistry and transport in groundwater. *Metal ions in biological systems* 2005, *44*, 145-169.

[26]   Smedley, P. L.; Kinniburgh, D. G., A review of the source, behaviour and distribution of arsenic in natural waters. *Applied Geochemistry* 2002, *17*, 517-568.

[27]   Hasegawa, H., Seasonal changes in methylarsenic distribution in Tosa Bay and Uranouchi Inlet. *Applied Organometallic Chemistry* 1996, *10*, 733-740.

[28]   Cullen, W. R.; Reimer, K. J., Arsenic speciation in the environment. *Chemistry Review* 1989, *89*, 713-764.

[29]   Andreae, M. O., Arsenic speciation in seawater and interstitial waters: The influence of biological-chemical interactions on the chemistry of a trace element. *Limnology and Oceanography* 1979, *24*, 440-452.

[30]   Navarro, M.; Sanchez, M.; Lóopez, H.; Lopez, M. C., Arsenic contamination levels in waters, soils, and sludges in southeast spain. *Bulletin of Environmental Contamination and Toxicology* 1993, *50*, 356-362.

[31]   Maher, W. A., Arsenic in coastal waters of South Australia. *Water Research* 1985, *19*, 933-934.
Cáceres, V. L.; Gruttner, D. E.; Contreras, N. R., Water recycling in arid regions: Chilean case. *Ambio* 1992, *21*, 138-144.

[32]   Azcue, J.; Nriagu, J., Impact of abandoned mine tailings on the arsenic concentrations in Moira Lake, Ontario. *Journal of Geochemical Exploration* 1995, *52*, 81-89.

[33]   Ashley, P. M.; Lottermoser, B. G., Arsenic contamination at the Mole River mine, northern New South Wales. *Australian Journal of Earth Sciences* 1999, *46*, 861-874.

[34]   Sharma, V. K.; Sohn, M., Aquatic arsenic: Toxicity, speciation, transformations, and remediation. *Environment International* 2009, *35*, 743-759.

[35]   Andreae, M. O., Organic compounds in the environment. In *Organometallic Compounds in the Environment: Principles and Reactions*, Craig, P. J., Ed. Longman: New York, 1986; pp 198-228.

[36]   Maher, W.; Butler, E., Arsenic in the marine environment. *Applied Organometallic Chemistry* 1988, *2*, 191-214.

[37]   Smedley, P. L.; Edmunds, W. M.; Pelig-Ba, K. B., Mobility of arsenic in groundwater in the Obuasi gold-mining area of Ghana: Some implications for human health. *Geological Society London Special Publications* 1996, *113*, 163.

[38]   Hasegawa, H.; Rahman, M. A.; Kitahara, K.; Itaya, Y.; Maki, T.; Ueda, K., Seasonal changes of arsenic speciation in lake waters in relation to eutrophication. *Science of the Total Environment* 2010, *408*, 1684-1690.

[39]   Hasegawa, H.; Rahman, M. A.; Matsuda, T.; Kitahara, T.; Maki, T.; Ueda, K., Effect of eutrophication on the distribution of arsenic species in eutrophic and mesotrophic lakes. *Science of the Total Environment* 2009, *407*, 1418-1425.

[40]   Howard, A. G.; Comber, S. D. W., The discovery of hidden arsenic species in coastal waters. *Appl. Organomet. Chem.* 1989, *3*, 509-514.

[41] Hasegawa, H.; Matsui, M.; Okamura, S.; Hojo, M.; Iwasaki, N.; Sohrin, Y., Arsenic speciation including 'hidden' arsenic in natural waters. *Applied Organometallic Chemistry* 1999, *13*, 113-119.

[42] De Bettencourt, A. M. M.; Andreae, M. O., Refractory arsenic species in estuarine waters. *Appl. Organomet. Chem.* 1991, *5*, 111-116.

[43] Bright, D. A.; Dodd, M.; Reimer, K. J., Arsenic in subArctic lakes influenced by gold mine effluent: the occurrence of organoarsenicals and 'hidden' arsenic. *Science of the Total Environment* 1996, *180*, 165-182.

[44] Koch, I.; Wang, L.; Ollson, C. A.; Cullen, W. R.; Reimer, K. J., The predominance of inorganic arsenic species in plants from yellowknife, northwest territories, Canada. *Environmental Science and Technology* 2000, *34*, 22-26.

[45] Lafabrie, C.; Major, K. M.; Major, C. S.; Cebrián, J., Arsenic and mercury bioaccumulation in the aquatic plant, *Vallisneria neotropicalis*. *Chemosphere* 2011, *82*, 1393-1400.

[46] Peng, K.; Luo, C.; Lou, L.; Li, X.; Shen, Z., Bioaccumulation of heavy metals by the aquatic plants *Potamogeton pectinatus* L. and *Potamogeton malaianus* Miq. and their potential use for contamination indicators and in wastewater treatment. *Science of the Total Environment* 2008, *392*, 22-29.

[47] Reuther, R., Arsenic introduced into a littoral freshwater model ecosystem. *Science of the Total Environment* 1992, *115*, 219-237.

[48] Foster, S.; Maher, W.; Taylor, A.; Krikowa, F.; Telford, K., Distribution and speciation of arsenic in temperate marine saltmarsh ecosystems. *Environmental Chemistry* 2005, *2*, 177-189.

[49] Henry, T. R. *Technical summary of information available on the bioaccumulation of arsenic in aquatic organisms*; EPA-822-R-03-032; Office of Science and Technology Office of Water, U.S. Environmental Protection Agency: Washington, DC, 2003; p 42.

[50] Caussy, D., Case studies of the impact of understanding bioavailability: Arsenic. *Ecotoxicology and Environmental Safety* 2003, *56*, 164-173.

[51] Silver, S.; Misra, T. K., Bacterial transformations of and resistances to heavy metals. *Basic life sciences* 1984, *28*, 23-46.

[52] Bottino, N. R.; Newman, R. D.; Cox, E. R.; Stockton, R.; Hoban, M.; Zingaro, R. A.; Irgolic, K. J., The effects of arsenate and arsenite on the growth and morphology of the marine unicellular algae *Tetraselmis chui* (Chlorophyta) and *Hymenomonas carterae* (Chrysophyta). *Journal of Experimental Marine Biology and Ecology* 1978, *33*, 153-168.

[53] Tu, C.; Ma, L. Q., Effects of arsenate and phosphate on their accumulation by an arsenic-hyperaccumulator *Pteris vittata* L. *Plant and Soil* 2003, *249*, 373-382.

[54] Wang, J.; Zhao, F.-J.; Meharg, A. A.; Raab, A.; Feldmann, J.; McGrath, S. P., Mechanisms of arsenic hyperaccumulation in *Pteris vittata*: Uptake kinetics, interactions with phosphate, and arsenic speciation. *Plant Physiology* 2002, *130*, 1552-1561.

[55] Robinson, B.; Kim, N.; Marchetti, M.; Moni, C.; Schroeter, L.; van den Dijssel, C.; Milne, G.; Clothier, B., Arsenic hyperaccumulation by aquatic macrophytes in the Taupo Volcanic Zone, New Zealand. *Environmental and Experimental Botany* 2006, *58*, 206-215.

[56] Chen, C. Y.; Folt, C. L., Bioaccumulation and diminution of arsenic and lead in a freshwater food web. *Environmental Science and Technology* 2000, *34*, 3878-3884.

[57] Mason, R. P.; Laporte, J. M.; Andres, S., Factors controlling the bioaccumulation of mercury, methylmercury, arsenic, selenium, and cadmium by freshwater invertebrates and fish. *Archives of Environmental Contamination and Toxicology* 2000, *38*, 283-297.

[58] Dushenko, W. T.; Bright, D. A.; Reimer, K. J., Arsenic bioaccumulation and toxicity in aquatic macrophytes exposed to gold-mine effluent: relationships with environmental partitioning, metal uptake and nutrients. *Aquatic Botany* 1995, *50*, 141-158.

[59] Kirby, J.; Maher, W.; Chariton, A.; Krikowa, F., Arsenic concentrations and speciation in a temperate mangrove ecosystem, NSW, Australia. *Applied Organometallic Chemistry* 2002, *16*, 192-201.

[60] Francesconi, K. A.; Goessler, W.; Panutrakul, S.; Irgolic, K. J., A novel arsenic containing riboside (arsenosugar) in three species of gastropod. *Science of the Total Environment* 1998, *221*, 139-148.

[61] USEPA *Methodology for Deriving Ambient Water Quality Criteria for the Protection of Human Health (2000)*; U.S. Environmental Protection Agency: Washington, D.C., 2000; p 185.

[62] Spehar, R. L.; Fiandt, J. T.; Anderson, R. L.; DeFoe, D. L., Comparative toxicity of arsenic compounds and their accumulation in invertebrates and fish. *Archives of Environmental Contamination and Toxicology* 1980, *9*, 53-63.

[63] Wagemann, R.; Snow, N. B.; Rosenberg, D. M.; Lutz, A., Arsenic in sediments, water and aquatic biota from lakes in the vicinity of Yellowknife, Northwest Territories, Canada. *Archives of Environmental Contamination and Toxicology* 1978, *7*, 169-191.

[64] Chen, C. Y.; Stemberger, R. S.; Klaue, B.; Blum, J. D.; Pickhardt, P. C.; Folt, C. L., Accumulation of heavy metals in food web components across a gradient of lakes. *Limnology and Oceanography* 2000, *45*, 1525-1536.

[65] Kaise, T.; Ogura, M.; Nozaki, T.; Saitoh, K.; Sakurai, T.; Matsubara, C.; Watanabe, C.; Hanaoka, K. i., Biomethylation of arsenic in an arsenic-rich freshwater environment. *Applied Organometallic Chemistry* 1997, *11*, 297-304.

[66] Giusti, L.; Zhang, H., Heavy metals and arsenic in sediments, mussels and marine water from Murano (Venice, Italy). *Environmental Geochemistry and Health* 2002, *24*, 47-65.

[67] Valette-Silver, N. J.; Riedel, G. F.; Crecelius, E. A.; Windom, H.; Smith, R. G.; Dolvin, S. S., Elevated arsenic concentrations in bivalves from the southeast coasts of the USA. *Marine Environmental Research* 1999, *48*, 311-333.

[68] Karadjova, I. B.; Slaveykova, V. I.; Tsalev, D. L., The biouptake and toxicity of arsenic species on the green microalga Chlorella salina in seawater. *Aquatic Toxicology* 2008, *87*, 264-271.

[69] Pawlik-Skowronska, B.; Pirszel, J.; Kalinowska, R.; Skowronski, T., Arsenic availability, toxicity and direct role of GSH and phytochelatins in As detoxification in the green alga *Stichococcus bacillaris*. *Aquatic Toxicology* 2004, *70*, 201-212.

[70] Yamaoka, Y.; Takimura, O.; Fuse, H.; Murakami, K., Effect of glutathione on arsenic accumulation by *Dunaliella salina*. *Applied Organometallic Chemistry* 1999, *13*, 89-94.

[71] Neff, J. M., Ecotoxicology of arsenic in the marine environment. *Environmental Toxicology and Chemistry* 1997, *16*, 917-927.

[72] Shi, H.; Shi, X.; Liu, K. J., Oxidative mechanism of arsenic toxicity and carcinogenesis. *Molecular and cellular biochemistry* 2004, *255*, 67-78.

[73] Levy, J. L.; Stauber, J. L.; Adams, M. S.; Maher, W. A.; Kirby, J. K.; Jolley, D. F., Toxicity, biotransformation, and mode of action of arsenic in two freshwater microalgae (*Chlorella* sp. and *Monoraphidium arcuatum*). *Environmental Toxicology and Chemistry* 2005, *24*, 2630-2639.

[74] Cullen, W. R.; Harrison, L. G.; Li, H.; Hewitt, G., Bioaccumulation and excretion of arsenic compounds by a marine unicellular alga, polyphysa peniculus. *Applied Organometallic Chemistry* 1994, *8*, 313-324.

[75] Takimura, O.; Fuse, H.; Murakami, K.; Kamimura, K.; Yamaoka, Y., Uptake and reduction of arsenate by *Dunaliella* sp. *Applied Organometallic Chemistry* 1996, *10*, 753-756.

[76] Taboada-de la Calzada, A.; Villa-Lojo, M. C.; Beceiro-González, E.; Alonso-Rodríguez, E.; Prada-Rodríguez, D., Accumulation of arsenic(III) by *Chlorella vulgaris*. *Applied Organometallic Chemistry* 1999, *13*, 159-162.

[77] Thursby, G. B.; Steele, R. L., Toxicity of arsenite and arsenate to the marine macroalga *Champia parvula* (Rhodophyta). *Environmental Toxicology and Chemistry* 1984, *3*, 391-397.

[78] Knauer, K.; Hemond, H., Accumulation and reduction of arsenate by the freshwater green alga *Chlorella* sp. (Chlorophyta). *Journal of Phycology* 2000, *36*, 506-509.

[79] Reish, D. J., Effects of metals and organic compounds on survival and bioaccumulation in two species of marine gammaridean amphipod, together with a summary of toxicological research on this group. *Journal of natural history* 1993, *27*, 781-794.

[80] Nelson, D. A.; Calabrese, A.; Nelson, B. A.; MacInnes, J. R.; Wenzloff, D. R., Biological effects of heavy metals on juvenile bay scallops, *Argopecten irradians*, in short-term exposures. *Bulletin of Environmental Contamination and Toxicology* 1976, *16*, 275-282.

[81] Krishnaja, A. P.; Rege, M. S.; Joshi, A. G., Toxic effects of certain heavy metals (Hg, Cd, Pb, As and Se) on the intertidal crab *Scylla serrata*. *Marine Environmental Research* 1987, *21*, 109-119.

[82] Madsen, K. N., Effects of arsenic on survival and metabolism of *Crangon crangon*. *Marine Biology* 1992, *113*, 37-44.

[83] Morel, F. M. M., *Principles of Aquatic Chemistry*. Wiley-Interscience: New York, 1983

[84] Pagenkopf, G. K., Gill surface interaction model for trace-metal toxicity to fishes: Role of complexation, pH, and water hardness. *Environmental Science and Technology* 1983, *17*, 342-347.

[85] Hassler, C. S.; Slaveykova, V. I.; Wilkinson, K. J., Some fundamental (and often overlooked) considerations underlying the free ion activity and biotic ligand models. *Environmental Toxicology and Chemistry* 2004, *23*, 283-291.

[86] Campbell, P. G. C., Interactions between trace metals and aquatic organisms: A critique of the free ion activity model. In *Metal Speciation and Bioavailability in Aquatic Systems*, Tessier, A.; Turner, D. R., Eds. John Wiley: New York, NY, USA, 1995; Vol. 3, pp 45-102.

[87] Tamaki, S.; Frankenberger, W. T. J., Environmental biochemistry of arsenic. *Reviews of Environmental Contamination and Toxicology* 1992, *124*, 79-110.

[88] Qin, J.; Lehr, C. R.; Yuan, C.; Le, X. C.; McDermott, T. R.; Rosen, B. P., Biotransformation of arsenic by a Yellowstone thermoacidophilic eukaryotic alga. *Proceedings of the National Academy of Sciences* 2009, *106*, 5213-5521.

[89] Seyler, P.; Martin, J. M., Distribution of arsenite and total dissolved arsenic in major French estuaries: dependence on biogeochemical processes and anthropogenic inputs. *Marine Chemistry* 1990, *29*, 277-294.

[90] Pettine, M.; Camusso, M.; Martinotti, W., Dissolved and particulate transport of arsenic and chromium in the Po River (Italy). *Science of the Total Environment* 1992, *119*, 253-280.

[91] Pettine, M.; Mastroianni, D.; Camusso, M.; Guzzi, L.; Martinotti, W., Distribution of As, Cr and V species in the Po-Adriatic mixing area, (Italy). *Marine Chemistry* 1997, *58*, 335-349.

[92] Lerda, D. E.; Prosperi, C. H., Water mutagenicity and toxicology in Rio Tercero (Cordoba, Argentina). *Water Research* 1996, *30*, 819-824.

[93] Nimick, D. A.; Moore, J. N.; Dalby, C. E.; Savka, M. W., The fate of geothermal arsenic in the Madison and Missouri Rivers, Montana and Wyoming. *Water Resources Research* 1998, *34*, 3051-3067.

[94] Robinson, B.; Outred, H.; Brooks, R.; Kirkman, J., The distribution and fate of arsenic in the Waikato River system, North Island, New Zealand. *Chemical Speciation and Bioavailability* 1995, *7*, 89-96.

[95] McLaren, S. J.; Kim, N. D., Evidence for a seasonal fluctuation of arsenic in New Zealand's longest river and the effect of treatment on concentrations in drinking water. *Environmental Pollution* 1995, *90*, 67-73.

[96] Williams, M.; Fordyce, F.; Paijitprapapon, A.; Charoenchaisri, P., Arsenic contamination in surface drainage and groundwater in part of the southeast Asian tin belt, Nakhon Si Thammarat Province, southern Thailand. *Environmental Geology* 1996, *27*, 16-33.

[97] Wilkie, J. A.; Hering, J. G., Rapid oxidation of geothermal arsenic(III) in streamwaters of the eastern Sierra Nevada. *Environmental Science and Technology* 1998, *32*, 657-662.

[98] Braman, R. S.; Foreback, C. C., Methylated forms of arsenic in the environment. *Science* 1973, *182*, 1247-1249.

[99] Abdullah, M. I.; Shiyu, Z.; Mosgren, K., Arsenic and selenium species in the oxic and anoxic waters of the Oslofjord, Norway. *Marine Pollution Bulletin* 1995, *31*, 116-126.

[100] Peterson, M. L.; Carpenter, R., Biogeochemical processes affecting total arsenic and arsenic species distributions in an intermittently anoxic fjord. *Marine Chemistry* 1983, *12*, 295-321.

[101] Seyler, P.; Martin, J. M., Arsenic and selenium in a pristine river-estuarine system: the Krka (Yugoslavia). *Marine Chemistry* 1991, *34*, 137-151.

[102] Howard, A. G.; Apte, S. C.; Comber, S. D. W.; Morris, R. J., Biogeochemical control of the summer distribution and speciation of arsenic in the Tamar estuary. *Estuarine, Coastal and Shelf Science* 1988, *27*, 427-443.

[103] Andreae, M. O.; Andreae, T. W., Dissolved arsenic species in the Schelde estuary and watershed, Belgium. *Estuarine, Coastal and Shelf Science* 1989, *29*, 421-433.

[104] Yusof, A. M.; Ikhsan, Z. B.; Wood, A. K. H., The speciation of arsenic in seawater and marine species. *Journal of Radioanalytical and Nuclear Chemistry* 1994, *179*, 277-283.

[105] Ishikawa, M.; Okoshi, K.; Kurosawa, M.; Kitao, K. In *Trace element analysis of seawater by PIXE*, 12th International Symposium on Application of Ion Beams in

Material Science, Hosei University, Tokio, Japan, 1987; Sebe, T.; Yamamoto, Y., Eds. Hosei University Press: Hosei University, Tokio, Japan, 1987; pp 445-456.

[106] Edmonds, J. S.; Shibata, Y.; Francesconi, K. A.; Rippingale, R. J.; Morita, M., Arsenic transformations in short marine food chains studied by HPLC-ICP MS. *Applied Organometallic Chemistry* 1997, *11*, 281-287.

[107] Rattanachongkiat, S.; Millward, G. E.; Foulkes, M. E., Determination of arsenic species in fish, crustacean and sediment samples from Thailand using high performance liquid chromatography (HPLC) coupled with inductively coupled plasma mass spectrometry (ICP-MS). *Journal of Environmental Monitoring* 2004, *6*, 254-261.

[108] Gallagher, P. A.; Shoemaker, J. A.; Wei, X.; Brockhoff-Schwegel, C. A.; Creed, J. T., Extraction and detection of arsenicals in seaweed via accelerated solvent extraction with ion chromatographic separation and ICP-MS detection. *Fresenius' Journal of Analytical Chemistry* 2001, *369*, 71-80.

[109] Ackley, K. L.; B'Hymer, C.; Sutton, K. L.; Caruso, J. A., Speciation of arsenic in fish tissue using microwave-assisted extraction followed by HPLC-ICP-MS. *Journal of Analytical Atomic Spectrometry* 1999, *14*, 845-850.

[110] Anderson, L. C. D.; Bruland, K. W., Biogeochemistry of arsenic in natural waters: the importance of methylated species. *Environmental Science and Technology* 1991, *25*, 420-427.

[111] Francesconi, K. A.; Pedersen, S. N.; Khokiattiwong, S.; Goessler, W.; Pavkov, M., A new arsenobetaine from marine organisms identified by liquid chromatography–mass spectrometry. *Chemical Communications* 2000, *2000*, 1083-1084.

[112] Benjamin, P. Y. L.; Michalik, P.; Porter, C. J., Identification and confirmation of arsenobetaine and arsenocholine in fish, lobster and shrimp by a combination of fast atom bombardment and tandem mass spectrometry. *Biomedical and Environmental Mass Spectrometry* 1987, *14*, 723-732.

[113] Lawrence, J. F.; Michalik, P.; Tam, G.; Conacher, H. B. S., Identification of arsenobetaine and arsenocholine in Canadian fish and shellfish by high-performance liquid chromatography with atomic absorption detection and confirmation by fast atom bombardment mass spectrometry. *Journal of Agricultural and Food Chemistry* 1986, *34*, 315-319.

[114] Mürer, A. J. L.; Abildtrup, A.; Poulsen, O. M.; Christensen, J. M., Effect of seafood consumption on the urinary level of total hydride-generating arsenic compounds. Instability of arsenobetaine and arsenocholine. *The Analyst* 1992, *117*, 677-680.

[115] Goessler, W.; Kuehnelt, D.; Schlagenhaufen, C.; Slejkovec, Z.; J. Irgolic, K., Arsenobetaine and other arsenic compounds in the National Research Council of Canada Certified Reference Materials DORM 1 and DORM 2. *Journal of Analytical Atomic Spectrometry* 1998, *13*, 183-187.

[116] Edmonds, J. S.; Francesconi, K. A., Organoarsenic compounds in the marine environment. In *Organometallic Compounds in the Environment*, Vraig, P. J., Ed. Wiley: West Succex, England, 2003; pp 196-222.

[117] Larsen, E. H., Speciation of dimethylarsinyl-riboside derivatives (arsenosugars) in marine reference materials by HPLC-ICP-MS. *Fresenius' Journal of Analytical Chemistry* 1995, *352*, 582-588.

[118] Madsen, A. D.; Goessler, W.; Pedersen, S. N.; Francesconi, K. A., Characterization of an algal extract by HPLC-ICP-MS and LC-electrospray MS for use in arsenosugar speciation studies. *Journal of Analytical Atomic Spectrometry* 2000, *15*, 657-662.

Reviewed by: Professor William Maher, Institute for Applied Ecology, University of Canberra, ACT 2601, Australia, Phone: (02) 6201 2531, Fax: (02) 6201 5305, Email: Bill.Maher@canberra.edu.au

SECTION II:
ARSENIC TOXICITY AND HUMAN HEALTH

In: Arsenic
Editor: Andrea Masotti

ISBN: 978-1-62081-320-1
© 2013 Nova Science Publishers, Inc.

Chapter 6

HUMAN HEALTH EFFECTS
OF CHRONIC ARSENIC TOXICITY

D. N. Guha Mazumder[*]

Department of Medicine and Gastroenterology
Institute of Post Graduate Medical Education and Research,
Kolkata, India

ABSTRACT

Chronic arsenic toxicity (arsenicosis), due to drinking of arsenic- contaminated ground water, is a major environmental health hazard throughout the world including India and Bangladesh. Lot Much of research on regarding the health effects of chronic arsenic toxicity in humans has been carried out in West Bengal during the last two decades. A review of literature on health-related issues including information available from West Bengal, Bangladesh, China, Chile, Argentina, Mexico and the USA has been made to characterize the problem. Scientific publications, journals, monographs and proceedings of conferences in regard to human health effects of chronic arsenic toxicity have been reviewed.

The symptoms of chronic arsenic toxicity (arsenicosis) are insidious in onset and are dependent on the magnitude of the dose and duration of its exposure. It needs to be mentioned that a few epidemiological studies in the USA reported that none of the population exposed to environmental arsenic showed any clinical manifestation of chronic arsenic toxicity. Further, there is a wide variation in the incidence of chronic arsenicosis in an affected population. Moreover, not even all members of an affected family show clinical effects. Pigmentation and keratosis are the specific skin lesions characteristic of chronic arsenic toxicity. However, in West Bengal, and Bangladesh, and in many other countries, it was found to produce various systemic manifestations such as chronic lung disease characterized by chronic bronchitis, chronic obstructive and/or restrictive pulmonary disease and bronchiectasis, liver disease like (e.g., non- cirrhotic portal fibrosis), polyneuropathy, peripheral vascular disease, hypertension, ischemic heart

[*] Director, DNGM Research Foundation; 37/C, Block B, New Alipore, Kolkata – 700 053, India; e-mail: guhamazumder@yahoo.com; and Professor & Head (Retd.); Department of Medicine and Gastroenterology; Institute of Post Graduate Medical Education & Research, Kolkata – 700 020, India.

disease, Cerebrovascular disease, non-pitting edema of feet/hands, conjunctival congestion, weakness and anemia.

During pregnancy, high concentrations of arsenic ≥200 µg/L during pregnancy were found to be associated with a six-fold increased risk for stillbirth. The deficiency of intellectual functioning has been reported in children born and brought up during periods of arsenic exposure. Severe keratosis and chronic lung disease are major causes of morbidity. Cancer of the skin, lungs and urinary bladder are significant cancers associated with this toxicity. Stoppage of discontinuing drinking of the arsenic-contaminated water and intaking of a nutritious diet are the main keys in the management of arsenicosis as specific chelation therapy. Early skin and bladder cancers, detectable by regular, active surveillance, are curable.

INTRODUCTION

Arsenic contamination of groundwater has been recognized as a great threat to water suppliesy and public health in many countries around in the world. The name term *arsenic* is derived from the Greek word *arsenikon*, meanings yellow orpiment. Arsenic compounds have been mined and used since ancient times. Currently, arsenic in drinking water has been recognized as a major public health problem in several regions of the world. Reports of arsenic contamination in ground water are emerging from more than 20 countries as and when cases of chronic arsenic toxicity are reported. Major countries affected are Bangladesh, India, Taiwan, China, Mexico, Argentina, Chile and the USA.

Chronic arsenic exposure in humans produces various non- specific symptoms because of its affection of it can affect multi-organ systems in the body. However, dermatological manifestations are quite distinctive and diagnostic of human arsenic toxicity. In a field guide for the detection, management and surveillance of arsenicosis cases, the the condition has been defined by WHO as:

'a chronic health condition arising from prolonged ingestion of arsenic above the safe dose for at leat six months, usually manifested by characteristic skin lesions of melanosis (hyperpigmentation) and keratosis, occurring alone or in combination, with or without involvement of internal organs' by WHO[1].

Their appearance usually follows a temporal progression, beginning with hyperpigmentation, and then progressing to palmar-planter keratosis which is dependent on the magnitude of the dose and duration of its exposure. Further, there is a wide variation in the incidence of chronic arsenicosis in an affected population. Not even all members of an affected family show clinical effects. The reason for such a variation of disease expression is an enigma. The specific skin lesions of chronic arsenic toxicity are characterized by pigmentation, depigmentation, keratosis and Bowen's disease.

The duration of the patient's arsenic exposure with the date of onset of symptoms onset does not follow a particular time frame. Arsenical skin lesions have been reported to occur in West Bengal, India after drinking arsenic contaminated water for one year or even less [2, 3]. In Taiwan, the youngest patient drinking contaminated water who developed hyperpigmentation was 3 years old [4]. Among the population exposed to arsenic in drinking water in the Antofagasta region of Chile, cases of cutaneous arsenicosis, including both hyperpigmentation and hyperkeratosis, have been described in children as young as 2 years of age [5, 6]. The mean arsenic dose in Antofagasta was estimated to be approximately 60 µg/kg per day for subgroups of children aged 3.13±3.33 years, but was approximately 20 µg/kg per day for subgroups in their teens and twenties and 6 µg/kg per day for a subgroup in their sixties, indicating an inverse relationship between daily arsenic dose rate/kg body weight and age [7]. Prolonged exposure to a non-lethal dose of 5 to 90 µg /kg-body weight /day has been suggested to cause arsenical skin lesions [1]. In a retrospective study of 262 adults treated with Fowler's solution, the minimal latency period for hyperkeratosis was found to be 2.5 years following the ingestion of approximately 2.2 g of arsenite [8]. Furthermore, development of hyperpigmentation was reported to occur within 6-12 months of the start of the arsenic treatment with arsenic at a dose of 4,750 µg /day. Hyperkeratosis appeared after approximately 3 years [9]. A history of chronic arsenic exposure for more than 6 months is recommended as one of the required diagnostic criteria for the clinical diagnosis of arsenicosis [1].

SKIN MANIFESTATIONS

Arsenical hyperpigmentation may appear as a diffused or generalisedized hyperpigmentation of the palm or body ("melanosis"), or appear as a finely freckled or spotted ("raindrop") pattern of pigmentation that is particularly pronounced on the trunk and or extremities and has a bilateral symmetrical distribution. Hyperpigmentation may sometimes be blotchy and involve mucous membranes such as the under surface of the tongue or buccal mucosa [10-16]. The raindrop appearance results from the presence of numerous rounded hyperpigmented or hypopigmented macules (typically 2-4 mm in diameter), widely dispersed against a tan-to-brown hyperpigmented background [12, 13], and and are localized or patchy pigmentation, particularly affecting skin folds [12, 17, 18]. Hypopigmentation, or so-called leukodermia, or leukomelanosis [13, 14], are is found in chronic arsenic- exposed people in whom the hypopigmented macules take a spotty, white appearance. Leucomelanosis appears to occur in an arsenicosis patient following their stoppage of discontinuance of drinking theing arsenic- contaminated water for some duration [19].

Table 1. Dermatological criteria and gradation of chronic arsenic toxicity

Scoring System

Pigmentation (Score)		
Mild (1)	Moderate (2)	Severe (3)
Diffuse Melanosis, mild spotty pigmentation, Leucomelanosis	Moderate spotty pigmentation	Blotchy pigmentation, pigmentation of under surface of tongue, buccal Mucosa
Keratosis (Score)		
Mild (1)	Moderate (2)	Severe (3)
Slight thickening, or minute papules (<2 mm) in palms and soles	Multiple raised keratosis papules (2 to 5mm) oin palms and soles with diffuse thickening	Diffuse severe thickening, large discreet or confluent keratotic elevations (>5mm), palms and soles (also dorsum of extremity and trunk)

Maximum total skin score = 6; [32].

Arsenical hyperkeratosis appears predominantly on the palms and the plantar aspects of the feet, although involvements of the dorsum of the extremities and the trunk have also been described. In the early stages, the involved skin might have an indurated, grit- like character that can be best appreciated by palpations; however, the lesions usually advance to form a raised, punctate, 2-4 mm wart- like keratosis that isare readily visible [12].

Occasionally keratotic lesions might be larger (0.5 to 1 cm) and have a nodular or horny appearance occurring onin the palms or dorsums of the feet. In severe cases, the hands and soles present with diffused verrucous lesions. Cracks and fissures may be severe oin the soles [4, 10-, 11, 12, 13, 15, 20-25]. A histological examination of the lesions typically reveals hyperkeratosis with or without parakeratosis, acanthosis, and enlargement of the rete ridges. In some cases, there might be evidence of cellular atypia, mitotic figures, in large vacuolated epidermal cells [10, 12, 26-28].

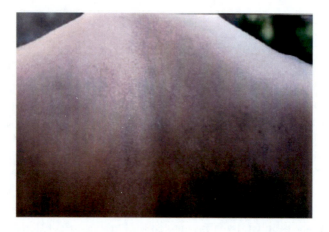

Figure 1. Diffuse Melanosis.

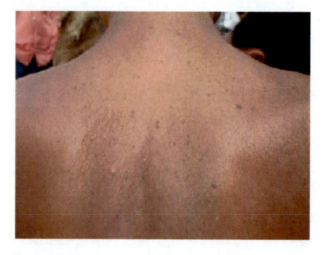

Figure 2. Mild Pigmentation (Spotty).

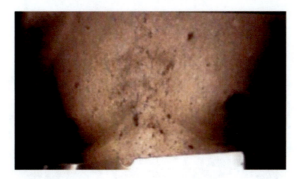

Figure 3. Moderate Pigmentation.

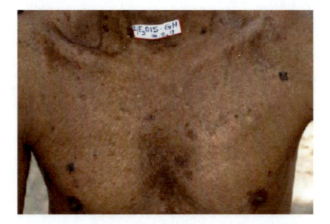

Figure 4. Severe Pigmentation (Blotchy).

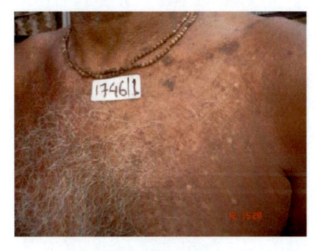

Figure 5. Pigmentation and Depigmentation (Leucomelanosis).

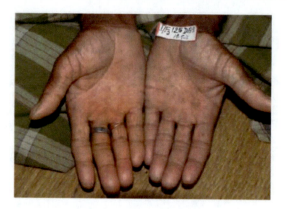

Figure 6. Mild Keratosi.

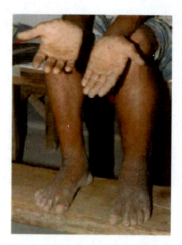

Figure 7. Moderate Keratosis with Non- Pitting Oedema of the legs.

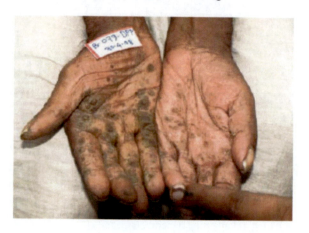

Figure 8. Severe Keratosis.

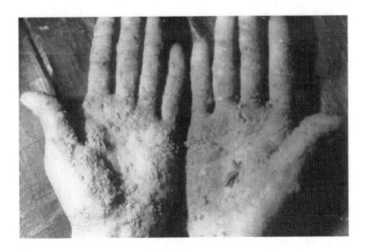

Figure 9. Verrucous keratosis of hand with keratotic horn.

Epidemiology

An epidemiological study was carried out to ascertain the prevalence of arsenical skin lesions in a population of 40,421 out of a total population of 103,154 in 37 villages of south west Taiwan where people were drinking arsenic- contaminated, artesian well water for more than 45 years. In the villages surveyed, the arsenic content of the well water examined ranged from, 10 to 1820 μg/L. The overall prevalence rates for keratosis and hyperpigmentation were 7.10 and 18.35 per 100 respectively. The male to female ratio for both the conditions was 1.1:1. The prevalence rates for hyperpigmentation steadily increased steadily with age for males, but for females a peak appeared inat the 50-59 age group 50-59 for females, followed by a gradual downturn in rates. The prevalence rates for keratosis for both sexes increased up to age 70 and then declined. However, arsenic levels in Taiwan were reported by villages. In this study, the youngest patient with hyperpigmentation in this study was 3 years old, and with keratosis was 4 years old [4]. The first population- based survey with individual arsenic exposure data from a drinking water source was carried out in West Bengal on 7,683 participants (4,093 female and 3,590 male) to assess the dose-response relationship and arsenical skin lesions [29]. The tube well arsenic concentration ranged from non-detectable to 3200μg/L. The overall prevalence rates for keratosis and hyperpigmentation were 3.64 and 8.82 per 100 respectively. Prevalence The prevalence of keratosis and hyperpigmentation was examined by water arsenic levels. A clear relationship was apparent between water levels of arsenic, and keratosis and hyperpigmentation. An age-adjusted prevalence of keratosis and hyperpigmentation was strongly related to water arsenic levels, rising from zero and 0.3 in the lowest exposure level (<50μg/L), to 8.3 and 11.5 per 100 respectively for females' drinking water containing >800 μg/L, and increasing from 0.2 and 0.4 per 100 in the lowest exposure category to 10.7 and 22.7 per 100 respectively for males in the highest exposure level (>800 μg /L). The prevalence of skin lesions was also examined by a daily dose per body weight (μg/Kg/day). This showed that men had roughly two to three times the prevalence of keratosis and pigmentation compared to that of those for women apparently ingesting the same dose of arsenic from drinking water. Subjects who were below 80% of the standard body weight for their age and sex had a 1.6 fold increase in the prevalence of keratosis, suggesting that malnutrition may play some role in increasing susceptibility [29]. A nested case control study was further carried out using a detailed lifetime (at least 20 years) exposure assessment among the above mentioned study population. Lifetime water arsenic exposure data could be obtained in a total of 192 cases having definite evidence of arsenical skin lesions and 213 age and sex matched controls. The lowest peak value of arsenic ingested by a confirmed case was 115 µg/L.

Strong dose response gradients with both peak and average arsenic water concentrations were also observed [30]. A cross- sectional study was conducted in Bangladesh in which arsenic- related skin lesions were assessed in 1,481 subjects (903 males and 578 females) in four rural villages with data for the arsenic exposure data (BDL to 2040µg/L) through drinking water (BDL to 2040µg/L) in the individuals .[31]. The crude prevalence of any arsenical hyperpigmentation or keratosis was 29/100. In the males in from the lowest exposure category (<150 µg/L), the age adjusted prevalence of any arsenic associated skin lesion was 18.6/100 and increased to 37.0/100 in at the highest exposure category (>1000 µg/L) (chi-square dose for trend p<0.0.001). The corresponding rates for females were 17.9/100 in from the lowest exposure category and 24.9 atin the highest exposure category (Test for trend, p<0.02). There was a significant trend for the prevalence rate, both in relation to exposure levels and to dose index (p < 0.05), regardless of sex. This study also shows a higher prevalence rate of arsenic skin lesions in males than females, with a clear dose-response relationship.

A cross- sectional study was further carried out in another arsenic affected district of West Bengal (Nadia) to ascertain the degree of skin lesion severity of skin lesion in among the arsenic affected population. Most of the people were poor and belonged to low socio-economic conditions with poor nutrition. Out of 10,469 participants examined, 1,616 (15.43%) patients showed clinical features of arsenicosis characterized by arsenical skin lesions. Mean The mean arsenic concentration in drinking water among these cases was 103.46 [standard deviation (SD) ± 153.28] µg/L, while whereas among the controls (no skin lesions) it was 73.18 (SD ± 115.10) µg/L. Duration The former's duration of arsenic exposure of the former was 12.47 (SD ± 7.39) years while whereas that of the later was 12.43 (SD ± 6.59) years. A scoring system has been adopted to classify the degrees of severity of associated with skin manifestations (Vide See table.1).

The skin score was mild in 87.5%, while moderate in 11.5% and severe in 0.87% of the people having that had arsenical skin lesions. The study showed that although a large number of people in the district of Nadia are showingpresented with arsenical skin lesions, in majority of cases these are were mild in a majority of the cases [32]. This information was helpful for undertaking an arsenic mitigation program, as providing arsenic- free drinking water in to the population could help in relieving the morbidity of associated with a majority of the arsenic-affected population in the state.

Oshikawa et al. (2001) [33] investigated the changes in skin lesion severity of skin lesions over a period of 10 years among an affected cohort in an area having that had arsenic-contaminated shallow wells due to tin mining activities in southern Thailand where interventions to reduce arsenic-contaminated water had been implemented. Over the 10-year period, drinking predominantly arsenic-free water had increased the probability of regression in subjects with mild-stage lesions, but not in those with more advanced-stage lesions.

RESPIRATORY DISEASE

A hospital- based study was carried out in Kolkata, West Bengal during 1995 to ascertain the pulmonary effects of chronic arsenic toxicity on 156 cases with arsenical skin lesions associated with drinking of arsenic-contaminated ground water (from a geogenic source: 50-3,200 µg/L in 136; and from pesticide contamination: 5,005-14,000 µg/L in 20 subjects). Symptoms of a chronic cough, with and without expectoration, were reported in 89 (57%) and 45 (28.5%) cases respectively. Lung function tests carried out on 17 patients showed features of restrictive lung disease in nine 9 (53%) and combined obstructive and restrictive lung disease in seven 7 (41%) cases [3, 16]. In another hospital-based study carried out in Kolkata on 107 arsenic- exposed (arsenic level in dinking water, mean ± SD : 550 ± 470 µg/L) and 52 control subjects (arsenic exposure <50 µg/L), 33 (30.8%) cases and 4 controls (7.6%) had symptoms of lung disease (p<0.01). Lung The lung function study showed obstructive, restrictive and mixed obstructive and restrictive patterns in 20 (68.9%), 1 (3.7%) and 8(27.6%) cases respectively. Bronchiectasis was diagnosed ion 3 cases on the basis of a chest X-Ray and HRCT (done on 4 cases) [34]. Pulmonary disease, characteristic of bronchitis, was also reported in 86 (23.7%) out of 360 people who were found to have arsenical skin lesions in a village in Bangladesh having an average arsenic level in their drinking water of 240µg/L [25].

Epidemiological Study

Symptoms of lung disease, characterisedized by a chronic cough, wereas reported from a cohort of 180 residents of Antafagosta, Chile exposed to drinking arsenic- contaminated water in 38.8% of 144 subjects with arsenical skin lesions compared to 3.1% of 36 subjects with normal skin [22]. In addition, the prevalence of the reported cough declined fom 38 % to 7% after an arsenic removal plant was installed in Antafagosta (p<0.001) [35]. Autopsies were conducted on five children, with evidences of arsenical skin lesions on their bodiesy, who died between 1968 and 1969 in Antafagosta. The lung tissue was examined in four of these children: with abnormalities were found in all four. Interstitial fibrosis was detected in two of the cases [5].

A large cross-sectional epidemiological survey was carried out during 1998-1999 on 6,864 participants (all non-smokers) in the district of South 24 Parganas, West Bengal, India to ascertain the relationship between chronic arsenic exposure and the occurrence of lung disease [36]. The participants were clinically examined and interviewed, and the arsenic content in their current drinking water source was measured (arsenic concentration ranged from non-detectable to 3,200µg/L). Among males and females, the prevalence of a cough, shortness of breath and chest sounds (crepitation and/or rhonchi) in the lungs rose with increasing arsenic concentrations in the drinking water. These respiratory effects were most pronounced in individuals with high arsenic water concentrations who had also had skin lesions. The prevalence odds ratio (POR) estimates were markedly increased for participants with arsenic- induced skin lesions who had high levels of arsenic in their drinking water source (≥500 µg/L) compared with individuals who had normal skin and was were exposed to low levels of arsenic (<50 µg/L). In participants with skin lesions, the age-adjusted POR estimates for coughing were 7.8 for females (95% CI:3.1-19.5) and 5.0 for males (95%CI:2.6-9.9); for chest sounds the POR for females was 9.6 (95% CI:4.0-22.9) and for males 6.9 (95%CI:3.1-15.0). The POR for shortness of breath in females was 23.2 (95% CI:5.8-92.8) and was in males 3.7 (95%CI:1.3-10.6). These results add to the evidence that longthe long-term ingestion of inorganic arsenic can cause respiratory effects [36].

A prevalence comparison study of respiratory effects among subjects with and without arsenic exposure through drinking water was conducted in Bangladesh. Exposed participants were recruited through a health awareness campaign program. Unexposed participants were randomly selected, from where tube wells were not contaminated with arsenic. A total of 218 individuals participated (94 exposed individuals exhibiting skin lesions; 124 unexposed individuals). The arsenic concentrations ranged from 136.0-1000.0 μg/L .Only non- smokers without any history of asthma or tuberculosis were recruited, and data on respiratory symptoms and signs were collected. The crude prevalence ratios for chronic bronchitis were found to be 1.6 (95% CI: 0.8-3.1) and 10.3 (95% CI:2.4-43.1) for males and females, respectively. These results also showd revealed respiratory effects on from long term arsenic exposure [37].

Furthermore, another population- based, cross- sectional study was carried out in another district of West Bengal (Nadia) to ascertain the incidence of respiratory disease. A total number of 2,297 households (having a population of 10,469) belonging to 37 arsenic-affected villages in all the 17 blocks of the district were selected by a multistage sampling method. The mean arsenic content in drinking water among the cases was (103.46; (SD ± 153.28 μg/L)) while whereas it was among the controls was (73.18; (SD ± 115.10 μg/L) among the controls). The duration of arsenic exposure of associated with the former was 12.47 (SD ± 7.39)) years while whereas that of the later was 12.43 (SD ± 6.59) years. Chronic lung disease characterisedized by chronic coughing and difficulty breathing difficulty was found in 207 (12.81%) out of 1,616 cases with arsenical skin lesions while whereas those symptoms were found it in 69 (0.78%) out of 8,853 participants without any such lesions (controls), (p<0.001) [32].

A case-control study was carried out during 2000-2003 to ascertain the relationship between respiratory symptoms, lung functioning and exposure to arsenic in drinking water in West Bengal among a cohort of 287 participants selected among a source population of 7,683 subjects surveyed for arsenic- related skin lesions during 1995-1996 [38]. Cases were selected that had having skin lesions and primary drinking water sources containinged arsenic at 50 μg/L -500μg/L, while whereas controls were selected who did not have arsenic-related skin lesions and whose main tube well-water source contained an arsenic concentration of <50 μg/L. For each case, one control matched on by age (within 5 years) and sex was randomly identified from all eligible non-cases. The study was confined to those participants of at least 20 years of age who completed the pulmonary function testing and for whom information on smoking was available. In this study, 132 and 155 subjects fulfilled the criteria of cases, and controls respectively. Respiratory symptoms were increased in men with arsenic- caused skin lesions (versus those without lesions), particularly "shortness of breath at night" [odds ratio (OR) = 2.8, 95% confidence interval (CI):1.1,7.6] and "morning cough" (OR=2.8,95% CI:1.2.6.6) in smokers and "shortness of breath ever" (OR=3.8,95% CI:0.7,20.6) in non-smokers.

Pronounced decrements in lung functioning were observed in males with skin lesions (both non-smokers and smokers) as compared to those without skin lesions. Male smokers had, on average, lower mean residual values for spirometric parameters than male non-smokers. The decreases in FEV1 and forced vital capacity (FVC) in male non-smokers with skin lesions as compared with non-smoking men without skin lesions were 157.3 mL (95% percent CI: - 24.7, 339.2) for FEV1 and 188.5 mL (95 percent% CI: 0.6,376.3) for FVC. In male smokers, the decreases were 271.1 mL (95 percent CI: 158.0, 384.2) for FEV1 and 304.1 mL (95 percent CI: 180.1, 428.1) for FVC.

Among women, the respective reductions were 63.2 mL (95% percent CI: -31.8, 158.2) for FEV1 and 101.5 mL (95% percent CI: -8.8, 211.8) for FVC. Decreases in FEV1 and FVC related to increased water arsenic concentrations were observed among men; the reduction in mean values from low exposure (arsenic level <100 μg/L) to high exposure (arsenic level ≥400 μg/L) were 194.7 mL (95% percent CI: 35.5, 353.9) for FEV1 and 83.8 mL (95% percent CI: - 93.8, 261.5) for FVC in non-smokers and 226.1 mL (95% percent CI: 45.2, 407.0) for FEV1 and 247.6 mL (95% percent CI: 58.3, 436.9) for FVC in smokers. Among women, the respective reductions were 28.5 mL (95 percent% CI: - 71.3, 128.2) for FEV1 and 7.5 mL (95 percent% CI: - 122.4, 137.5) for FCV.

In the multivariate linear regression analyses stratified by sex and adjusted for age, height, and smoking, lung function was significantly decreased for from signs of arsenic-related skin lesions among men, with a reduction in FEV1 of 256.2 mL (95 percent% CI : 113.9, 398.4, p value: <0.001) and in FVC of 287.8 mL (95 percent CI : 134.9, 440.8, p value: <0.001)

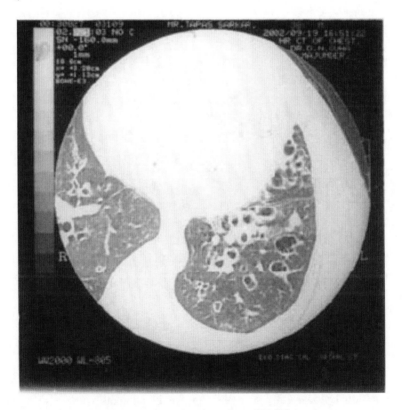

Figure 10. Bronchiectasis of a Lung in an Arsenicosis patient (HRCT).

To further investigate the effects of ingested arsenic on flow, the FEV$_1$/FVC ratio and the forced expiratory flow between 25 and 75% percent of forced vital capacity (FEF$_{25-75}$) were investigated, and significant reductions related to the presence of skin lesions consistent with the findings for FEV$_1$ and FVC in men were found. Reductions were also observed related to smoking in FEV$_1$ (156.4 mL; 95% percent CI: -3.2, 316.0, p value: 0.055) and FVC (119.7mL: 52.0, 291.4, p value: 0.2), but the effect size was smaller than for the presence of skin lesions. Using arsenic levels in water as a measure of exposure instead of skin lesions, significant decreases were found in FEV$_1$ of 45.0 mL (95 percent % CI : 6.2, 83.9, p value: 0.02) and in FVC of 41.4 mL (95 percent % CI : -0.7, 83.5, p value: 0.054) in men per100.0 µg/L increase of arsenic. Potential confounders such as weight, type of house, education and occupation were assessed in the multivariate models but did not change the estimates for skin lesions or arsenic in water. Interestingly, in women, estimates for skin lesions or arsenic in water did not indicate a strong relationship with lung functioning. In this study, consumption of arsenic- contaminated water was found to be associated with respiratory symptoms and reduced lung functioning in men, particularly with in those with skin lesions [38].

To ascertain the incidence of bronchiectasis in the population, 108 subjects with arsenic-caused skin lesions and 150 subjects without skin lesions were studied from a population survey of over 7,000 people of an endemic region of West Bengal done during 1995-96 [39]. The median highest median level of arsenic in drinking water was 330µg/L [standard deviation (SD) ±881µg/L] in subjects with skin lesions compared with 28µg/L [(SD) ± 147µg/L] in those without such lesions.

Subsets of both the groups who reported a chronic cough (more than 3 months per year for at least 2 years) were referred to a tertiary referral center in Kolkata for a high resolution computed tomography (HRCT). These scans were read by two radiologists, independently and without knowledge of presence or absence of skin lesions. The severity of bronchiectasis severity in each lobe was ranked on a scale of 0 to 4 using a modification of the system described by Lynch et al. (1999) [40]. Briefly, a five-point grading system was employed: 0 = no bronchiectasis; 1 = mild bronchiectasis (non-tapering cylindrical internal bronchial diameter 1.5-3 times the diameter of the accompanying artery); 2 = moderate bronchiectasis (non-tapering cylindrical internal bronchial diameter more than 3 times the diameter of the accompanying artery); 3 = varicose bronchiectasis; and 4 = cystic bronchiectasis. The lungs were divided into 6 lobes (considering the lingual as a separate lobe), and a score was assigned to each lobe. A single bronchiectasis severity score was assigned to each subject by summing the bronchiectasis scores from each lobe of each lung.

Thirty- three (31%) subjects with skin lesions and 18 (12%) subjects without lesions reported a chronic cough for more than 3 years (OR = 3.2; CI = 1.7 -6.1). Of these, twenty-seven27 subjects with skin lesions and 11 of the subjects without lesions agreed to travel to Kolkata for a HRCT. Overall, the participation rate was 82% in subjects with skin lesions and 61% in subjects without skin lesions. For those subjects who underwent the HRCT, the average bronchiectasis severity score was 3.4 (SD ± 3.6) in 27 subjects with skin lesions and 0.9 (SD ± 1.6) in 11 subjects without lesions. In subjects who reported a chronic cough, CT evidence of bronchiectasis was found in 18(67%) participants with skin lesions and 3 (27%) subjects without skin lesions. Only one (9%) of the 11 subjects without skin lesions had a bronchiectasis severity score greater than 2, while 14 (52%) of the 27 skin lesion subjects had bronchiectasis severity scores greater than 2.

The unadjusted odds ratio for bronchiectasis was 10 (95% CI = 2.9-35) in subjects with arsenic-caused skin lesions compared with subjects having no lesions. The corresponding adjusted odds ratio was 10 (2.7-37). The adjusted odds ratio was 13 (2.6-62) in men and 6.1 (0.6-62) in women.

This study was the first investigation of HRCT findings in a population exposed to high levels of arsenic in drinking water. In this study, the authors found that persons people with arsenic-caused skin lesions have a 10-fold higher rate of bronchiectasis s being revealed on the HRCT. Given the large magnitude of the relative estimated risk estimate they identified, this association is not likely to be due to chance. The highly characteristic skin lesions diagnosed in this study are known to result from the consumption of arsenic-contaminated drinking water. The findings, therefore, provide evidence that the long-term ingestion of arsenic results in increased risks of non-malignant pulmonary disease, in particular, bronchiectasis [41].

Mortality Studies

In a black-foot disease prone area of Taiwan (arsenic in drinking water: median=780µ/L), an eclogical study of mortality showed that the Standardized Mortality Ratio (SMR) for "bronchitis" significantly increased significantly relative to a nearby reference population (SMR=1.53; 95% CI=1.30-1.80) during the period from 1971 to 1994. The authors observed that it was unlikely that the differences in the rate of smoking would account for the increased bronchitis mortalitiesy [42].

Furthermore, a retrospective study on mortality due to bronchiectasis, carried out for the period (between 1988-2000) in Antafagosta, Chile, suggested that the exposure of arsenic in through drinking water during early childhood or *in utero* has pronounced pulmonary effects, greatly increasing subsequent mortality in young adults from the condition. For the birth cohort born just before the high exposure period (1950-1957) and exposed in early childhood, the standardized mortality ratio (SMR) for bronchiectasis was 12.4 (95% CI, 3.3-31.7;p<0.001) while whereas those born during the high exposure period (1958-1970) with probable exposure *in utero* and early childhood, the corresponding SMR was 46.2 (95% CI,21.1-87.7;p<0.001) [43].

Lung Disease and Biomarkears

Recently a study on the serum level of the Clara cell protein, (CC16),- a novel biomarker for lung disease, has been studied in arsenic- exposed people and correlated with well arsenic, total urinary arsenic, and urinary As methylation indices in Bangladesh [44]. The study was conducted on 241 non- smoking individuals having that had arsenic exposure (0.1–761µg/L) from drinking water.

An inverse association was observed between urinary arsenic and serum CC16 among persons people with skin lesions (β = −0.13, p = 0.01). A positive association was observed between the secondary methylation index in urinary arsenic and CC16 levels (β = 0.12, p = 0.05) in the overall study population; the association was stronger among people without skin lesions (β = 0.18, p = 0.04), indicating that an increased methylation capability may be protective against arsenic-induced respiratory damage. The authors suggested that serum CC16 may be a useful biomarker of pulmonary epithelial damage in individuals with arsenical skin lesions. As the clinical significance of early epithelial changes in lung disease detected by serum CC16 are is yet to be fully determined, how far the results of this study could be of clinical relevance remains to be ascertained.

GASTROINTESTINAL AND LIVER DISEASE

Gastrointestinal Disease

Chronic arsenic toxicity has been reported to produce various gastrointestinal (GI) symptoms. Mild gastrointestinal symptoms were reported in 76% of subjects exposed to environmental arsenic exposure at Torku, Japan [45]. Gastroenteritis was reported in a study of 1,447 cases of chronic arsenicosis caused by drinking arsenic- contaminated water (50-1,800 μg /L) in the Inner Mongolian Autonomous region of China [46]. Out of 248 patients suffering from chronic arsenicosis following from drinking arsenic- contaminated water (50 – 14,200μg/L) in West Bengal, India, symptoms of dyspepsia was were present in 60 out of 156 (38.4%) of the cases studied [16]. Many investigators variously reported symptoms like such as nausea, diarrhoea, anorexia and abdominal pain in cases of chronic arsenic toxicity [15, 21-24]. However, in an epidemiological study carried out in the affected population, there was no difference in the incidence of abdomen pain abdomen among people drinking arsenic contaminated water and the control population (27.84% vs 31.81%,p>0.05) [47].

Liver Disease

Exposure of to inorganic arsenic compounds has been found to be associated with the development of chronic pathological changes in the liver. Several authors reported cases of liver damage following the treatment of patients with arsenic as Fowler's solution [17, 48, 49]. All of these patients developed features of portal hypertension with signs of liver fibrosis. Typical cutaneous signs of long-term arsenic exposure were also observed in some of the patients. Moreover, there been case reports on liver cirrhosis following medication with inorganic arsenic compounds [5, 50].

Portal hypertension associated with periportal fibrosis was reported in nine patients who were found to have high arsenic levels in their liver in Chandigarh, India. Two of those patients had been found to be drinking arsenic- contaminated water (549 and 360 μg /L) [51]. From a population- based study in West Bengal, hepatomegaly was reported in 62 out of 67 members of families who drank arsenic- contaminated water (200-2000 μg /L), while whereas this was only the case only in six out of 96 people who took safe water (<50 μg /L) from the same area [15]. Thirteen arsenic- exposed patients who had hepatomegaly were further investigated in a Kolkata hospital in Kolkata. All showed various degrees of portal zone expansion and fibrosis on liver histology. Four of the five patients who had splenomegaly showed evidence of increased intra splenic pressure (30-36 cm saline), suggesting portal hypertension. Splenoportography done in those cases showed evidence of intrahepatic portal vein obstruction. Although routine liver function tests were normal in all these cases, the Bromosulphathlion retention test done in three patients were abnormal. The arsenic level in the liver tissue (estimated by Neutron Activation Analysis) was found to be elevated in 10 out of those 13 cases (As levels: Cases- 0.5 to 6 mg/kg; control- 0.10 ± 0.04 mg/kg) [15]. In another study from West Bengal, hepatomegaly was found in 190 out of 248 cases of chronic arsenicosis investigated in the same hospital. Arsenic levels in drinking water varied from 50 µg/L to 3,200 µg/L. Evidence of portal zone fibrosis on liver histology was found in 63 out of 69 cases of hepatomegaly. Liver functions tests carried out on 93 such patients showed evidence of elevation of ed ALT, AST and ALP in 25.8%, 6.3% and 29% of cases respectively. Serum globulin was found to be high (>3.5 gm/dl) in 19 (20.7%) cases [52].

Liver function tests were studied forin arsenic- exposed subjects in three towns of Leguneara, Mexico. Significant elevation of ed bilirubin and alkaline phosphatase levels wereas observed amongst people living in the town having the highest (239 ± 88 µg/L) arsenic exposure category compared to those living in the lowest (14 ± 3.1 µg/L) exposed group. Nonetheless, sSerum transaminases and albumin levels were however normal in all of the groups [53].

A cross- sectional epidemiological study was carried out on 7,683 people in an arsenic- affected district of West Bengal. Out of these, 3,467 and 4,216 people consumed water contaminated with arsenic below and above 50 µg /L, respectively. The prevalence of hepatomegaly was significantly higher in arsenic- exposed people (10.2%) compared to controls (2.99%, p< 0.001). The incidence of hepatomegaly was found to have a linear relationship proportional to the increasing exposure of arsenic in drinking water in both sexes (p < 0.001) [47].

Liver enlargement was also reported in people drinking of arsenic- contaminated water by other workers from West Bengal, Inner Mongolia and Bangladesh [13, 25, 46, 54].

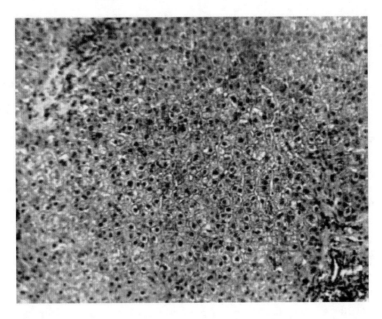

Figure 11. Liver histology showing non-cirrhotic portal fibrosis with expansion of the portal zone in the case of chronic arsenic toxicity.

Experimental Model for Liver Fibrosis

In an experimental study, BALB/C mice were given water contaminated with arsenic (3,200 µg /L) *ad libitum* for 15 months, the animals being sacrificed at 3- months intervals. In the experimental animals, the progressive reduction of hepatic glutathione and enzymes of the anti- oxidative defencedefense system were found to be associated with lipid peroxidation. Liver histology showed fatty infiltration at 12 months and hepatic fibrosis at 15 months [55]. In another experimental study, increasing the dose and duration of arsenic exposure in mice were found to cause a progressive increase of oxy-stress associated with an elevation of the cytokines, IL-6 and TNF- ἀ, with increasing collagen content and arsenic levels in liver [56].

All of these studies show that the prolonged drinking of arsenic- contaminated water is associated with hepatomegaly. Predominant lesions of hepatic fibrosis appears to be caused by arsenic- induced oxy-stress.

CARDIOVASCULAR DISEASE

Severe cardiovascular manifestations, associated shock and hypotension, cardiac arrhythmia, or congestive heart failure occur in acute or sub- acute arsenic poisoning following the acute ingestion of inorganic arsenic in large doses. The arrhythmias may be a prolongation of the Q-T interval or ventricular Tachycardia [19]. In a study of three cases of acute arsenic poisoning in the USA, a prolongation of the Q-T interval and an abnormality of the T-wave wereas reported [57]. The abnormalities returned to normal in two cases in whom a repeat ECG was done after two months. The ECG abnormalities were not related to any electrolyte disturbance.

Various cardiovascular manifestations had been reported from south-western Taiwan, China following the chronic drinking of arsenic- contaminated water from artesian wells. The iIngestion of arsenic- contaminated water from these wells from the early 1900s to the early 1960s was reported to be associated with the development of more than 1,000 cases of black foot disease (BFD), a form of severe peripheral vascular disease [19]. Most of the reports of chronic arsenic toxicity from drinking water studies involving these BFD patients had shown a significant positive correlation between arsenic in drinking water and death rates from cardiovascular diseases (*vide infra*). However, a systematic review of the epidemiological evidence on arsenic exposure and cardiovascular disease showed that methodological limitations limited the interpretation of the moderate- to- strong association between high arsenic exposure and cardiovascular outcomes in Taiwan. In other populations and in occupational settings, the evidence was inconclusive [58].

Recently, a study was carried out to assess the cardiac effects of arsenic in a human population chronically exposed to arsenic from artesian wells in residents of Ba Men, Inner Mongolia by investigating QT interval alterations. A total of 313 participants with a mean arsenic exposure of 15 years were divided into three arsenic exposure groups: low ($\leq$ 21 µg/L), medium (100–300 µg/L), and high (430–690 µg/L). ECGs were obtained on all study subjects. A value of the QTc (corrected QT) interval $\geq$ 0.45 seconds was considered to be prolonged. The prevalence rates of QT prolongation and water arsenic concentrations showed a dose-dependent relationship (p = 0.001). The prevalence rates of QTc prolongation were 3.9, 11.1 and 20.6% for low, medium, and high arsenic exposure, respectively.

QTc prolongation was also associated with sex (p < 0.0001), but not age (p = 0.486) or smoking (p = 0.1018). Females were more susceptible to QT prolongation than males. The investigators demonstrated a significant association between chronic arsenic exposure and QT interval prolongation in a human population [59].

An ecological study carried out in Spain showed an increased cardiovascular mortality rate at the municipal level with mild- to- moderate elevated arsenic concentrations in drinking water. Standardized mortality ratios (SMRs) for cardiovascular (361,750 deaths), coronary (113,000 deaths), and cerebrovascular (103,590 deaths) disease were analyseanalyzed for 1,721 municipalities from where data of arsenic concentrations in drinking water were available for the period 1999-2003, covering 24.8 million people. Mean municipal drinking water arsenic concentrations ranged from <1 to 118 µg/L. Compared to the overall Spanish population, sex- and age-adjusted mortality rates for cardiovascular (SMR 1.10), coronary (SMR 1.18), and cerebrovascular (SMR 1.04) diseases were increased in municipalities with arsenic concentrations in drinking water > 10 µg/L.

Compared to municipalities with arsenic concentrations < 1 µg/L, fully adjusted cardiovascular mortality rates were increased by 2.2% (-0.9% to 5.5%) and 2.6% (-2.0% to 7.5%) in municipalities with arsenic concentrations between 1-10 and >10 µg/L, respectively (P-value for trend 0.032). The corresponding figures were 5.2% (-0.8% to 9.8%) and 1.5% (-4.5% to 7.9%) for coronary heart disease mortality, and 0.3% (-4.1% to 4.9%) and 1.7% (-4.9% to 8.8%) for cerebrovascular disease mortality [60].

Further A further association between arsenic exposure and increased mortality from cardiovascular disease was reported from a prospective cohort study with arsenic exposure measured in drinking water from wells and urine in 11,746 men and women in Bangladesh. Death occurred in 198 people from diseases of the circulatory system, accounting for 43% of the population's total mortalitiesy in the population. The mortality rate for cardiovascular disease was 214.3 per 100, 000 person years in people drinking water containing <12.0 µg/L arsenic, compared with 271.1 per 100, 000 person years in people drinking water with ≥12.0 µg/L arsenic. There was a dose-response relation between exposure to arsenic in well water assessed at the baseline, and mortality from ischaemic heart disease and other heart diseases; the hazard ratios in increasing quarters of the arsenic concentration in well water (0.1-12.0, 12.1-62.0, 62.1-148.0, and 148.1 864.0 µg/L) were 1.00 (reference), 1.22 (0.65 to 2.32), 1.35 (0.71 to 2.57), and 1.92 (1.07 to 3.43) (P=0.0019 for trend), respectively, after adjustment for potential confounders including age, sex, smoking status, educational attainment, body mass index (BMI), and changes in urinary arsenic concentration changes since the baseline [61]. The Bangladesh study demonstrated a dose- response relation for mortality due to cardiovascular diseases with a moderate level of arsenic exposure. One of the limitations of the study was that the data might not be generalisedized as the study population consisted of only lean married men and women with low BMIs.

Peripheral Vascular Disease (Blackfoot Disease)

Occurrence. The occurrence of Blackfoot disease (BFD), a form of peripheral vascular disease following the drinking of arsenic- contaminated, artesian well water has been extensively studied in the southwestern region of Taiwan. Arsenic has been well documented as one of the major risk factors for BFD, a unique peripheral arterial disease characterized by severe systemic arteriosclerosis as well as dry gangrene and spontaneous amputations of affected extremities at its end stages. Histologically, black foot disease (BFD) can be divided into two reaction groups, arteriosclerosis obliterans and thromboangitisobliterans, particularly affecting small vessels [62]. Clinically, the disease begins with patients' subjective complaints of coldness or numbness in the extremities (usually in the feet) and intermittent claudication, progressing over the course of several years to ulceration, gangrene, and spontaneous amputation [63]. The prevalence of BFD has been reported to be 8.9 per 1,000 among 40,421 inhabitants studied at in Taiwan [4]. The villages surveyed were arbitrarily, divided according to the arsenic content in the well water in to low (below 300μg/L), mid (300-600 μg/L) and high (above 600 μg/L) groups. The prevalence of BFD revealed a clear-cut ascendency gradient from low to mid to high groups for both sexes and the three different age groups studied [64]. That a humic substance plays an etiologic role in BFD has not been substantiated by epidemiologic or animal studies. A causal role for arsenic in the induction of BFD offers the best explanation for these observations in Taiwan [65].

A close relation between long-term arsenic exposure and peripheral vascular disease in blackfoot disease- endemic villages in Taiwan was observed, even after stopping the consumption of artesian well water. The correlation between previous arsenic exposure and peripheral vascular disease after stopping the consumption of high-arsenic artesian well water for more than two decades in blackfoot disease- endemic villages in Taiwan has been done on a total of 582 adults (263 men and 319 women, aged 52.6 +/- 10.6 years) living in arsenic endemic villages of Taiwan. Doppler ultrasound measurements of systolic pressures were done on bilateral ankles (posterior tibial and dorsal pedal) and brachial arteries, and as well as an estimation for long-term arsenic exposure estimation. The diagnosis of peripheral vascular disease was based on an ankle-brachial index (the ratio between ankle and brachial systolic pressures) <0.90 on either side. A multiple logistic regression analysis was used to assess the association between peripheral vascular disease and arsenic exposure. A dose-response relation was observed between the prevalence of peripheral vascular disease and the long-term arsenic exposure. The odds ratios (95% confidence intervals) after adjustment for age, sex, body mass index, cigarette smoking, serum cholesterol and triglyceride levels, diabetes mellitus and hypertension were 2.77 (0.84-9.14), and 4.28 (1.26-14.54) for those who had a cumulative arsenic exposure of 0.1-19.9 and > or = 20.0 mg/l-years, respectively, compared with those who were not exposed [66].

Peripheral vascular disorders with varying degrees of severity and incidence, including Raynaud's syndrome and acrocyanosis, have also been reported among people drinking arsenic- contaminated water in other countries [5, 16, 22, 25, 46, 67, 68].

Atherosclerosis and Macro Vascular Macrovascular Disease

The Athereogenecity of high arsenic- contaminated artisan well water was studied in Taiwan [69]. The life table method used to analyseanalyze the mortalitiesy of 789 BFD patients, followed for 15 years, showed a significantly higher mortality rate from cardiovascular, peripheral vascular disease among BFD patients as compared with the general population in Taiwan or residents in the BFD- endemic area. That a humic substance plays an etiologic role in BFD has not been substantiated by epidemiologic or animal studies. A causal role for arsenic in the induction of BFD offers the best explanation for the observations in

Taiwan [65]. The extent of carotid atherogenecity was studied on 463 residents of arsenic-endemic areas of Taiwan by duplex ultrasonography. Long- term arsenic exposure was found to be significantly associated with a prevalence of Carotid atherosclerosis in a dose- response relationship. A significant biological gradient was found after adjustment for age, sex, hypertension, diabetes mellitus, and addiction of smoking and alcohol addictions, and the serum level of lipids. The multivariate age adjusted odds ratio for atherosclerosiswas 3.1 (95% CI1.37.4) for those who had a cumulative arsenic exposure of $\geq$20 mg/L-years, compared to those without exposure to arsenic from drinking artesian well water [70].

The relation between the genetic polymorphisms of GSTP1 and p53, and the risk of carotid atherosclerosis following arsenic exposure was studied in an arsenic- contaminated region of Taiwan. Arsenic exposure through drinking well water and evidence of carotid atherosclerosis were determined ion 605 residents (289 men and 316 women). Carotid atherosclerosis was diagnosed by either a carotid artery intima-media thickness (IMT) of >1.0 mm, a plaque score of > or =1, or stenosis of >50%. A significant age- and gender-adjusted odds ratio of 3.3 for the development of carotid atherosclerosis was observed among the high-arsenic exposure group who drank well water containing arsenic at levels >50 μg/L. The high-arsenic exposure group with GSTP1 variant genotypes of Ile/Val and Val/Val, and with the p53 variant genotypes of Arg/Pro and Pro/Pro had 6.0- and 3.1-fold higher risks of carotid atherosclerosis, respectively. A multivariate-adjusted odds ratio of 3.4 for the risk of carotid atherosclerosis among study subjects with the two variant genotypes of GSTP1 and p53 was also found [71].

In a cross- sectional study on a population of 10,910 in Bangladesh, there was a positive association between low-to-moderate levels of arsenic exposure from drinking water and high pulse pressure [72], an indicator of increased arterial stiffness which is associated with an increased risk of atherosclerosis [73, 74].

Long-term exposure to inorganic arsenic from artesian well drinking well water with carotid atherosclerosis in the a Blackfoot Disease (BFD)-hyperendemic area in Taiwan was further studied on 304 adults (158 men and 146 women) .in regards to the percentage of arsenic species, primary methylation index [PMI = MMA(V) / (As(III) + As(V)] and secondary methylation index [SMI = DMA(V) / MMA(V)] as indicators of arsenic methylation capacity. Results showed that carotid atherosclerosis cases had a significantly greater percentage of MMA(V) [%MMA(V)] and a lower percentage of DMA [%DMA (V)] compared to controls. The authors conclude that individuals with a greater exposure to arsenic and a lower capacity to methylate inorganic arsenic may be at a higher risk forto carotid atherosclerosis [75].

The atherogenicity of arsenic could be associated with its effects of hypercoagulability, endothelial cell injury, smooth muscle cell proliferation, somatic mutation, oxidative stress, and apoptosis. However, its interaction with some trace elements and its association with hypertension and diabetes mellitus could also explain part of its higher risk of developing atherosclerosis. Although humic substances have also been suggested as a possible cause of BFD, epidemiologic studies are required to confirm its etiologic role [76].

Recently, the animal model for arsenic- induced atherosclerosis and liver sinusoidal endothelial cell dysfunction has been developed. Initial studies in these models show that arsenic exposure accelerates and exacerbates atherosclerosis in apolipoprotein E-knockout mice. Findings from mechanistic studies indicate that arsenic causes inflammation of the vascular tissues and active oxidative signaling [77].

Microvascular Disease

An angiographic study of 12 patients with cutaneous arsenic poisoning from Salta Province of Argentina (Torres Soruco et al., 1991, as cited in Engel et al., 1994) [65, 78] showed evidences of microangiopathies in all patients, half of whom were under 25 years of age. In another study, a baseline and heat- induced cutaneous perfusions of the big toe were assessed in 45 non-smoker residents of the blackfoot -disease- endemic area of Taiwan without any history of cardiac and peripheral vascular disease and 51 similar age matched controls using laser Doppler flowmetry. Both basal and heat-induced cutaneous perfusion was substantially lower in formerly exposed subjects, suggesting microcirculatory changes [79]. However, no report of presence of neither cutaneous sign of arsenicosis, nor magnitude of past or present arsenic exposure was provided.

Arsenic has been reported to increase the risk of non-insulin-dependent diabetes mellitus and its related micro- and macro vascular diseases. The prevalence of non-insulin-dependent diabetes and related vascular diseases waswere ascertained by the age and sex among residents in arsenicosis-endemic and non-endemic areas in Taiwan from records of medical reimbursement claims of the National Health Insurance Database for 1999-2000. The study included 66,667 residents living in endemic areas and 639,667 in non-endemic areas, all ≥ 25 years of age, and the age- and gender-adjusted prevalence of non-insulin-dependent diabetes to in the Taiwan's general population in Taiwan, was 7.5% (95% confidence interval, CI: 7.4-7.7%) in the arsenicosis -endemic areas and 3.5% (CI:3.5-3.6%) in the non-endemic areas. Among both diabetics and non-diabetics, a higher prevalence of micro-vascular and macrovascular diseases was observed in arsenicosis –endemic areas rather than in the non-endemic areas. The a Age- and gender-adjusted prevalence of microvascular disease in endemic and non-endemic areas was 20.0% and 6.0%, respectively, for diabetics, and 8.6% and 1.0%, respectively, for non-diabetics. The corresponding prevalence of macrovascular disease was 25.3% and 13.7% for diabetics, and 12.3% and 5.5% for non-diabetics [80]. As the prevalence of microvascular disease was estimated from records of medical reimbursement claims of health insurance, it might have been overestimated if the patients were more likely to visit clinicians and included in the database than were the unaffected people.

An increased prevalence of microvascular diseases, including neurological and renal disorders, were reported to be associated with arsenic ingestion. Microvascular diseases in relation to arsenic exposure levels haves been studied on a total of 28, 499 subjects living in the study area of Taiwan, information being obtained through their medical records from the National Health Insurance database in 1999–2000. The arsenic concentrations of artesian well water in the villages of the study area were utilized as indices of previous ingestion levels. Both stratified analysis and unconditional logistic regression were used to mainly examine mainly neurological and renal disease in relation to the arsenic exposure, taking into account the diabetices status. The age-adjusted and gender-adjusted prevalence of microvascular diseases was 7.51% [95% confidence interval (CI) 7.50–7.51] for an arsenic level of 100µg/L, and then increased from 6.59% (6.59– 6.60) for the arsenic concentrations of 100–290µg/L to 8.02% (8.02–8.03) and 11.82% (11.81–11.83) for those of 300–590µg/L mg/L and ≥600µg/L mg/L in non-diabetic subjects. For diabetic patients, the prevalence was 16.41% (95% CI 16.37–16.45), 15.85% (15.8–15.9), 21.69% (21.6–21.8), and 28.31% (28.2–28.4) for arsenic levels <100, 100–0.290, 300–590, and ≥ 600µg/L, respectively. The prevalence of microvascular diseases significantly increased significantly with arsenic exposure, especially at higher levels, and the relationship is stronger in diabetics than in non-diabetic subjects. The results for neurological disease are very similar, and the patterns are the same for renal disease [81]. In the report the authors acknowledged that the precise level of arsenic exposure at the individual level, as a consequence of wellwater consumption, was rather difficult to assess without individual lifetime histories. The consequences of exposure misclassification might have some limitation in the observed relationship of arsenic exposure and microvascular disease.

Ischemic Heart Disease

Increased mortality rates from ischemic heart disease (IHD) has been reported among arsenic- exposed residents in BFD- endemic regions of Taiwan [19]. Mortality rates from ischemic heart disease (IHD) among residents in 60 villages of the area in Taiwan with endemic arsenicos are from 1973 through 1986 was analyzed to examine their association with arsenic concentrations in drinking water. Based on 1,355,915 person years and 217 IHD deaths, the cumulative ISHD mortalities from birth to age 79 years were 3.4%, 3.5%, 4.7% and 6.6%, respectively, for residents who lived in villages in which the median arsenic concentrations in drinking water were <0.1, 0.1 to 0.34, 0.35 to 0.59 and $\geq$ 0.6 mg/L. A cohort of 263 patients affected with BFD and 2,293 non-BFD residents in the endemic area of arseniasis were recruited and followed up for an average period of 5.0 years. There was a monotonous biological gradient relationship between cumulative arsenic exposure through drinking artesian well water and IHD mortality.

The relative risks were 2.5, 4.0, and 6.5 respectively, for those who had a cumulative arsenic exposure of 0.1 to 9.9, 10.0 to 19.9, and $\geq$ 20.0 mg/L years compared with those without the arsenic exposure after adjustment for age, sex, cigarette smoking, body mass index, serum cholesterol and triglyceride levels, and disease status for hypertension and diabetes through proportional hazards regression analysis [82].

Arsenic-related ischemic heart disease (IHD) has been suggested to have a pathogenic mechanism partially different from that of IHD, unrelated to long-term arsenic exposure to arsenic. A study in Taiwan reported an increase in the risk of arsenic-related IHD associated with hypertension and an elevated body mass index, but not with the serum lipid profile, cigarette smoking and alcohol drinking.

A significant biological gradient between the risk of ischemic heart disease (IHD) and the duration of intake of arsenic- contaminated artesian well water was reported from in Taiwan. Furthermore, a community-based health survey in an arsenic- endemic region was carried out to find out any association between arsenic-related IHD and the serum antioxidant micronutrient level. A total of 74 patients affected with IHD, who were diagnosed through both electrocardiography and a Rose questionnaire interview, and 193 age-sex matched healthy controls were selected for the examination of serum levels of micronutrients by a high performance liquid chromatography (HPLC). A significant reverse dose-response relationship with arsenic-related IHD was observed for the serum levels of alpha- and beta-carotene, but not for serum levels of retinol, lycopene and alpha-tocopherol. A multivariate analysis showed a synergistic interaction on arsenic-related IHD between duration of consumption of artesian well water and low serum carotene level. An increased risk of arsenic-related IHD was also associated with hypertension and elevated body mass index, but not with serum lipid profile, cigarette smoking and alcohol drinking. The authors suggested that arsenic-related IHD has a pathogenic mechanism which is at least partially different from that of IHD unrelated to long-term exposure to arsenic [83].

The occurrence of ischemic heart disease (IHD) was further ascertained by electrocardiography in a community- based study on a total of 462 subjects living in the blackfoot disease-endemic villages of Taiwan. A diagnosis of IHD was made by evaluating the resting electrocardiograms with the Minnesota code. The history of arsenic exposure was estimated through information obtained from a personal interview according to a structured questionnaire and the arsenic content in the artesian well water of the villages. Cumulative arsenic exposure (CAE) was calculated as the sum of the products multiplying the arsenic concentration in artesian well water (mg/l) by the duration of drinking the water (years) in consecutive periods of when living in the different villages. Among the subjects, 78 cases (16.9%) were diagnosed as having IHD. The prevalence rates of IHD for the age groups of 30-39, 40-49, 50-59, and >/=60 years were 4.9, 7.5, 16.8, and 30.7%, respectively (P<0.001). For those with a CAE of 0, 0.1-14.9 and >/=15 mg/l-years, the prevalence rates of IHD were 5.2, 10.9 and 24.1%, respectively (P<0.001). The odds ratios (95% confidence intervals) for IHD were 1.60 (0.48, 5.34), and 3.60 (1.11, 11.65), respectively, for those with a CAE of 0.1-14.9 and >/=15.0 mg/l-years, when compared with those lacking drinking water exposure to arsenic after a multivariate adjustment. The authors concluded that IHD is associated with long-term arsenic 'exposure [84]. However, the limitation of the paper was that cumulative arsenic exposure was measured in group levels by village. Further prospective studiesy with individual arsenic data and in other geographic regions are needed to establish the cause-effect relationship.

Hypertension

An increased prevalence of hypertension was observed in 6.2% patients affected with arsenic- induced skin lesions amongst 144 cases compared to none without skin lesions in 36 subjects in Antafagesta, Chile [22]. An epidemiological study reported an increased prevalence of hypertension among residents in anthe endemic area of B black Ffoot disease and a dose-response relationship between ingested inorganic arsenic and prevalence of hypertension [85]. The investigators studied a total of 382 men and 516 women residing in arsenic- hyperendemic areas in Taiwan. They observed 1.5 fold increases in age- and sex-adjusted prevalence of hypertension compared with residents in non-endemic areas. The higher the cumulative arsenic exposure the higher was the prevalence of hypertension. The dose response relation remained significant after adjustment for age, sex, diabetes mellitus, protein-urea, body mass index and serum triglyceride levels.

A cross sectional study for the determination of hypertension was done on 1,595 arsenic-exposed and 114 unexposed (aged more than 30 years) people in four villages in Bangladesh. The prevalence ratio for hypertension was assessed after adjustment for age, sex and body mass index. The Mantel-Haenszel-adjusted prevalence ratio for hypertension increased with an increase in arsenic in the drinking water. Exposure categories were assessed as <500 µg/L, 500 to 1000 µg/L and > 1000 µg/L. Corresponding to the exposure categoryrie, and using the "unexposed" as a reference, the prevalence ratios for hypertension adjusted for age, sex, and body mass index were 1.2, 2.2 and 2.5 in relation to arsenic exposure in micrograms per liter, indicating a significant dose- response relationship (p<0.001). A similar significant dose response relationship was also observed with a cumulative dose of arsenic exposure [86]. However, potential limitations of the study include a failure to identify all eligible participants and the fact that a direct measurement of exposure is preferred to a recall of exposure. In another study from Bangladesh, though the investigators found arsenic exposure was positively associated with systolic hypertension and high pulse pressure, no apparent associations were observed between the time-weighted well arsenic concentration (TWA) and general or diastolic hypertension [72]

Cerebrovascular Disease

Increased incidence of cerebrovascular disease in chronic arsenicosis cases has been reported in case studies from Japan and Inner Mongolia [45, 46].

The increased prevalence of cerebrovascular disease and ingestion of inorganic arsenic in drinking water was reported in a cross- sectional study in Taiwan. A total of 8,102 men and women from 3,901 households were recruited forin this study. The status of cerebrovascular disease of study subjects was identified through home visits, personal interviews and by a review of hospital medical records according to the WHO criteria. Information on well water consumption of well water, socio-demographic characteristics, cigarette smoking, and alcohol consumption habits, as well as personal and family disease history of disease, was also obtained. A significant dose-response relationship was observed between the arsenic concentration in well water and prevalence of cerebrovascular disease after an adjustement for age, sex, hypertension, diabetes mellitus, cigarette smoking and alcohol consumption. The biological gradient was more prominent for cerebral infarction, showing multivariate-adjusted odds ratios of 1.0, 3.4, 4.5, and 6.9, respectively, for those who consumed well water with an arsenic content of 0, 0.1 to 50.0, 50.1 to 299.9, and >300 µg/L [87]. However, there were limitations in the methodology described in the study. Subjects were recruited from a registry of all adult residents within 18 villages, but the criteria for village selection, methods of recruitment, and percentage participation were not stated.

Other studies from the region had not produced similar findings. Increased mortality was not observed due to cerebrovascular accidents when mortality and population data were obtained from the local household registration offices and Taiwan 's Provincial Department of Health during 1973-1986 and correlated with arsenic levels in well water determined in 1964-1966 in 42 study villages in Taiwan [88]. Another study from Taiwan reported that elevation in mortality due to cerebrovascular disease in arsenic- affected regions was not significant compared to cardio vascular disease [89].

However, a recent study from Taiwan again showed that the exposure to arsenic in drinking water was associated with a higher risk of cerebrovascular disease (CVD). A study of (CVD) mortality was carried out with the national death registry data from two areas in Taiwan from 1971 to 2005. The arsenic levels in the drinking water in the BFD area were generally higher than those in the unexposed area. Standardized mortality ratios (SMRs) adjusted for gender and age were calculated using the whole population of Taiwan and the population of the unexposed area as the reference population respectively. The standardized mortality rate of CVD in Taiwan decreased from 2.46/100 in 1971 to 0.63/100 in 2005. The exposed group had higher SMRs of CVD in comparison with the reference populations, with SMRs from 1.06 to 1.09 in men and 1.12 to 1.14 in women. The BFD- endemic area had higher mortality rates of CVD than the unexposed area, with SMR = 1.05 in men and SMR = 1.04 in women [90].

Recently, an increased mortalitiesy wereas reported due to cerebrovascular disease with a low dose range of arsenic in the USA's ground water in USA. A standardized mortality ratio (SMR) analysis was conducted in a contiguous six county study area of south eastern Michigan to investigate the relationship between moderate arsenic levels and several selected disease outcomes. Arsenic data wasere compiled from 9,251 well water samples tested by the Michigan Department of Environmental Quality from 1983 through 2002. The six county study area had a population-weighted mean arsenic concentration of 11.00 µg/L and a population-weighted median of 7.58 µg/L. Michigan Resident Death Files data wasere amassed for 1979 through 1997, and sex-specific SMR analyses were conducted with an indirect adjustment for age and race; 99% confidence intervals (CI) were reported. Elevated mortality rates were observed for both males (M) and females (F) for cerebrovascular disease (M SMR, 1.19; CI, 1.14–1.25; F SMR, 1.19; CI, 1.15– 1.23) .[91].

NEUROLOGICAL DISEASE

There are many reports of occurrence of peripheral neuropathy occurring due to chronic arsenic exposure of arsenic through drinking water [13, 23, 25, 45, 46, 92]. In a hospital-based study in west Bengal, an electro-physiological investigation on eight arsenic- exposed subjects (arsenic level in drinking water: 200-2000µg/L) showed sensory predominant distal polyneuropathy in eight patients with arsenical skin lesions [93]. In a further study on 156 patients of chronic arsenicosis associated with arsenical skin lesion and a history of prolonged drinking of arsenic contaminated water (50 – 14200µg/L), peripheral neuritis characterized by paraesthesia (e.g., tingling, numbness, limb weakness, etc.) was present in 74 (47.4%) cases. An objective evaluation of neuronal involvement, done in 29 patients, showed an abnormal electromyography (EMG) in 10 (30.8%) and altered nerve conduction velocity and EMG in 11 (38%) cases having chronic arsenic exposure [3, 16]. In another electrophysiological study carried out on 88 arsenicosis patients of arsenicosis in West Bengal, sensory neuropathy was found in 24 (27.3%), motor neuropathy in 13 (14.7%) and abnormal EMG in 5 (5.7%) cases [94]. Abnormal EMG findings, suggestive mostly of sensory neuropathy, wereas reported in 10 out of 32 subjects exposed to drinking arsenic- contaminated well water (range 60 to 1,400 µg/L) in Canada [95]. Increased arsenic exposure, as measured by cumulative and urinary arsenic measurements, was found to be associated with subclinical sensory neuropathy as determined by an elevation of vibration tactile sensitivity in the toe oin 137 subjects in Bangladesh .[96].

Peripheral neuropathy was clinically diagnosed in 21 (52.5%) out of 40 cases of chronic arsenic exposure through the drinking water (202-1654 µg/L) in a village in Bihar, India having arsenical skin lesions and elevated arsenic levels of arsenic in their urine, nails and hair. Predominant symptoms were limb pain and paresthesia (e.g., burning) in a stocking and glove distribution, numbness, hyperpathea, distal hypesthesia (reduced perception of sensation of pin prick sensations, vibration sense, joint sense), diminished joint pain and calf tenderness [97].

Epidemiological Study

A cross- sectional study of nerve conduction velocity was carried out at in Taiwan on 130 students aged 12-24 years who were exposed to arsenic from drinking arsenic- contaminated water. After adjustments for t of gender and height, a significant odds ratio of 2.9 (95% Confidence Interval.CI,1.1-7.5) was observed for the development of the slow conduction velocity of sural sensory action potential (SAP) among the study subjects with a cumulative arsenic exposure of > 100 mg. The authors concluded that chronic arsenic exposure might induce peripheral neuropathy [98].

Increased prevalence of peripheral neuritis was reported in a cross- sectional study from an arsenic- endemic district (Nadia) of West Bengal. Out of a population of 10,469 arsenic-exposed participants selected from all 17 affected blocks of the district by multistage sampling, 1,616 (15.43%) cases had arsenical skin lesions while 8,853 control subjects had no skin lesions. The mean arsenic content in drinking water among the cases was (103.46; (SD ± 153.28 µg/L) while among the controls it was (73.18 (SD ± 115.10 µg/L). Peripheral neuropathy, characterisedized by tingling, numbness and limb pain, was found in 257 (15.90%) cases while whereas it was found in only 136 (1.54%) controls (p<0.001) [32].

Peripheral neuritis, sleep disturbances, weakness, and cognitive and memory impairment have been reported in residents of Byan College Station, Texas that were exposed to arsenic from air and water from arsenic trioxide used to produce defoliants from an Atochem plant [92]. Headaches haves been reported to occur in people drinking arsenic- contaminated water in Mexico [23] and in West Bengal [16].

Irritability, lack of concentration, depression, sleep disorders, headache and vertigo were reported in arsenicosis people showing features of neuropathy in West Bengal [94].

HEMATOLOGICAL EFFECTS

Anemia, leucopenia and thrombopenia have been reported in acute and chronic arsenic poisoning [19]. A characteristic pattern of anaemia, leucopenia and thrombocytopenia were found in 55 individuals exposed to arsenic in drinking water in Niigata Prefecture in Japan for approximately 5 years, half of the subjects having arsenical skin lesions [99]. In a hospital based study in Kolkata,West Bengal, anaemia was found in all the 13 patients exposed to arsenic- contamiated ground water (0.2-2 mg/L) and affected with arsenical skin lesions and liver disease (non cirrhotic portal fibrosis) [15].

A further study on 156 people exposed to arsenic contaminated water (0.05-14.2 mg/L) showed incidence of anemia in 47.4% of cases [16]. However, no association of anemia was found in people drinking well water (mean 0.22mg/L) in Alaska [100] and in two towns of Utah (arsenic exposure 0.18 mg/L and 0.27 mg/L) [101].

DIABETES MELLITUS

The association between ingested inorganic arsenic and prevalence of diabetes mellitus was studied ion 891 adults residing in arsenic- contaminated (arsenic level in artesian well water, 700 to 930 µg/L) villages in Southern Taiwan for 30 years or more. The cumulative lifetime ingestion of arsenic in drinking water could be determined for 718 subjects aged over 30 years old. The status of diabetes mellitus was determined by an oral glucose tolerance test and a history of diabetes regularly treated with sulfonylurea or insulin. They observed a dose-response relation between cumulative arsenic exposure and prevalence of diabetes mellitus prevalence. The relation remained significant after adjustment for age, sex, body mass index and activity level at work by a multiple logistic regression analysis giving a multivariate adjusted odds ratio of 6.61 (95% CI, 0.86-51.0) and 10.05 (95% CI, 1.30-77.9), respectively, for those who had a cumulative arsenic exposure of 0.1-15.0 mg/L year and greater than 15.0 mg/L year compared with those who were unexposed. However, the study had a few limitations, one being the prevalence of diabetes observed in 108 unexposed subjects as only 0.9%, much lower than the 5.1% background prevalence cited by the authors [102].

Increased prevalence of diabetes mellitus due to drinking arsenic- contaminated ground water was reported among 163 subjects with arsenical keratosis living in 7 villages in Bangladesh having ground water contamination when compared with the prevalence of diabetes among 854 subjects recruited from an region uncontaminated by arsenic uncontaminated region [103]. A diagnosis of diabetes was made from the history of symptoms, previously diagnosed disease, glycosuria and blood glucose after an oral glucose challenge. Using the control population's age- and sex- adjusted prevalence as a reference (1.0), the prevalence ratios for diabetes mellitus were calculated to be 2.6, 3.9 and 8.8 for time-weighted-average exposure categories of less than 500µg/L, 500-1000µg/L and greater than 1000µg/L. The lack of a comprehensive, systematic long-term sampling of the water supplies in the study area is a limitation of the study because directly measured individual exposure data over time would have been desirable. However, these results suggest that chronic arsenic exposure may induce diabetes mellitus in humans.

A high mortality rate, primarily attributed primarily to diabetes mellitus, was reported from four arsenic- contaminated (arsenic levels in artesian well water, 250-1,400 µg/L) four townships of Taiwan from the early 1900s until the mid-to- late1970s. The observed mortalitiesy between 1971 and 1994 were as compared with age- and sex- specific expected mortalitiesy based on data from (1) a local reference group similar to a study group derived from nearby counties and (2) all of Taiwan [42].

A prospective cohort follow- up study was carried out on 446 subjects who agreed to participate in an investigation for to detection of diabetes mellitus by estimating fasting blood sugar and a glucose tolerance test in the years 1991 and 1993. The participants were drawn from 632 non- diabetic subjects, enrolled in the year 1970 from three villages in Taiwan having history of arsenic exposure (700 - 930 µg/L) from artesian well water. The incidence of diabetes mellitus in the study population was calculated as the total number of incident cases divided by the sum of follow-up person-times in all subjects. Data on each subject included age, sex, body-mass index and an index of lifetime cumulative arsenic exposure (CAE). CAE (in units of milligrams per liter-years) was calculated as the product of the median arsenic concentration of the well water in every village that a subject inhabited at some point in his or her life multiplied by the length of time they consumed well water in that village. Incidences for diabetes mellitus in the study population were compared with those reported for a demographically similar population that was contemporaneously studied contemporaneously. During the follow- up period, that included 1,499.5 person-years, 41 out of 446 subjects developed diabetes mellitus (all non-insulin-dependent diabetes mellitus). The incidence for new cases was particularly increased in subjects 55 years of age or older (50.8 per 1,000 person-years). The relative risk for developing diabetes mellitus among those more than 17 mg/L-years CAE compared with those with less than 17mg/L CAE was 2.1(995% CI, 1.1-4.2), adjusted for age, sex and body-mass index in a multivariate Cox proportional hazard model. When considered as a continuous variable, the CAE was associated with an adjusted relative risk of developing diabetes mellitus of 1.03 for every 1mg/L years of exposure (p<0.05) [104].

Although earlier reports from Taiwan and Bangladesh showed an association between diabetes mellitus following exposure to relatively high concentrations (> 500 µg/L) of arsenic in drinking water, the risk of diabetes following exposure to lower levels of As (< 300 µg/L) has not been conclusively demonstrated. A population-based cross-sectional study was done on 11,319 participants using baseline data in from the Health Effects of Arsenic Longitudinal Study in Araihazar, Bangladesh, to evaluate the associations of well water arsenic and total urinary arsenic concentrations, and the prevalence of diabetes mellitus and glycosuria. The investigators also assessed the concentrations of well water arsenic, total urinary arsenic, and urinary arsenic metabolites in relation to blood glycosylated haemoglobin (HbA1c) levels in subsets of the study population. More than 90% of the cohort members were exposed to < 300µg/L of arsenic in drinking water.

The authors reported no association between arsenic exposure and the prevalence of diabetes, and no association between arsenic exposure and prevalence of glycosuria. The adjusted odds ratios for diabetes were 1.00 (referent), 1.35 [95% confidence interval (CI), 0.90–2.02], 1.24 (0.82–1.87), 0.96 (0.62–1.49), and 1.11 (0.73–1.69) in relation to quintiles of time-weighted water arsenic concentrations of 0.1–8, 8–41, 41–91, 92–176, and ≥ 177 µg/L, respectively, and 1.00 (referent), 1.29 (0.87–1.91), 1.05 (0.69–1.59), 0.94 (0.61–1.44), and 0.93 (0.59–1.45) in relation to quintiles of urinary arsenic concentrations of 1–36, 37–66, 67–114, 115–204, and ≥ 205 µg/L, respectively. No association between arsenic exposure and prevalence of glycosuria and no evidence of an association between well water arsenic, total urinary arsenic, or the composition of urinary arsenic metabolites and HbA1c levels were observed in the study [105].

The study population in this report was generally lean and belonged to low socioeconomic and nutritional statuses. The findings, therefore, may not be generalizable to other study populations, given the possible different distribution of risk factors for diabetes that may influence the effect of arsenic exposure.

PREGNANCY OUTCOME

Associations between arsenic concentrations in drinking water and adverse pregnancy effects included spontaneous abortion [19, 106]. Investigators in Hungary reported a correlation between increased concentrations of arsenic in well water (60-270µg/L) and increased incidences of spontaneous abortion and perinatal death without an increase in premature birth from 1980-1992 [107]. Another evaluation of demographic data from 1970-1987 indicated that the frequency of still births and spontaneous abortions were as increased in a population drinking from wells with arsenic concentrations exceeding 100µg/L in Hungary [108, 109].

In an ecological study carried out in Chile, stillbirths (rate ratio 1.7; 95 percent CI: 1.5, 1.9), neonatal and postneonatal infant mortality rates were found to be increased in the high arsenic exposure city of Antofagasta as compared with the low exposure city, Valparaiso [110].

A cross-sectional study from Bangladesh compared rates of spontaneous abortions, stillbirths and preterm delivery between 96 women in one village who were exposed to ≥100 µg/L of arsenic to rates in 96 women in another village who were exposed to less than 20 µg/L, and showed two to three times higher rates among the exposed women [111]. Another study conducted in Bangladesh showed an increased stillbirth risk for stillbirth for women with a current arsenic level of ≥100µg/L, although the risk estimates were smaller (OR= 2.5; 95 percent CI: 1.5, 5.9). The authors further reported significant effects for spontaneous abortions (OR=2.5; 95 percent CI: 1.5, 4.4) [112]. No information on high exposure levels of 200 µg/L and or more were separately considered separately in this study. Both the Bangladeshi studies reported a relation to the overall duration of women's exposure without taking into account exposure during the actual time period of their pregnancies.

In West Bengàl, India, a retrospective study of pregnancy outcomes and infant mortality were done among 202 married women selected from a source population of 7,683 between 2001 and 2003. Reproductive histories were ascertained by structured interviews. Arsenic exposure during each pregnancy was assessed based on all water sources used, involving measurements from 409 wells. Odds ratios for spontaneous abortions, stillbirth, neonatal and infant mortality were estimated with logistic regressions based on the method of generalized estimating equations. High concentrations of arsenic ≥200 µg/L during pregnancy were associated with a six-fold increased risk for stillbirth after adjusting for potential confounders (odds ratios [OR] = 6.25; 95% confidence interval [CI]: 1.59, 24.6, p=0.009). Arsenic-related skin lesions were found in 12 women who had a substantially increased risk of stillbirth (OR=13.1, 95% CI: 3.17, 54.0, p=0.002). The odds ratio for neonatal death was 2.03 (95% CI: 0.57, 7.24). No association was found between arsenic exposure and spontaneous abortion (OR = 0.90; 95%CI 0.36, 2.26) or overall infant mortality (OR =1.18, 95% CI: 0.38, 3.64). This study adds to the limited evidence that exposure to high concentrations of arsenic during pregnancy increases the risk of stillbirth. However, there was no indication of increased rates of spontaneous abortion and overall infant mortality [113].

EYE DISEASE

In a hospital- based study in West Bengal, eye symptoms characterisedized by a burning of the eye, and conjunctival congestion was found in 69 (42.2%) out of 156 patients of with chronic arsenicosis associated with arsenical skin lesions and a history of prolonged drinking of arsenic contaminated water [16]. Conjunctivitis associated with chronic arsenic ingestion had also been reported from Bangladesh in 57(15.7%) out of 363 arsenic exposed to arsenic

[25]. Increased incidence of conjunctivitis was reported from a study of with a total of 1,482 arsenicosis patients, who were identified through household screening, living in 6 of 496 upazilas (sub-districts) of Bangladesh, who were identified through household screening. The investigators observed increased duration and older age to be significantly associated with an increased occurrence of conjunctivitis [114]. Papillary conjunctivitis was reported in 2 members of a family with dermatological manifestations and a history of drinking arsenic-contaminated water for 15 years in Bangladesh. Arsenic levels estimated in the nails and hairs of both patients were very high. A histopathological examination of conjunctival tissue confirmed the inflammatory response of a papillary type; however, an arsenic estimation in conjunctival tissue was not possible [115].

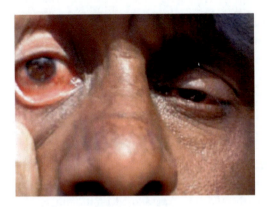

Figure 12. An arsenicosis patient showing conjunctivitis.

Pterygium

Pterygium is a fibrovascular growth of the bulbar conjunctiva and underlying subconjunctival tissue that may cause blindness. Chronic exposure to arsenic in drinking water was reported to be associated with the occurrence of pterygium in Taiwan. Eye examinations and a questionnaire interview of 223 participants from three exposure villages and 160 from four comparison villages were carried out, and photographs were taken. The subsequent pterygium status grading of pterygium status was done by an ophthalmologist. After adjusting for age, sex, working under sunlight, and working in sandy environments, the authors found that a cumulative arsenic exposure of 0.1–15.0 mg/L-year and ≥ 15.1 mg/L-year were associated with increased risks of developing pterygium. The adjusted odds ratios were 2.04 [95% confidence interval (CI), 1.04–3.99] and 2.88 (95% CI, 1.42–5.83), respectively [116].

Cataract

An increasing prevalence of posterior subcapsular cataracts with the increased in exposure to ingested arsenic was reported from Taiwan. The study was carried out on a total of 349 residents living in arseniasis- hyperendemic villages of southwestern Taiwan with by recording the cumulative arsenic exposure and determiningation of different types of lental opacity. The cataract surgery prevalence was 10% for the age group of 50 or more years. Cortical opacity was most common (35%), where asile nuclear and posterior subcapsular opacities were observed in 24% and 22% of subjects, respectively. Diabetes mellitus was a significant risk factor for all types of cataracts. Occupational sunlight exposure was associated with cortical and posterior subcapsular opacities in a dose-response relationship. The cumulative exposure to arsenic from artesian well water and the duration of consuming artesian well water were associated with an increased risk of all types of lens opacity. But However, statistically significant dose-response relations with the cumulative arsenic exposure and the duration of consuming artesian well water were only observed only for posterior subcapsular opacity (P=0.014 and P=0.023, respectively) after adjustments for age, sex, diabetes status, and occupational sunlight exposure [117].

MISCELLANEOUS

Erectile Dysfunction

Erectile dysfunction (ED) has a profound impact on the quality of life of many men. Few studies have so far been done so far to assess the relationship between arsenic exposure and ED. Reports from Taiwan suggested that chronic arsenic exposure has a negative impact on erectile functioning. A study had been conducted on 177 males ≥ 50 years of age through health examinations conducted in three hospitals in Taiwan through an assessment of the risk factors associated with ED, such as aging, sex hormone levels, hypertension, cardiovascular diseases, and diabetes mellitus. The investigators used a questionnaire (International Index of Erectile Function-5) to measure the level of erectile function and determined Sex hormones, including total testosterone and sex hormone–binding globulin radioimmunoassays. Another standardized questionnaire was used to collect background and behavioral information (e.g., cigarette smoking; alcohol, tea, or coffee drinking; and physical activity). The prevalence of ED was found to be greater in the arsenic-endemic area (83.3%) than in the non–arsenic-endemic area (66.7%). Subjects with arsenic exposure > 50 μg/L had a significantly higher risk of developing ED than those with exposure ≤ 50 μg/L, after adjusting for age, cigarette smoking, diabetes mellitus, hypertension, and cardiovascular disease [odds ratio (OR) = 3.4, (95% CI, 1.1–10.3)]. Results also showed that the risk of developing severe ED was drastically enhanced by arsenic exposure [OR = 7.5, (95% CI, 1.8–30.9)], after adjusting for free testosterone and traditional risk factors of ED [118].

Non- Pitting Oedema of Limbs

Oedema of the legs was described as early as 1983 from Mexico in a higher number (18 out of 296; 6.1%) of arsenic- exposed cases from drinking water (>50μg/L) compared to control unexposed subjects (4 out of 318;1.3%, p<0.01) .[23]. However, whether the oedema was pitting or not was not mentioned in the paper. In a hospital- based study, non-pitting oedema was observed in 23 (9.3%) cases out of 248 patients of chronic arsenicosis associated with arsenical skin lesions and a history of prolonged drinking of arsenic contaminated water (50 – 14200µg/L) in West Bengal, India [52, 119]. Non- pitting oedema of the limbs has also been reported in 10 (2.5%) cases out of 363 cases of arsenicosis in Bangladesh with a history of drinking arsenic- contaminated (82-1371 µg/L) water [25].

Further, increased incidences of non-pitting oedema of the limbs (hand and/or leg) wereas reported in a population- based cross- sectional study carried out in all of the 17 arsenic- affected blocks of a district in West Bengal (Nadia). The non-pitting oedema of limbs was found in 4 out of 1,616 (0.25%) cases (arsenic level in drinking water: Mean, 103.46; SD ± 153.28 µg/L) with arsenical skin lesions whereasile it was only found in 2 out of 8,853 (0.02%) control subjects without skin lesions (arsenic level in drinking water: Mean,73.18; SD ± 115.10 µg/L) (p<0.01) [32].

Proteinuria

Proteinuria has been recognized as a marker for an increased risk of chronic renal disease. A study in Bangladesh suggested adverse effects of arsenic exposure from drinking dug well water on the risk of proteinuria, and the effects are modifiable by recent changes in arsenic exposure. Proteinuria was detected by urinary dipstick tests at the baseline and at 2-year intervals on 11,122 participants in the Health Effects of Arsenic Longitudinal Study (HEALS).

At the baseline, well arsenic was positively related to the prevalence of proteinuria; prevalence odds ratios (PORs) for proteinuria in increasing quintiles of well arsenic (</=7, 8-39, 40-91, 92-179 and 180-864 µg/L) were 1.00 (ref), POR 0.99 [95% confidence interval (CI) 0.771.27], POR 1.23 (95% CI, 0.97-1.57), POR 1.50 (95% CI, 1.18-1.89) and POR 1.59 (95% CI 1.26-2.00) (P for trend <0.01). Hazard ratios for incidence of proteinuria were POR 0.83 (95% CI 0.67-1.03) and POR 0.91 (95% CI 0.74-1.12) for participants with a decreasing level of >70 and 17-70 µg/L in urinary arsenic over time, respectively, and were POR 1.17 (95% CI 0.97-1.42) and POR 1.42 (95% CI 1.16 1.73) for participants with an increasing level of 16-68 and >68 µg/L in urinary As over time, respectively, compared with the group with relatively little changes in urinary arsenic as the reference group (urinary As -16 to 15 µg/L) [120].

Weakness and Fatigue

Generalized weakness and fatigue have been reported in chronically exposed arsenic exposed people following drinking arsenic- contaminated water by many workers [13, 15, 16, 21, 93, 119, 120].

A large cross-sectional epidemiological survey was carried out during 1998-1999 on 6,864 participants (all non-smokers) in the district of South 24 Parganas, West Bengal, to ascertain the prevalence of weakness in the arsenic- exposed (arsenic in drinking water, 0.3-3400 µg/L) population. The age adjusted prevalence of weakness strongly increased strongly with arsenic concentrations (lowest,< 50 µg/L and highest, ≥ 800 µg/L) in both sexes (from1.7 per100 to 11.9 per 100 among women, $p < 0.0001$, and from 0.9 to 9.5 per 100 among men, $p<0.0001$), [36].

ARSENICOSIS IN CHILDREN

Skin abnormalities of chronic arsenic toxicity occur in children similar to those as in adults. In Taiwan, the youngest patient drinking arsenic- contaminated water who developed hyperpigmentation was 3 years old [(4)]. Arsenical skin lesions were reported in 144 school children in Antofagasta, Chile during a cross-sectional survey in 1976. The investigators further reported that a chronic cough was complained of by 38.8% of children with skin lesions compared to 3.1% of children with normal skin [(22].)

Less a fewer number of children was reported to have arsenical skin lesions compared to adults in the Indo-Bangladesh subcontinent following drinking of arsenic- contaminated water (below detection level to 3200 µg/L). In an epidemiological study carried in West Bengal , 9 (1.7%) out of 536 girls and 12 (1.9%) out of 613 boys younger thanbelow the age of 9 years old had pigmentation due to exposure of high levels of arsenic in water. The number of subjects with keratosis were 1 (0.2%) and 3 (0.48%) in girls and boys respectively. [(29]). In another study in the region, 114(3.7%) out of 6,695 children younger thanbelow 11 years old in West Bengal and 298 (6.11%) out of 4,877 children younger thanbelow 11 years old in Bangladesh had evidences of arsenical skin disease due to drinking of arsenic- contaminated water [(121]). However, the incidence of arsenical skin lesions in children was found to be much higher (12.2%) in Inner Mongolia, China when studied in a population of 728 subjects younger thanbelow 19 years old who were exposed to arsenic [(122]). In another study, out of 97 subjects, belonging to 40 families of a small village of Cambodia drinking arsenic-contaminated well water, 27 children below the age of 16 years old were clinically examined and 10 (37%) had arsenical skin lesions [(123]). From the reports available in the literature, it appears that children are similarly affected as adults due to chronic arsenic exposure, though the incidence of arsenic related skin manifestation varies in different countries.

A study report on children's intellectual functioning in the arsenic exposed region of Thailand was available for review [124]. Chronic arsenic exposure assessed by hair concentrations was related to developmental retardation as judged by IQs measured by using the Wechsler Intelligence Scale Test for children. A multiple classification analysis was conducted with data from 529 children aged 6-9 years old who had lived in Ronpiboon district since birth. The percentage of children in the average IQ group decreased remarkably from 56.8 to 40.0 as the arsenic level increased. After adjusting for confounders, they observed a statistically significant relationship that arsenic could explain 14% of variance in the children's IQ. The extent of arsenic exposure was difficult to assess. Hair concentrations were found to have an average of 2.42 mg/kg (range: 0.48-26.94 mg/kg) of arsenic, whereas normal was quoted as less than 1mg/kg.

The possible influence of long-term arsenic exposure on the development of cognitive functioning among adolescents was studied in Taiwan. Forty-nine junior school students drinking arsenic-containing well water and 60 controls matched with age, sex, education, body height, body weight, body mass index, and socioeconomic status were compared. The arsenic exposed group was divided into high and low exposure on the basis of mean cumulative arsenic exposure. Four neurobehavioral tests, including continuous performance test (CPT), symbol digit (SD), pattern memory (PM) and switching attention (SA), were applied. A strong correlation between age and education caused a collinearity in the multiple regression model ($r=0.84$, $P<0.0001$). Pattern memory and switching attention were significantly affected by long-term cumulative exposure to arsenic after adjusting for education and sex [125].

A cross-sectional study on intellectual functioning was done on 201 children, 10 years of age, from an arsenic exposed village in Bangladesh. Water As and manganese concentrations were tested from the water of the tube wells ofof each child's home. Children's intellectual functioning tests were done from the Wechsler Intelligence Scale for Children, version III, and assessed by summing weighted items across domains to create Verbal, Performance, and Full-Scale raw scores. Exposure to arsenic from drinking water was associated with reduced intellectual functioning after adjustments for sociodemographic covariates and water Mn. Water As was associated with reduced intellectual functioning, in a dose-response manner [126].

A cross-sectional study of intellectual development was done in West Bengal on 351 children aged 5-15 years in families selected from a surveyed source population of 7,683 people [(12)] in West Bengal, India. Intellectual functioning was assessed based on 6 subsets from the Weschsler Intelligence Scale for children (WISC), the total Sentence recall test, the Colored Progressive Matrices (CPM) test and a Pegboard test. Information on socio-demographic factors was collected and height and weight were measured. Arsenic levels in urine samples collected from the participants and water samples consumed by them were measured by AAS. Urinary arsenic concentrations, stratified into tertiles, showed an inverse trend with the vocabulary test scores, the object assembly test scores and the picture completion test scores, adjusted for potential confounders. This corresponds to a relative reduction of the mean scores related to exposure in the upper tertile in the vocabulary tests of 12.6%, in the object assembly test of 20.6% and in the picture completion test of 12.4%. Reduction in intellectual functioning scores, particularly the vocabulary and picture completion test scores, were associated with increased urine arsenic concentrations, but not with various measures of water concentrations [127].

Thus, results of various studies show that chronic arsenic toxicity due to drinking of arsenic- contaminated water causes significant morbidity in children characterized by skin pigmentation, keratosis and chronic lung disease. Arsenicosis has a socio economic effect on children. Children affected by pigmentation and/or keratosis due to chronic arsenic toxicity are being discouraged from attending schools for fear of being ridiculed.

ARSENICOSIS AND CANCER

The evidence of carcinogenicity in humans from exposure to arsenic is based on epidemiological studies of cancer in relation to arsenic in drinking water.

Skin Cancer

Skin cancer is a commonly observed malignancy related to drinking of arsenic-contaminated water [128-131]. The working group of the International Agency for Research on Cancer [128] has incriminated arsenic as a carcinogen for skin cancer on the basis of an evaluation of data obtained from ecological studies from Taiwan, Mexico, Chile and the

USA; cohort studies from Taiwan; and a case-control study from tThe USA. Numerous cases of skin cancer have been documented from communities with arsenic- contaminated drinking water. The largest prevalence study was conducted on a population of 40,421 in arsenic-endemic areas of Taiwan where 238 cases of skin cancer had been clinically detected. Prevalence rates for inhabitants residing in low-(<300µg/L), medium-(300-600µg/L), and high-(>600µg/L) arsenic areas represented over an eight-fold difference from the highest to the lowest category [4]. The age- standardized mortality rates for skin cancer increased in the three categories for both genders [69]. In an ecological study, Smith et al., (1998) compared sex- and site-specific mortalitiesy for the years1989-93 in Region II of Chile with national mortality rates. The standardized mortality ratio (SMR) for skin cancer was 7.7 (95% CI, 4.7-11.9) among men and 3.2 (95% CI, 1.3-6.6) among women [132].

Malignant arsenical skin lesions may be Bowen's disease (intraepithelial carcinoma, or carcinoma in situ), basal-cell carcinoma, or squamous-cell carcinoma. Skin cancer might arise in the hyperkeratotic areas or might appear on non-keratotic areas of the trunk, extremities, or hands [11, 20]. Bowen's disease appears as sharply demarcated round plaque or has an irregular polycyclic lenticular configuration. The lesions are usually erythematous, pigmented, crustated, fissured and keratotic. Some may be nodular, ulcerated or eroded. The diameter of the lesions may varry from 0.8 to 3.5 cm. Clinically, arsenic- induced Bowen's Disease can be distinguished from non-arsenical Bowen's disease by its multiple and recrudescent lesions and predominant occurrence on the sun-protected areas of the skin [(64].) Histologically, with Bowen's disease, a skin biopsy in Bowen's disease will show a wind blown appearance of the epithelial cells, with multinucleated giant cells, vaculated cells, dyskeratosis, and abnormal mitotic figures at all levels of the epidermis. Arsenic- related basal cell carcinoma appears to be a deeply ulcerative or superficial type as those of an ordinary type of basal cell carcinoma. Histologically, the cells have scanty and ill-defined cytoplasms. Nuclear atypy and giant cells are not ordinarily found. The size and gross appearance of epidermoid carcinoma varied greatly, some forming fungating masses measuring up to 5 cm by 5 cm. and some causing large crater-like ulcers with elevated margins measuring up to 6 cm in diameter. Histologically, the grade of differentiation of epidermoid carcinoma varies [11].

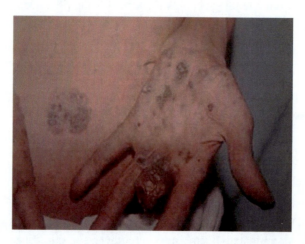

Figure 13. Bowens Disease in the body, Squamous cancer in the finger (Same patient).

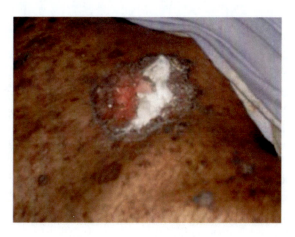

Figure 14. Squmaous cell cancer in the body.

Urinary Bladder Cancer

The working group of IARC (2004) [128] evaluated ecological studies in Taiwan, Chile, Argentina and Australia; cohort studies from Taiwan, Japan and USA; and case- control studies in Taiwan, The USA and Finland and found evidence of an increased risk for urinary bladder cancer associated with arsenic in drinking water. The first report of bladder cancer associated with drinking arsenic- contaminated water was published from the Province of Cordoba in Argentina where 11% of cancer deaths were caused by this cancer [129]. An ecological mortality study for bladder cancer was conducted in Chile in Region II (an arsenic-endemic region) and in Region VIII (a non- arsenic area) for the period 1950-92 and the SMR was found to be 10.2 (95% CI, 8.6-12.2) [130]. In Taiwan, the evidence for anof increased occurance of bladder cancer due to arsenic was supported by case-control and cohort studies within the exposed communities that demonstrated evidence of a dose-response relationship with levels of arsenic in drinking water [131]. There was also evidence of increased risks of bladder cancer from a small cohort study in Japan of peoplersons drinking from wells that had been highly contaminated with arsenic wastes from a factory. The findings of epidemiological studies are consistent with a strong association of arsenicosis with to bladder cancer.

Lung Cancer

On the basis of ecological studies using mortality data in Taiwan, Chile, Argentina and Australia; cohort studies in Taiwan, Japan and the USA; and case-control studies in Taiwan and Chile, a strong association of lung cancer has been observed in populations with high arsenic exposure [128]. In an ecological study, an increased mortality from lung cancer was observed in men and women in 1968-82 in an area endemic for Blackfoot disease in Taiwan [131]. There was an exposure-response relationship between the SMR of lung cancer and the prevalence of Blackfoot disease. Elevated SMRs (about 3) were observed for lung cancer for in both sexes in Region II of Chile using the national rate as a standard [132].

Other Cancers

Identified from death certificates, an iIncreased risk of liver cancer has been reported in several studies from Taiwan, Japan and Chile identified from death certificates. There is limitation of interpretation of these findings because of questionable accuracy regarding the diagnosis of liver cancer on death certificates and the potentially confounding or modifying effects of a hepatitis virus infection or other factors. Ecological studies in Taiwan, Chile, Argentina and Australia and cohort studies from Taiwan and the USA involving populations with high long- term exposure to arsenic found increased risks for kidney cancer. Relative risk estimates for kidney cancer were generally lower than those for urinary bladder cancer, and no studies have reported dose-response relationships on the basis of individual exposure data. Excess mortality from prostate cancer was found in southwest Taiwan. Inconsistent findings were reported for other cancers [128].

DIAGNOSIS

Although chronic arsenic toxicity produces varied systemic manifestations, as well as skin cancer of skin and different internal organs, dermal manifestations such as pigmentation and keratosis are diagnostic of chronic arsenicosis. For this reason, WHO's field guide of diagnostic algorithms of arsenicosis of WHO (2005) [1] is based on the presence or absence of characteristic dermatological manifestations of chronic arsenic toxicity.

A clinically confirmed case of arsenicosis is a "probable case with pigmentation (diffused, spotty or blotchy) and/or keratosis (thickening or nodularity or verrucousity) of the skin, distributed bilaterally symmetrically in the body", in whom the presence of other arsenicosis simulating skin lesions has been ruled out by a differential in-depth skin examination by either a trained dermatologist or an arsenic expert . A "clinically and laboratory confirmed case" is a "clinically confirmed case" in whom the arsenic test is also positive by the recommended laboratory criteria recommended. The laboratory criteria for establishing the exposure history of arsenicosis cases are : i) consumption of drinking water with an arsenic concentration in excess of prevailing national standards for that country for at least six months. (Country Standard in Asia Pacific Region is 50µg/L while WHO Standard is 0.01 µg /L), and ii) an elevated concentration of arsenic in hair (> 1 mg/Kg of hair) or in nail clippings (> 1.5 mg/Kg of nail) [1].

Biomarkers with Special Focus on Diagnosis

On the basis of arsenic metabolism data, the most important biomarkers of internal exposure are the urinary excretions of the element and its concentration in hair and nails. Blood concentrations are generally too low and transient to be of use. The use of arsenic measurements in hair and nails as indices of absorbed doses appears limited. Efforts are needed to develop a standardized procedure to solve the problem of externally contaminatingon of samples. The relationship between arsenic intake through water and urinary excretion of inorganic arsenic and of MMA and DMA appears better.

The concentration of total arsenic in urine has often been used as an indicator of recent arsenic exposure, because urine is the main route of excretion of associated with most arsenic species [133].

As a parameter for the oral intake of arsenic via drinking water in steady-state condition, urinary levels of arsenic (seafood arsenic excluded) has been reported to be diagnostic by several authors from different countries. Despite possible ethnic and environmental differences, reported results display a quite a satisfactory consistency. Most strikingly, an increased excretion rate is observed when the water arsenic concentration reaches 100–200 µg/L [134]. The half-lifetime of inorganic arsenic in humans is about 4 days. Average background concentrations of arsenic in urine are generally below 10µg/L. A urine sample showing more than 50µg/L may be taken as evidence of recent exposure provided the subject has not consumed seafood during the previous four days [1]. Although high arsenic excretion in urine is indicative of continued arsenic exposure, this is not always diagnostic of chronic arsenic toxicity. In West Bengal, India, a significant number of people (9 out of 17) who were drinking arsenic-contaminated water and had high urinary arsenic excretions did not show cutaneous manifestations of chronic arsenic toxicity. On the other hand, many (33 out of 40) of the chronically arsenic-exposed people showing arsenical skin lesions did not have high urinary arsenic excretions [135]. These results might be explained by the fact that all those who wereare drinking arsenic-contaminated water at a point of time may not be showingshow clinical features of chronic arsenic toxicity, whereas others who might have consumed arsenic-contaminated water for a long time in the past, but have since stopped drinking arsenic contaminated water, might still have clinical expressions of arsenic toxicity.

A study from Bangladesh reported that urinary arsenic may be a stronger predictor of skin lesions than arsenic in drinking water in the population. Arsenic levels in the drinking water source and urinary arsenic value had been correlated with incidence of arsenical skin lesions in residents of three villages in Bangladesh. Current arsenic levels in current the drinking water source wereas found to be <50µg/L in 13 (36.1%) out of 36 subjects having arsenical skin lesions among the 167 people studied. The risk of skin lesions in relation to the exposure estimates based on urinary arsenic was elevated more than 3-fold, with the odds ratios for the highest versus the lowest quartiles of 3.6 (95% CI:1.2-12.1) for total urinary total arsenic.

The risks for skin lesions in relation to exposure estimates based on arsenic in drinking water with the odds ratios for the highest versus lowest quartiles of exposure being 1.7 (95% CI: 0.6-5.1) for current drinking water arsenic and 2.3 (95% CI:0.7-7.6) for the cumulative arsenic index [136].

Arsenic is normally found in higher concentrations in human hair and nails than in other parts of the body. This has been explained by the high content of keratin in these tissues [137]. In people with no known exposure to arsenic, the concentration of arsenic in hair is generally 0.02–0.2 mg/Kg [138]. In one study in West Bengal, arsenic content in hair (analyseanalyzed by nutron activation analysis) in patients drinking arsenic- contaminated water (220-2000 µg/L), was found to vary from 1.4 to 20 mg/Kg [15]. External contamination of the hair by arsenic must be excluded in order to use hair arsenic concentrations to assess toxicity.

Normal arsenic values in nails appear to range from 0.02 to 0.5 mg/Kg [139, 140]. Arsenic content in nails (estimated by nutron activation analysis) in arsenic- (220-2000 µg/L) exposed people (220-2000 µg/L) in West Bengal was found to vary between 16 to 66 mg/ Kg [15].

There is no correlation between arsenic levels in hair and nails and clinical features of chronic arsenic toxicity. In a village in West Bengal, all the 17 people drinking arsenic-contaminated water had raisedraised hair and nail arsenic levels, but only 8 had cutaneous lesions [3]. Further, out of 40 people with arsenical skin lesions in another village of West Bengal with a history of drinking arsenic-contaminated water, normal arsenic levels of arsenic in hair and nails wereas found in 31 and 26 cases, respectively [135].

Several autopsy studies in smelter workers have linked exposure to inhaled arsenic with persistence of arsenic in the lungs [141, 142]. In one study, exposed workers had lung arsenic concentrations that were 6 times higher than in controls (47 µg/Kg tissue vsversus 8 µg/kg tissue).

These increases were not consistently seen in the kidney or in the liver, and the elevation in the lung did not significantly decline, significantly even as the time from retirement to death increased, suggesting a long half-life [141]. Other evidence indicates that ingested arsenic reaches the lungs. A fatal poisoning following arsenic ingestion by a 3-year-old boy resulted in an arsenic concentration in the lungs of 7,550 µg/Kg [143]. In another fatal case, the arsenic lung concentration was 2,750 µg/Kg [144].

The maximum arsenic content of the liver in people with hepatomegaly, drinking arsenic-contaminated water in West Bengal, India, was 6 mg/kg (measured by neutron activation analysis), although the arsenic was undetectable in 6 out of the 21 case samples tested (mean of case samples, 1.39 ± 0.3 mg/Kg; mean of control samples, 0.016 ± 0.04 mg/kg) (Guha Mazumder et al., 1988). There was no correlation between the arsenic content in biological tissues (liver, hair and nails) and the arsenic dose taken by the patients [3].

MANAGEMENT OF CHRONIC ARSENIC TOXICITY

Chronic arsenicosis leads to irreversible damage in several vital organs, and arsenic is an established carcinogen. Although there is no significant morbidity for milder forms of the disease, mortality is high in severe cases. Despite the magnitude of this potentially fatal toxicity, there is no effective therapy for theis disease. Complications of moderate and severe form of arsenicosis may not be prevented even after stopping of drinking of the arsenic-contaminated water.

However, people should be advised advised to stop drinking arsenic- contaminated water or avoid exposure to arsenic from any other source. To determine the effect of providing safe water to affected people, a cohort of 24 patients with chronic arsenicosis were re-examined after drinking arsenic-free water (<10 µg/L) for a period varying from 2 to 10 years (13 patients 10 years, 11 patients 2–5 years) in West Bengal. These people had been drinking arsenic-contaminated water (130–2,000 µg/L) for 4–15 years. A partial improvement of pigmentation and keratosis were observed in 45% and 46% of patients, respectively. However, liver enlargement was persistent in 86% of cases. The most distressing observation was the new appearance of signs associated with of chronic lung disease (cough, shortness of breath and chest signs) in 41.6% of cases. There was a slight reduction of clinical symptoms associated with neuropathy [145]. Study reports are available on changes associated with the severity of skin lesions amongst an affected cohort of arsenicosis patients in Southern Thailand where interventions to reduce arsenic- contaminated water had been implemented. Over a 10- year period, both the regression and progression of lesions occurred, although the majority of the subjects followed- up remained the same. Drinking predominantly arsenic-free water increased the probability of regression in subjects with mild stage lesions but not in those with more advanced stage lesions. By contrast, high arsenic content in the household well water, even though it was not used for drinking, decreased the probability of lesion regression among the subjects in more advanced stages but not among milder stage cases. Irrespective of initial stage a period of absence from the affected area increased the likelihood of lesion regression [33]. Another cohort follow-up study was carried out on 1,074 people (arsenic exposed people 623, control population 451) in 2000, five years after the original clinical examination was done on the same population at South 24 Pparganas, West Bengal. Out of 199 people with skin lesions among the arsenic exposed population who were consuming safe water during the previous 5 years, the skin lesions cleared or decreased in 49.7% of people. However, out of 306 people who did not previously have such lesions previously, new skin lesions appeared in 32 (10.5%) [146]. Skin lesions were reported to improve, to some extent, in cases of arsenicosis in Inner Mongolia, China, after drinking low arsenic- containing water for one year. However, after five years, a follow- up study showed no more significant improvement of skin lesions, while whereas the potential risk of for arsenic- induced cancers after cutting off high arsenic exposure was still uncertain and indefinite [147].

From the results of the few studies as described above, it becomes apparent that a significant improvement of mild and moderate dermatological manifestations occurs in many cases of arsenicosis after continuously drinking of arsenic- free water. However, symptoms of severe keratosis and systemic manifestations of arsenicosis may persist in spite of stoppage ofstopping the consumption of arsenic-contaminated water. Further, there remains the potential risk of arsenic- induced cancer in these cases. Hence, there is further need for effective therapeutic intervention for the treatment of chronic arsenicosis.

Chelation therapy for chronic arsenic toxicity is thought to be the specific therapy for relief of to relieve systemic clinical manifestations and reducetion of arsenic stores in the body, thereby reducing any subsequent cancer risk. However, a study evaluating the efficacy of specific chelation therapy with DMSA (Dimercaptosuccinic Acid) for patients suffering from chronic arsenic toxicity has not yielded better efficacy than control subjects treated with a placebo. Twenty-one consecutive patients with chronic arsenicosis were randomized into two groups. Eleven patients (10 males, ages 25.5 ± 8.0 years) received DMSA at 1,400 mg/day (1,000 mg/m^2) in four divided doses in the first week and then 1,050 mg/day (750 mg/m^2) in three divided doses during the next 2 weeks. The same dose schedule was repeated after 3 weeks during which no drug was administered. The other 10 patients (all males, ages 32.2 ± 9.7 years) were given placebo capsules (resembling DMSA) byin the same schedule. The patients were blinded about the nature of treatment being given. The patients included in the study were selected from the arsenic clinic on the basis of their history of drinking arsenic-contaminated water (≥ 0.05 mg/l) for 2 years or more, and clinical symptoms and signs of chronic arsenicosis. The symptoms and signs of patients were evaluated by a scoring system before and after treatment. Therapy with DMSA did not cause any significant clinical improvement compared with patients treated with the placebo. The clinical score improved after therapy with DMSA, but a similar improvement was observed in patients treated with the placebo [28]. In another study, the efficacy of DMPS (Dimercapto propane sulphonate) treatment of DMPS (Dimercapto propane sulphonate) was tested in a single-blind placebo-controlled trial in patients suffering from chronic arsenic toxicity in West Bengal. The trial design was similar to that carried out in the DMSA trial described above. DMPS was given in a dose of 100-mg capsules 4 times a day for a course of 7 days for four courses, with a 1-week drug-free period between each course. Eleven patients (9 males and 2 females) received the drug, whereas 10 patients (5 males, 5 females) received placebo capsules. Therapy with DMPS caused a significant improvement in the clinical condition of chronic arsenicosis patients as evidenced by the significant reduction in the total clinical score from 8.90 ± 2.84 to 3.27 ± 1.73 ($P < 0.0001$).

Exposure cessation alone with placebo treatment also reduced the clinical score (from 8.50 ± 1.96 to 5.40 ± 2.12; $P < 0.003$), but the post-treatment total clinical score of DMPS-treated patients (3.27 ± 1.73) was significantly lower than that of placebo-treated patients (5.40 ± 2.12; $P < 0.01$). The most significant improvement was noted in the clinical scores for weakness, pigmentation and lung disease. No difference was noted between groups in the skin histology before and after treatment. Increased urinary excretions of arsenic during the period of therapy are the possible cause of this improvement. No DMPS-related adverse effects were noted [119].

Improvement of symptoms of arsenicosis patients have been reported to occur following the use of antioxidants like Vitamin A, C and E in Bangladesh. The effectiveness of managing chronic arsenicosis was evaluated by administering a vitamin A, C and E combination regimen. Forty-three patients with chronic arsenicosis had been given the following regimen for 6 months: a) withdrawal of exposure; b) vitamin A (50 000 IU on alternate days); c) vitamin E (200 mg daily); d) v itamin C (250 mg twice daily) and topical application of a keratolytic agent (for arsenical keratosis only). An iImprovement of melanosis and keratosis was observed in 90.9% and 86.4% of patients, respectively, from among 22 patients who had used safe water and had regularly taken the regimen regularly [148]. However, the characteristics of skin lesions for evaluation of ing the severity of arsenicosis were not described in the methodology. Further, no neither a placebo- controlled trial with those vitamins has been carried out nor has the toxicity of long term use of vitamine A been ascertained.

Supportive treatment could help in reducing many symptoms of these patients. Treatment in a hospital with a good, nutritious diet has been found to reduce symptom scores in subsets of placebo- treated arsenicosis patients during the course of DMSA and DMPS trials. [28, 119].

Presently the most prevailing practice of symptomic treatment of keratosis is to locally apply locally 5-10% of salicylic acid and 10-20% of a urea- based ointment on keratotic skin lesions [1]. A higher dose (20% salicylic acid) may be used in severe cases of arsenical keratosis. Although a specific treatment for chronic arsenic toxicity has not yet been fully established, supportive and symptomatic treatment could help in reducing many patient symptoms of the patients. Arsenic- induced cancers could be cured if detected early. Hence, a good cancer surveillance programme in a chronically arsenic- exposed population is essential for preventing cancer- related deaths. Mass communication measures should be undertaken in the arsenic- endemic areas highlighting that people should get their drinking water source tested for arsenic and stop its consumptiont if found contaminated.

REFERENCES

[1]   WHO. *A Field Guide for Detection, Management and Surveillance of Arsenicosis Cases* Technical Publication No. 30, ed. Caussy D, WHO Regional Office for South East Asia, New Delhi, (2005).

[2]   R. Garai, A. K. Chakraborty, S. B. Dey, K. C. Saha, *J. Indian Med. Assoc.*, 82, 34 (1984).

[3]   D. N. Guha Mazumder, J. Das Gupta, A. Santra, A. Pal, A. Ghose, S. Sarkar, Chattopadhaya, D. Chakraborti, In, C.O. Abernathy, R.L. Calderon, and W.R. Chappell, eds. *Arsenic Exposure and Health Effect.* London: Chapman and Hall, 112 (1997).

[4]   W. P. Tseng, H. M. Chu, S. W. How, J. M. Fong, C. S. Lin, S. Yeh, *J. Natl. Cancer Inst.* 40, 453 (1968).

[5]   H. G. Rosenberg, *Arch. Pathol.* 97(6), 360 (1974).

[6]   R. Zaldivar, A. Guillie, *Zentralbl. Bakteriol. Parasitenkd. Infektionskr. Hyg. Abt. Orig. Reihe B.* 165, 226 (1977). (Cited by Cebrian 1983)

[7]   R. Zaldivar, G. L. Ghai, *Zentralbl. Bakteriol. Abt. 1 Orig. B.* 170, 402 (1980). (Cited by NRC 1999)

[8]   U. Fierz, *Dermatologica,* 131, 41 (1965) (in German) (Cited by NRC 1999).

[9]   H. Rattner, M. Dorne, *Arch. Dermatol. Syphilol.*, 48, 458 (1943), (Cited by NRC 1999).

[10] M. D. Black, *Pharm. J.*, (Dec. 9), 593 (1967).

[11] S. Yeh, *Hum. Pathol.*, 4, 469 (1973).

[12] C. H. Tay, *Australas. J. Dermatol.*, 15(3), 121 (1974).

[13] K. C. Saha, *Indian J Dermatol.*, 29(4), 37 (1984).

[14] K. C. Saha, *Ind. J. Dermatol.*, 40, 1 (1995).

[15] D. N. Guha Mazumder, A. K. Chakraborty, A. Ghosh, J. Das Gupta, D.P. Chakraborty, S.B. Dey, N. Chattopadhaya, *Bull. Wld Health Org.*, 66, 499 (1988).

[16] D. N. Guha Mazumder, J. Das Gupta, A. Santra, A. Pal, A. Ghose, S. Sarkar, *J. Indian Med. Assocn.*, 96(1), 4-7 and 18 (1998).

[17] I. M. Szuler, C. N. Williams, J. T. Hindmarsh, H. Park-Dinesoy, *Can. Med. Assoc. J.*, 120, 168 (1979).

[18] H. Luchtrath, *J. Cancer Res. Clin. Oncol.*, 105, 173 (1983).

[19] NRC (Natioanl Research Council) *Arsenic in drinking* Academic Press, (1999). *Water* Washington DC : National.

[20] S. G. Sommers, Mc. Manus, *Cancer.* 6, 347 (1953)

[21] R. Zaldivar, *Beitr. Pathol.* 151, 384 (1974). (Cited by Cebrian 1983)

[22] J. M. Borgono, P. Vicent, H. Venturino, A. Infante, *Environ. Health Perspect.* 19,103 (1977).

[23] M. E. Cebrian, A. Albores, M. Aguilar, E. Blakely, *Hum. Toxicol.,* 2,121 (1983).

[24] S. A. Ahmad, D. Bandaranayake, A.W. Khan, S.A. Hadi, G. Uddein, Ma. Halim, *Interntl. J. of Env. Heath Res.* 7, 271 (1997).

[25] S. A. Ahmad, M. H. S. U. Sayed, S. A. Hadi, M. H. Faruquee, M. A. Jali, R. Ahmed, A.W. Khan, *Int. J. Environ. Health Res.*, 9, 187 (1999).

[26] K. V. Ratnam, M. J. Espy, S. A. Muller, T. F. Smith, W. P. Su, *J. Am. Acad. Dermatol.*, 27, 120 (1992).

[27] G. Alain, J. Tousignant, E. Rozenfarb, *Int. J. Dermatol.,* 32, 899 (1993).

[28] D. N. Guha Mazumder, U. C. Ghoshal, J. Saha, A. Santra, B. K. De, A. Chatterjee, S. Dutta, C. R. Angle, J. A. Centeno, *J. Toxicol. Clin. Toxicol.*, 36, 683 (1998c).

[29] D. N. Guha Mazumder, R. Haque, N. Ghosh, B.K. De, A. Santra, D. Chakraborty, A.H. Smith, *Int. J. Epidemiol.* 27(5), 871 Oct. (1998).

[30] R. Haque, D. N. Guha Mazumder, S. Samanta, N. Ghosh, D. Kalman, M.M. Smith, S. Mitra, A. Santra, S. Lahiri, S. Das, B.K. De, A.H. Smith, *Epidemiology.*, 14, 174 (2003).

[31] M. Tondel, M. Rahman, A. Magnuson, O. A. Chowdhury, M.H. Faruquee, S.A. Ahmad, *Environ. Health Perspect.*, 107, 727 (1999).

[32] D. N. Guha Mazumder, A. Ghosh, K. K. Majumdar, N. Ghosh, C. Saha, R.N. Guha Mazumder, *Ind. J. of Community Medicine.* 35, 331 (2010).

[33] S. Oshikawa, A. Geater, V. Chongsuvivatwong, T. Piampongsan, D. Chakraborti, G. Samanta, B. Mandel, N. Hotta, Y. Kojo, H. Hironaka, *Environmental Sciences*, 85, 435 (2001).

[34] B. K. De, D. Majumdar, S. Sen, S. Guru, S. Kundu, *J. Assoc. of Physcn. India,* 52 : 395 (2004).

[35] R. Zaldivar, G.L. Ghai, *Zentralbl. Bakteriol.*, 170, 409 (1980). ,(Cited from NRC 1999).

[36] D. N. Guha Mazumder, R. Haque, N. Ghosh, B. K. Dey, A. Santra, D. Chakraborti, A.H. Smith, *International Journal of Epidemiology*, 29, 1047 (2000).

[37] A. H. Milton, Z. Hasan, A. Rahman, M. Rahman, *J. Occup. Health.* 43, 136 (2001).

[38] S. Ondine, O. S. von Ehrenstein, D. N. GuhaMazumder, Y. Yuan, S. Samanta, J. Balmes, A. Sil, N. Ghosh, H. M. Smith, R. Haque, R. Purushothamam, S. Lahiri, S. Das, A. H. Smith, *India. Am. J. Epidemiology*, 162, 533 (2005).

[39] D. N. Guha Mazumder, R. Haque, N. Ghosh, B. K. De, A. Santra, D. Chakraborty, A. H. Smith, *India. Int. J. Epidemiol.* 27, 871 (1998).

[40] D. Lynch, J. Newell, V. Hale, D. Dyer, K. Corkery , N.L. Fox, P. Gerend, R. Fick, *AJR Am. J. Roetgenol.*, 173, 53 (1999).

[41] D. N. Guha Mazumder, C. Steinmaus, P. Bhattacharya, O.S. von Ehrenstein, N. Ghosh, M. Gotway, A. Sil, J. R. Balmes, R. Haque, M. M. Hira-Smith, A. H. Smith, *Epidemiology*, 16, 760 (2005).

[42] S. M. Tsai, T. N. Wang, Y. C. Ko, *Arch. Environ. Health*, 54(3), 186 (1999).

[43] A. H. Smith, G. Marshall, Y. Yuan, C. Ferreccio, J. Liaw, O. S. Von Ehrenstein, C. Steinmaus, M. N. Bates, S. Selvin, *Environ. Hlth. Perspect.*, 114, 1293 (2006).

[44] F. Parvez, Y. Chen, P. W. Brandt-Rauf, A. Bernard, X. Dumont, V. Slavkovich, M. Argos, J. D'Armiento, R. Foronjy, M. R. Hasan, H. E. M. M. Eunus, J. H. Graziano, H. Ahsan, *Environ. Health Perspect.*, 116, 190 (2008).

[45] N. Hotta, Nippon Taishitsugaku Zasshi [*Jpn. J. Const. Med.*] 53(1/2), 49 (1989).

[46] H. Z. Ma, Y. J. Xia, K. G. Wu, T. Z. Sun, J. L. Mumford, In: Chappell, W. R., Abernathy, C. O. and Calderon, R.L., eds, *Arsenic Exposure and Health Effects,* Amsterdam, Elsevier Science, 127 (1999).

[47] D. N. Guha Mazumder, N. Ghosh, B. K. De, A. Santra, S. Das, S. Lahiri, R. Haque, A. H. Smith, D. Chakraborti, In *Arsenic Exposure and health effects IV*, (Eds C. O. Abernathy, R. L. Calderon and W. R. Chappell) Oxford, UK: Elsevier Science, 153 (2001a).

[48] J. S. Morris, M. Schmid, S. Newman, P. J. Scheuer, S. Sherlock, *Gastroenterology*, 66, 86 (1974).

[49] F. Nevens, W. Van Steenbergen, R. Sciot, V. Desmet, J. D. Groote, *J. Hepatology*, 11, 80 (1990).

[50] M. Franklin, W. Bean, R. C. Harden, Fowler's solution as an etiologic agent in cirrhosis. *Am. J. Med. Sci.* 1950; 589-596.

[51] D. V. Datta, S. K. Mitra, P. N. Chhuttani, R. N. Chakravarti, *Gut.*, 20, 378 (1979).

[52] A. Santra, J. Das Gupta, B. K. De, B. Roy, D. N. Guha Mazumder, *Ind. J. Gastroenterology*, 18(4), 152 (1999).

[53] A. Hernandez-Zevala, L. M. Del Razo, G. G. Garcia-Vargas, V. H. Borza, M. E. Cebrian, *Toxicol. Lett.*, 99(2),79 (1998).

[54] A. K. Chakraborty, K. C. Saha, *Indian J. Med. Res.*, 85, 326 (1987).

[55] A. Santra, A. Maiti, S. Das, S. Lahiri, S. K. Charkaborty, D. N. Guha Mazumder, *Clin. Toxicol.*, 38, 395 (2000).

[56] S. Das, A. Santra, S. Lahiri and D. N. GuhaMazumder, *Toxicology and Applied Pharmacology.* 204, 18 (2005).

[57] S. F. Glazener, Joseph G. Ellis, K. P. Johnson, *California Medicine*, 109, 158 (1968).

[58] A. Navas-Acien, A. R. Sharreet, E. K. Silbergeld, B. S. Schwartz, K. E. Nachman, T.A. Burke, E. Guallar, *American Journal of Epidemiology*, 162, 11, 1037 (2005).

[59] J. L. Mumford, K. Wu, Y. Xia, R. Kwok, Z. Yang, J. Foster, W. E. Sanders, *Environ Health Perspect.*, 115, 690 (2007).

[60] M. M. Medrano, R. Moix, R. Pastor-Barriuso, M. Palau, J. Damian, R. Ramis, J. L. Del Barrio, A. Navas-Acien, *Environ. Res.*, 110, 448 (2010).

[61] Y. Chen, J. H. Graziano, F. Parvez, Liu, V. Slavkovich, T. Kalra, M. Argos, T. Islam, A. Ahmed, M. Rakibuz-Zaman, R. Hasan, G. Sarwar, D. Levy, A. van Geen, H. Ahsan, *BMJ*, 5, 342 (2011).

[62] H. S. Yu, H. M Sheu, S. S. Ko, L. C. Chiang, C. H. Chien, S. M. Lin, B. R. Tserng, C. S. Chen, *J. Dermatol.* 11, 361 (1984).

[63] W. P. Tseng, W. Y. Chen, J. L. Sung, J. S. Chen, *Memoirs College Med.* Natl Taiwan Univ. 7, 1 (1961).

[64] W. P. Tseng, *Environ. Health Perspect.*, 19, 109 (1977).

[65] R. Engel, A. Smith, *Arch. Environ. Health,* 49, 418 (1994).

[66] C. H. Tseng, C. K. Chong, C. J. Chen, T. Y. Tai, *Atherosclerosis.* 120, 125 (1996).

[67] M. G. Alam, G. Allinson, F. Stagnitti, A. Tanaka, M. Westbrooke, *Int. J. Environ. Health Res.*, 12, 235 (2002).

[68] H. M. Anawar, J. Akai, K. M. Mostofa, S. Safiullah, S. M. Tareq, *Environ. Int.*, 27, 597 (2002).

[69] C. J. Chen, M. M. Wu, S. S. Lee, J. D. Wang, S. H. Cheng, H. Y. Wu, *Arteriosclerosis,* 8,452 (1988).

[70] C. H. Wang, J. S. Jeng, P. K. Yip, C. L. Chen, L. I. Hsu, Y. M. Hsueh, H. Y. Chiou, M. M. Wu, C. J. Chen, *Circulation*, 105, 1804 (2002).

[71] Y. H. Wang, M. M. Wu, C. T. Hong, L. M. Lien, Y. C. Hsieh, H. P. Tseng, S. F. Chang, C. L. Su, H. Y. Chiou, C. J. Chen, *Atherosclerosis.* 192, 305 (2007).

[72] Y. Chen, P. Factor-Litvak, G. R. Howe, J. H. Graziano, P. Brandt-Rauf, F. Parvez , A. van Geen , H. Ahsan, *Am. J. Epidemiol.*, 165, 541 (2007)

[73] A. M. Dart, B. A. Kingwell, *J. Am. Coll. Cardiol.*, 37, 975 (2001).

[74] M. E. Safar, B. I. Levy, H. Struijkar-Boudier, *Circulation.*, 107, 2864 (2003).

[75] Y. L. Huang, Y. M. Hsueh, Y. K. Huang, P. K. Yip, M. H. Yang, C. J. Chen, *Science of the Total Environment*, 407, 2608 (2009).

[76] C. H. Tseng, *Journal of Environmental Science and Health*, Part C. 23, 55 (2005).

[77] J. C. States, S. Srivastava, Y. Chen, A. Brachowsky, *Toxicol. Sc.*, 107, 312 (2009).

[78] C. A. Torres Soruco, R. E. Bagini, M. A. Salvador, *La Semana Med.*, 175, 35 (1991). (Cited from NRC 1999)

[79] C. H. Tseng, C. H. Chong, C. J. Chen, T. Y. Tai, *Atherosclerosis.* 120, 125 (1995).

[80] S. L. Wang, J. M. Chiou, C. J. Chen, C. H. Tseng, W. L. Chou, C. C. Wang, T. N. Wu, L. W. Chang, *Environ. Health Perspect.* 111, 155 (2003).

[81] J. M. Chiou, S. L. Wang, C. J. Chen, C. R. Deng, W. Lin, T. Y. Tai, *International J. of Epidemiol.* 34, 93 (2005).

[82] J. C. Chen, H. Y. Chiou, M. H. Chiang, L. J. Lin, T. Y. Tai, *Arteriosclerosis*, 16, 504 (1996).

[83] Y. M. Hsueh, W. L. Wu, Y. L. Huang, H. Y. Chiou, C. H. Tseng, C. J. Chen, *Atherosclerosis.* 141, 248 (1998).

[84] C. H. Tseng, C. K. Chong, C. P. Tseng, Y. M. Hsueh, H. Y. Chiou, C. C. Tseng, C. J. Chen, *Toxicol. Letters.* 137, 15 (2003).

[85] C. J. Chen, Y. M. Hsueh, M. S. Li, M. P. Shyu, S. Y. Chen, M. M. Wu, T. L. Kuo, T. Y. Tai, *Hyperension.* 25, 53 (1995).

[86] M. Rahman, M. Tondel, S.A. Ahmad, I.A. Chowdhury, M.H. Faruquee, O. Axelson, *Hypertention*, 33(1), 74 (1999).

[87] H. Y. Chiou, Y. M. Hsueh, Li. Hsieh, S. F. Chang, Y. H. Hsu, C. J. Chen, *Mutat. Res.*, 386, 197 (1997).

[88] M. M. Wu, T. L. Kuo, Y. H. Hwang, C. J. Chen, *Am. J. Epid.*, 130, 1123 (1989).

[89] Tsi, T. N. Wand, Y. C. Ko, *Arch. of Env. Hlth.*, 54, 186 (1999).

[90] T. J. Cheng, D. S. Ke, Y. J. Wang, H. R. Guo. *Epidemiology.* 20, S59 (2009).

[91] J. R. Meliker, L. W. Robert, L. C. Lorraine, J. O. Nriagu, *Environmental Health.* 6, 4 (2007).

[92] K. H. Kilburn, Eds. C. O. Abernathy, R. L. Calderon and W. R. Chappell, *Arsenic Exposure and health effects,* London. Chapman and Hall, 14, 159 (1997).

[93] D. Basu, J. Dasgupta, A. Mukherjee, D. N. Guha Mazumder, *JANEI.*, 1, 45 (1996).

[94] S. C. Mukherjee, M. M. Rahman, U. K. Chowdhury, M. K. Sengupta, D. Lodh, C. R. Chanda, K. C. Saha, D. Chakraborti, *J. Eenviron, Sci. Health.* A 38, 165 (2003).

[95] J. T. Hindmarsh, O. R. McLetchine, L. P. M. Heffernan, O. A. Hayne, H. A. A. Ellenberger, R. R. McCurdy, H. J. Thiebaux, *J. Analytical Toxicol.*, 11, 270 (1977).

[96] D. M. Hafeman, H. Ahsan, E. D. Lueis, A. B. Siddique, V. Slavkovich, Z. Cheng, A. van Geen, J. H. Graziano, *J. Occup. Env. Med.*, 47, 778 (2005).

[97] D. Chakraborti, S. C. Mukherjee, S. Pati, M. K. Sengupta, M. M. Rahaman, U. K. Chowdhury, D. Lodh, C. R. Chanda, A. Chakraborti, G. K. Basu, *Environ. Health Perspect.*, 111,1194 (2003).

[98] H. P. Tseng, Y. H. Wang, M. M. Wu, H. W. The, H. Y. Chiou, C. J. Chen, *J. Health Popul. Nutrn.*, 24, 182 (2006).

[99] H. Terada, T. Sasagawa, H. Saito, H. Shirata, T. Sekiya, *Acta Med. Biol.*, 9, 279 (1962).

[100] J. M. Harrington, J. Middaugh, D. L. Morse, J. Housworth, *Am. J. Epidemol.*, 108, 377 (1978)

[101] J. W. Southwick, A. E. Western, M. M. Beck, T. Whitley, R. Isaac, J. Petajan and C.D. Hansen, *Arsenic: Industrial Biomedical Environmental Perspectives.*, (Eds Lederer W. and Fensterheim, R.) New York, Van Nostrand Reinhold, 210 (1983) (Cited from NRC 1999).

[102] M. S. Lai, Y. M. Hsueh, C. J. Chen, M. P. Shyu, S. Y. Chen, T. L. Kuo, M. M. Wu, T. Y. Tai, *Am. J. Epidemiol.*, 139, 484 (1994).

[103] M. Rahman, M. Tondel, S. A. Ahmad, O. Axelson, *Am. J. Epidemiol.*, 148, 198 (1998).

[104] C. H. Tseng, T. Y. Tai, C. K. Chong, C. P. Tseng, M. S. Lai, B. J. Lin, H. Y. Chiou, Y. M. Hsueh, K. H. Hsu, C. J. Chen, *Environ. Health Perspect.*, 108, 847 (2000).

[105] Y. Chen, H. Ahsan, V. Slavkovich, G. L. Peltier, R. T. Gluskin, F. Parvez, X. Liu, J. H. Graziano, *Environ. Health Perspect.*, 118, 1299 (2010).

[106] A. Aschengrau, S. Zierler, A. Cohen, *Arch. Environ. Health*, 44, 283 (1989).

[107] I. Dési, *Geogr. Med.*, 22, 45 (1992), (Cited from NRC 1999).

[108] M. B'rzs'nyi, A. Berecsky, P. Rudnai, M. Csanády, A. Horvath, *Arch. Toxicol.*, 66, 77 (1992). (Cited from NRC 1999)

[109] P. Rudnai, M. Csanády, E. Sárkány, V. Kiss, G. Mucsi, [abstract]. *International Seminar on Arsenic Exposure: Health Effects, Remediation Methods in Treatment Costs.* Universidad de Chile, Oct. 8-10, 1996. (Cited from NRC 1999)

[110] C. Hopenhayn-Rich, S. Browning, I. Hertz-Picciotto, C. Ferreccio, C. Peralta, H. Gibb, *Environ Health Perspect.*, 108, 667 (2000).

[111] S. A. Ahmad, M. H. S. U. Sayed, S. Barua, M. H. Khan, M. Faruquee, A. Jalil, S. A. Hadi, H. K. Talukder. *Environ Health Perspect.*, 109, 629 (2001).

[112] A. H. Milton, W. Smith, B. Rahman, Z. Hasan, U. Kulsum, K. Dear, M. Rakibuddin, A. Ali, *Epidemiology*, 16, 82 (2005).

[113] O. S. von Ehrenstein, D. N. Guha Mazumder, M. Hira-Smith, N. Ghose, Y. Yan, G. Windham, A. Ghosh, R. Haque, S. Lahiri, D. Kalman, S. Das and A. H. Smith, *Am. J. Epidemiol.*, 163, 662 (2006).

[114] M. K. Hossain, M. M. Khan, M. A. Alam, A. K. Chowdhury, H. M. Delwar, A. M. Feroze, K. Kobayashi, F. Sakauchi, M. Mori, *Toxicol. Appl. Pharmacol.*, 208, 78 (2005).

[115] K. Baidya, A. Raj, L. Mondal, G. Bhaduri, A. Todani, *J. Ocul. Pharmacol.Ther.*, 22, 208 (2006).

[116] W. Lin, S.-L. Wang, H. J. Wu, K. H. Chang, P. Yeh, C. J. Chen, H. R. Guo, *Environ Health Perspect.*, 116, 952 (2008).

[117] L. C. See, H. Y. Chiou, J. S. Lee, Y. M. Hsueh, S. M. Lin, M. C. Tu, M. L. Yang, C. J. Chen, *J. Environ. Sci. Health A Tox. Hazard Subst. Environ. Eng.*, 42, 1843 (2007).

[118] F. I. Hsieh, T. S. Hwang, Y. C. Hsieh, H. C. Lo, C. T. Su, H. S. Hsu, H. Y. Chiou, C. J. Chen, *Environ. Health Perspect.*, 116, 532 (2008).

[119] D. N. Guha Mazumder, B. K. De, A. Santra, N. Ghosh, S. Das, S. Lahiri, T. Das, *Clinical. Toxicol.*, 39, 665 (2001b).

[120] Y.Chen, F.Parvez, M. Liu, G.R. Pesola, M.V. Gamble, V, Islam T. Slavkovich, A. Ahmed, R. Hasan, JH. Graziano, H. Ahsan. *Int. J. Epidemiol.*,40,828(2011).

[121] M. M. Rahaman, U. K. Chowdhury, S. C. Mukherjee, B. K. Mondal, K. Paul, D. Lodh, B. K. Biswas, C. R. Chanda, G. K. Basu, K. C. Saha, S. Roy, R. Das, S. K. Palit, Q. Quamaruzzaman, D. Chakraborti, *Clinical Toxicol.* 39, 683 (2001).

[122] X. Guo, Y. Fujino, S. Kaneko, K. Wu, Y. Xia, T. Yoshimura, *Mol. Cellular Biochem*, 222, 137 (2001).

[123] D. N. Guha Mazumder, K. K. Majumdar, S. C. Santra, Hero Kol, Chan Vicheth. *J. Environmental Science and Health*, Part-A, 44,1 (2009).

[124] U. Siripitayakunkait, P. Vishudhiphan, M. Pradipasen, T. Vorapongsathron, *Arsenic Exposure and Health Effects*. In: W. R. Chappell, C. O. Abernathy, R. L. Calderon, eds. Oxford, UK: Elsiver Science Ltd, 141-149 (1999).

[125] S. Y. Tsai, H. Y. Chou, H. W. The, C. M. Chen, C. J. Chen, *NeuroToxicology*, 24, 747 (2003).

[126] G. A. Wasserman, X. Liu, F. Parvez, H. Ahsan, P. Factor-Litvak, A. van Geen, V. Slavkovich, N. J. LoIacono, Z. Cheng, I. Hussain, H. Momotaj, J. H. Graziano, *Environ. Health Perspect.*, 112, 1329 (2004).

[127] O. S. von Ehrenstein, S. Poddar, Y. Yuan, D. N. Guha Mazumder, B. Eskanazi, A. Basu, M. Hira-Smith, N. Ghosh, S. Lahiri, R. Haque, A. Ghosh, D. Kalman, S. Das, A. H. Smith, *Epidemiology*, 18, 44 (2007).

[128] IARC (International Agency of Research on Cancer) *Monographs on the Evaluation of Carcinogenic risk to Humans Some drinking-water Disinfectants and contaminants, including Arsenic.* Lyon, France: WHO. 84, 227 (2004).

[129] R. M. Bergoglio, *Pren. Med. Argent.* 51, 994 (1964). (in Spanish) (Cited by IARC 2004).

[130] M. I. Rivara, M. Cebrian, G. Corey, M. Hernandez, I. Romieu, *Toxicol Ind Health*, 13, 321 (1997).

[131] C. J. Chen, Y. C. Chuang, T. M. Lin, H. Y. Wu. *Cancer Res*, 45, 5895 (1985).

[132] A. H. Smith, M. Goycolea, R. Haque, M. L. Biggs, *Am. J. Epidemiol.*, 147,660 (1998).

[133] M. Vahter, *Appl. Organomet. Chem.* 8, 175 (1994). (Cited by NRC 1999)

[134] J. P. Buchet, P. Hoet,V.Haufroid, D. Lison, *Arsenic Exposure and Health Effects.* In: W. R. Chappell, C. O. Abernathy, R. L. Calderon, eds. Oxford, UK: Elsiver Science Ltd, 141-149 (1999).

[135] T. R. Chowdhury, B. K. Mandal, G. Samanta, G. K. Basu, P. P. Chowdhury, C. R. Chanda, N. K. Karan, D. Lodh, R. K. Dhar, D. Das, K. C. Saha, D. Chakraborti, In: C. O. Abernathy, R. L. Calderon and W. R. Chappell., eds. *Arsenic Exposure and health effects*, London. Chapman and Hall 1997; 93-111.

[136] H. Ahsan, M. Perrin, A. Rahman, F. Parvez, M. Stute, Y. Zeng, A. H. Milton, P. B. Rauf, A. van Geen, J. Graziano, *J. Occup. Environ. Med.*, 42, 1195 (2000).

[137] H. A. Shapiro. *J. forensic Med.*, 14, 65 (1967).

[138] J. L. Valentine, H. K Kang, G. Spivey, *Environ. Res.*, 20, 24 (1979).

[139] A. P. S. Narang, L. S. Chawls, S. B. Khurana. Levels of arsenic in Indian opium eaters. *Drug Alcohol Depend*, 20, 149 (1987).

[140] Y. Takagi *Bull. Environ. Contam. Toxicol.* 41, 690 (1988).

[141] D. Brune, G. Nordberg, P. O. Wester. *Sci. total Environ.*, 16(1), 13 (1980).

[142] L. Gerhardsson, D. Brune, G. F. Nordberg, P. O. Wester, *Science of the Total Environment*, 74, 97 (1988).

[143] J. J. Saady, R. V. Blanke, A. Polkis. *Journal of Analytical Toxicology*, 13, 310 (1989).

[144] G. Quatrehomme, O. Ricq, P. Lapalus, Y. Jacomet, A. Ollier, *Journal of Forensic Sciences*, 37, 1163 (1992).

[145] D. N. Guha Mazumder, B. K. De, A. Santra, J. Dasgupta, N. Ghosh, B. K. Roy, U. C. Ghosal, J. Saha, A. Chatterjee, S. Dutta, R. Haque, A. H. Smith, D. Chakraborti, C.

Angle, J. A. Centino, In: W. R. Chappell C. O. Abernathy, R. L. Calderon, eds. *Arsenic Exposure and Health Effects*, London, UK: Elsevier, 335-347, 1999.

[146] D. N. Guha Mazumder, N. Ghosh, K. Mazumder, A. Santra, S. Lahiri, S. Das, A. Basu, and A. H. Smith. In: W. R. Chappell, C. O. Abernathy, R. L. Calderon and D. J. Thomas., eds. *Arsenic Exposure and Health Effects* V, Oxford, UK: Elsevier Science, 381-389, 2003b.

[147] G. Sun, X. Li, J. Pi, Y. Sun, B. Li, Y. Jin, Y. Xu, *J. Health Popul. Nutr.* 24, 176 (2006).

[148] S. A. Ahmad, M. H. Faruquee, M. H. S. U. Sayed, M. H. Khan, M. A. Jalil, R. Ahmed, S. A. Hadi, *J. of Prev. and Soc. Med.*, 17(1), 19 (1998).

In: Arsenic
Editor: Andrea Masotti

ISBN: 978-1-62081-320-1
© 2013 Nova Science Publishers, Inc.

Chapter 7

OCCURRENCE OF INORGANIC ARSENIC IN RICE-BASED INFANT FOODS: SOIL-RICE-INFANT RELATIONSHIPS

A. Ramírez-Gandolfo[1], P. I. Haris[2], S. Munera[1], C. Castaño-Iglesias[3], F. Burló[1] and A. A. Carbonell-Barrachina[1,]*

[1]Universidad Miguel Hernández. Departamento Tecnología Agroalimentaria.
Ctra. de Beniel, Alicante, Spain
[2]Faculty of Health and Life Sciences. De Montfort University.
Hawthorn Building. The Gateway. Leicester, UK
[3]Universidad Miguel Hernández, Departamento de Farmacología,
Pediatría y Química Orgánica. Sant Joan, Alicante, Spain

ABSTRACT

About 100 million rural people in Asia (mainly Bangladesh and India) are exposed to arsenic-polluted drinking water and agricultural products in what it is considered the biggest mass poisoning case in the human history. This problem that seemed so distant from the European countries is perhaps closer than expected due to the high consumption of rice-based foods by babies and infants. It is well demonstrated that elevated contents of inorganic arsenic (iAs) are found in infant products, such as rice milk and baby rice, being marketed in European countries. Workable solutions to limit arsenic in paddy rice by breeding rice cultivars with low arsenic accumulation are being sought but still are not fully implemented. Meanwhile, simple recommendations for processing and cooking rice as preliminary unit operations in the manufacturing of infant foods will significantly help in reducing arsenic exposure in European infants. For instance, avoiding the use of organic brown rice (high concentrations of arsenic are found in rice bran) and cooking rice using high volumes of arsenic-free water may be easy and cheap ways of reducing arsenic exposure in infants.

---

[*] E-mail address: angel.carbonell@umh.es.

INTRODUCTION

The word arsenic (As) has made its way through history on the toxic properties of a number of its compounds and the strength of its killing properties [1]. Fortunately there are great differences in the toxicity of different compounds, and the species that are most commonly found in soils, sediments, plants, and edible vegetal products are not the most toxic. The inorganic arsenic forms (iAs) are more toxic, both acute and chronic, than the organic ones (oAs). Furthermore, iAs is non-threshold, class 1 carcinogen; this classification was based on the induction of primary skin cancer, as well as the induction of lung and urinary bladder cancer.

Consequently, it is very important to produce As speciation data and not only data on total As (tAs) in foods, including infant products. The Joint FAO/WHO Expert Committee on Food Additives (JECFA) established a Provisional Tolerable Weekly Intake (PTWI) for iAs of 0.015 mg/kg of bodyweight/week in 1988. EFSA has recently concluded that this PTWI is no longer appropriate as data had shown that iAs causes cancer of the lung and urinary bladder in addition to skin, and that a range of adverse health effects had been reported at exposures lower than those reviewed by the JECFA [2].

A proposed model of a local soil arsenic system is shown in Figure 1. In this cycle arsenic pesticides and soil parental materials are the main inputs [3]. The most important translocation factors are adsorption by soil (1), precipitation as insoluble materials (2), oxidation/reduction reactions (3), leaching or runoff (4), volatilization (5) after biomethylation (6), and uptake by vegetation and incorporation into the food chain (7). The most obvious pathway linking contaminated soils and man is food chain (Figure 2), but potable surface and ground waters, respired and ingested dust and tobacco fumes may also play significant roles.

Food chain relationships are critical as contaminants tend to become more concentrated as they move up the food-chain (e.g. soil $\Rightarrow$ plant $\Rightarrow$ herbivore $\Rightarrow$ carnivore). The most likely dietary sources of arsenic, with the exception of fish and shellfish [4, 5] are crops and vegetables grown in contaminated soil [6-8].

Some additional arsenic may be found in milk and meat products derived from animals grazed on contaminated land [9]. However, the peculiarity of rice farming, flooding soils for periods of time long enough to drastically reduce their redox potential will make a huge difference compared to other cereals [10]. In anaerobic paddy rice soil systems arsenic is more mobile than in aerobic wheat soils. Transfer of arsenic from soil to grain was one order of magnitude greater in rice than in wheat or barley mainly due to high shoot/soil ratios for rice, ~0.8, compared to 0.2 and 0.1 for barley and wheat, respectively [10].

The adsorption and retention of arsenic by soils control its persistence, activity, movement, transformation, and ecological effects. Arsenic adsorption is related to the pH, chemical and physical properties (e.g. amount of sesquioxides, clay, and exchangeable Ca^{2+} and Mg^{2+}), and cation exchange capacity of soils, and to the amount of arsenic in the soils. The influence of redox on arsenic solubility in soils was found [11-14] to be governed by (i) the dissolution of Fe-oxyhydroxides and concurrent release of coprecipitated arsenate, and (ii) reduction of arsenate to arsenite.

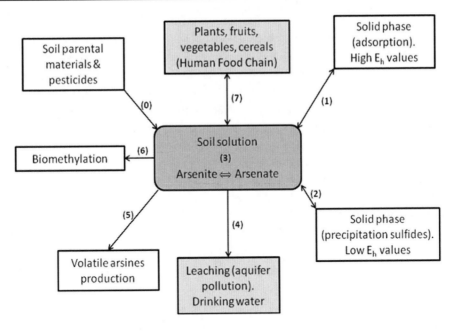

Figure 1. Arsenic cycle in the soil system.

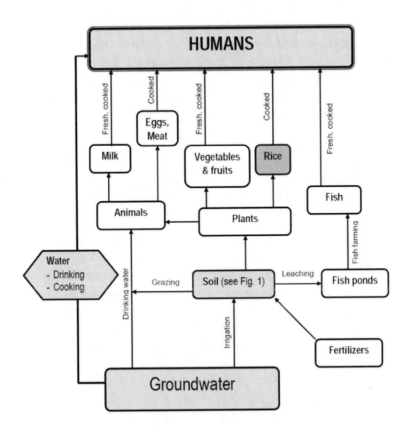

Figure 2. Possible pathways of arsenic ingestion in rural populations of India and Bangladesh [8].

In soils of rice-producing areas of the USA elevated arsenic concentrations are most commonly associated with MAA, a herbicide widely used for post-emergence weed control in cotton [15, 16] as a direct spray. Methylarsonic acid residues can cause severe damage to succeeding rotational crops, such as rice as was the case in fields of Louisiana [15, 16]. On the other hand, in Asia (Bangladesh and India) most of the arsenic found in rice grain is present under inorganic species because the main source of arsenic is the groundwater. In world-wide rice-producing areas, arsenic is considered the probable cause for straighthead, a physiological disease of flooded rice that results in blank florets, distorted palea and lemma and, in extreme cases, failure of panicles to form [15-17]. The affected panicles are erect rather than bent and have few filled florets [18]. Consequently, there is a clear relationship between rice and arsenic; relationship that is exclusive of rice and does not apply to other cereal, such as wheat or corn.

WHY THIS ASSOCIATION BETWEEN RICE AND ARSENIC?

Arsenicals were widely used in agriculture as pesticides or plant defoliants for many years. Now the use of both inorganic and organic arsenicals is legally forbidden or it is reduced as much as possible. However, a legacy of contaminated orchard soils has been left behind due to the extensive use of inorganic forms in the past; this is of great importance because residues from the application of these compounds can produce phytotoxic effects long after application has ceased. Besides, it is also important to highlight that the most popular fertilizers N-P-K will incorporate trace amounts of arsenic to agricultural soils as contaminants due to the high chemical similarity between phosphorus and arsenic.

Soils have been referred as heavy metals and trace elements "sinks"; however, what they really do is to convert a short-term problem, through adsorption onto clays and Fe and Mn oxides or precipitation of insoluble sulfides under oxidizing or reducing redox conditions, respectively, in a long-term risk. The biogeochemical conditions of soils containing trace amounts of arsenic could someday change and the most toxic arsenic species could become more available. Therefore, precaution must always be taken when working with arsenic-polluted-soils even if the sink mechanism of soils and sediments seems to be solving current pollution hazards.

This precaution must be even higher in the case of paddy rice fields because of their very specific farming conditions; flooding the soils will almost simultaneously solubilize ferric iron oxides, which are the main compounds responsible for the arsenic adsorption to soils, and will reduce arsenate [As(V)] to the most toxic species arsenite [As(III)].

As mentioned previously, the reducing conditions under which rice is grown will induce the reduction and subsequent dissolution of the ferric iron oxides and hydroxides to soluble ferrous iron. Later pentavalent arsenate will be reduced to the more mobile and toxic trivalent arsenite (Figure 3).

An anomalously high soil concentration due to contamination generally implies greater availability to plants. But, the bioavailability of arsenic is controlled by a large number of factors, including the source and form of the element, soil pH, redox and drainage conditions as well as the type and amount of organic matter present [19].

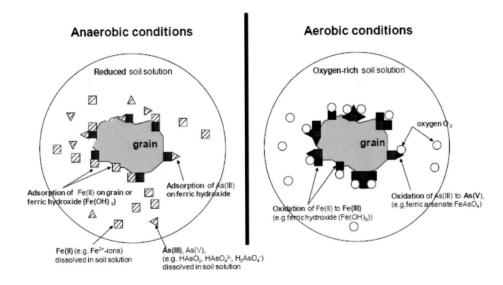

Figure 3. Effect of soil redox potential on the geochemistry of arsenic on soil particles.

PARTICULARITIES OF RICE PLANTS

In general the highest residues of arsenic are found in roots, with intermediate contents being found in green-vegetables, with edible seeds and fruits containing the lowest levels of arsenic. In this way, arsenic contents tend to be restricted during the transport within the plant system. Consequently, arsenic contents in cereals, including rice, are expected to be low. Williams et al. [10] demonstrated that this statement is not true for rice. These authors reported that in rice the median shoot/soil transfer factors were nearly 50 times higher than those in wheat and barley. However, median grain/shoot transfer factors for wheat and barley were 4 times higher than those of rice. Additionally, several authors have clearly demonstrated that methylated oAs are rapidly moved from the roots into shoot tissues in different plants, including turnip [20], tomato [6] and beans [21]. The differences in these transfer factors among rice, wheat and barley are probably due to differences in arsenic speciation and dynamics in anaerobic rice soils compared to aerobic soils fro barley and wheat. In summary, transfer of arsenic from soil to grain was an order of magnitude greater in rice than in wheat and barley, despite lower rates of shoot-to-grain transfer in rice.

ARSENIC POLLUTION — A GLOBAL ISSUE

Contamination of groundwater with arsenic and the impact of this contamination on humans have been reported in 23 countries, but the magnitude of this problem is especially severe in Bangladesh and India [7]. As a consequence of long periods of exposure to high levels of arsenic intake from drinking water and cooked rice, people suffer from damage to the skin, kidneys, brain, heart, and circulation; bladder and lung cancers are the major killers [8]. About 100 million rural people in Asia are exposed to arsenic-polluted drinking water and agricultural items in what is considered the worst mass poisoning event in the human history.

This problem that seemed until a couple of years very distant from people living in the European Union (EU) is now closer than expected. The main exposure route to iAs in the EU is dietary [2], and it is of particular importance for young infants because of their high food consumption per body mass. The national exposures to iAs from food and drinking water in the European Union have been estimated to range between 0.1 and 0.6 µg/kg body weight (bw) per day for average consumers [2].

The main food subclasses contributing to the iAs exposure were cereal grains and cereal based products, food for special dietary uses, bottled water, coffee and beer, rice grains and rice based products, fish and vegetables. That early-life exposure to iAs via baby rice may increase the risk for adverse health effects was first reported by Meharg et al. [22]. Their main conclusion was that when baby iAs intake from rice products was considered, median consumption (0.2 µg/kg/day) was higher than drinking water maximum exposures predicted for adults in these regions when water intake was expressed on a bodyweight basis. The need for legislation on As in baby food in order to prevent excessive As exposure in infants consuming rice-based products has been highlighted [23].

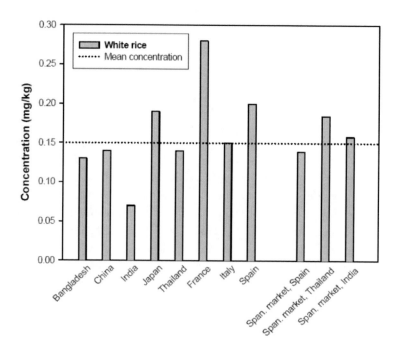

Figure 4. Total arsenic contents in white rice produced and marketed in different countries [24, 52].

High As accumulation in rice occurs due to paddy cultivation, i.e. anaerobic conditions, where soil As is highly available for plant uptake [8]. European rice has high tAs and iAs contents as depicted in Figure 4, in which the highest contents were found in rice samples from France, Italy and Spain together with samples of Japan. Besides, Burló et al. [24] has shown that samples from different geographical origin are being marketed worldwide. As an example these authors have reported arsenic contents in rice samples from Spain, Thailand and India marketed in Spain.

Recent studies showed high iAs in rice products marketed worldwide. For instance, Signes-Pastor et al. [25] studied As speciation in Japanese rice drinks and condiments sold in the UK. These authors concluded that a) rice based products displayed higher iAs contents than those from barley and millet; b) most of the tAs in the rice products was iAs (63-83 %), and c) high consumers of Japanese products could be at serious risk.

EXPOSURE TO ARSENIC WITHIN THE EUROPEAN UNION

By modeling the dose-response data from key epidemiological studies and selecting a benchmark response of 1% extra risk, a range of benchmark dose lower confidence limit ($BMDL_{01}$) values between 0.3 and 8 µg/kg b.w. per day were identified for cancers of the lung, skin and bladder, as well as skin lesions [2].

The EFSA [2] identified the food subclasses of cereal grains and cereal based products to be the dominant pathway of exposure to "iAs" in the EU (Figure 5). It must be noted that within this food category, due to its high tAs amount, rice grain is one of the major contributors to the iAs forms (Figure 6). Thus, rice and rice-based products drastically contribute to iAs and are the dominant pathway of iAs exposure in the EU. Other contributors to iAs exposure are cereal and cereal products, food for special dietary uses, coffee and beer, fish and fish products, and other vegetables and vegetable products [2]. As rice-based products are often used in weaning foods for infants, exposure of infants to As is of great importance and its assessment is a priority.

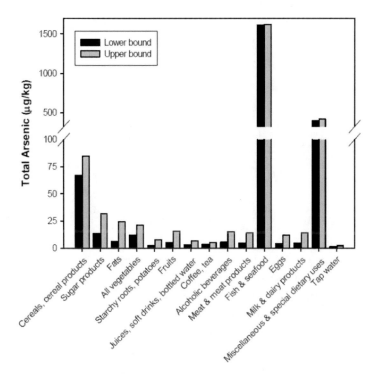

Figure 5. Total arsenic contents in the main food categories of European consumers [2].

Carbonell-Barrachina et al. [26] have estimated the intake of iAs in Spanish infants of 4, 6, 8 and 12 months of age with the celiac disease (force to consume mainly rice-based products) and control infants at mean values of 0.26, 0.27, 0.40 and 0.41 μg/kg/day and 0.05, 0.16, 0.25 and 0.26 μg/kg/day, respectively. These results clearly show that infants with celiac disease are at high risk regarding exposure to inorganic arsenic.

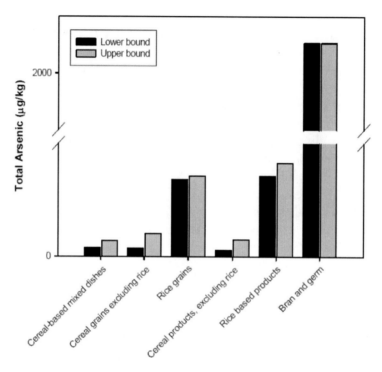

Figure 6. Total arsenic contents in products from the different subcategories within the food category "cereal and cereal products" [2].

ARSENIC IN RICE-BASED INFANT PRODUCTS MARKETED IN THE EUROPEAN UNION

Most children are weaned using, initially, pure rice porridge, a precooked, dried and milled product [22, 27]; milled rice is a dominant carbohydrate source to weaning babies up to 1 year of age due to its blandness, material properties, low allergenic and nutritional value. As the child develops, this porridge is used for the basis of more complex meals, by mixing it initially (from 6 months) with puréed fruits or vegetables and later (from 8 months) with meat (mainly chicken) and fish (mainly hake), either home-made mixtures with baby rice or pre-prepared commercial products. Rice biscuits are used during teething. Many common food items eaten by infants are also rice-based such as cereals and biscuits [22, 27]. This dependence on rice is exacerbated in infants with food intolerances. Rice milk is often used as an animal milk substitute for those with milk intolerance [27]; while rice based products are the staple for those with gluten intolerances [28]. Children from families that follow typical Southeast Asian rice based diets will have further enhanced exposures. Even though there is no data for infants, the Bangladeshi community living in the UK has a *ca.* 30-fold higher rice consumption than white Caucasians; this trend seems to be quite realistic also for infants and will imply a very important exposition to rice and thus to iAs in this population [27].

Recent studies have proved that high iAs are also present in rice products commercialised in the EU, for instance Japanese rice drinks and condiments [25]. *Rice milk*, an alternative to cow milk for lactose intolerance sufferers, has been recently analysed in UK [28]. Results proved that all samples analysed in a supermarket survey would fail the EU limit (10 µg tAs L^{-1}), with up to 3 times this concentration recorded, while out of the subset that had As species determined, 80 % had iAs levels above 10 µg L^{-1}. Later Meharg et al. [22] and Carbonell-Barrachina et al. [26] studied the contents of iAs in *baby rice* and *baby cereals* from UK and Spain.

The iAs contents were significantly higher in gluten-free rice than in cereals mixtures with gluten, placing infant with celiac disease at high risk. All rice-based products displayed a high iAs content, with values being >60% of the total As content and the remainder being dimethylarsinic acid (DMA). A clear positive correlation was found between the iAs content and the rice percentage in Spanish infant foods, with iAs increasing as the rice percentage increased (Figure 7).

Finally, it is important to highlight that not just pure and supplemented infant rice are of concern, because a wide range of other rice based products that are fed to babies, such as crackers, biscuits, crisped and puffed rice cereals, pasta, noodles, puddings, plain polished and whole grain rice are made primarily with, or are formulated with, rice.

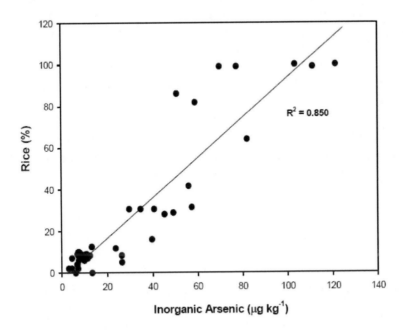

Figure 7. Relationship between inorganic arsenic contents and rice percentage in Spanish infant foods [26].

Because of high food consumption rates per body mass, children are at very higher risk of exposure of dietary contaminants, including As, than adults from the EU [2, 22, 27]. The UK Food Standards Agency (UK FSA) has already issued advice that children under 4.5 years do not take rice milk [29], as typical consumption patterns would lead to iAs intakes exceeding WHO Permissible Maximum Tolerable Daily Intakes (PMTDI). EU baby rice itself is elevated in arsenic and leads to a high dietary exposure [22]. Similarly, rice based products pose a risk to young children due to high iAs content [28, 30]. The EFSA arsenic review concluded that iAs levels in EU diets are at a level consistent with the possibility of risk to consumers, that rice was one of the most important sources of iAs, and that young children had the highest exposures with respect to the general population.

Very recently Ljung et al. [31] studied the contents of essential (Ca, Mg, Fe, Zn, Cu, Mn, Mo and Se) and toxic (As, Cd, Sb, Pb and U) elements in infant formula from Sweden and reported elevated tAs concentrations in rice-based products. These authors concluded that contents of toxic elements such as As, Cd and Pb have to be kept at an absolute minimum in food products intended for infant consumption.

STRATEGIES TO REDUCE ARSENIC OCCURRENCE IN RICE-BASED INFANT FOODS

The most important conclusion of the Scientific Opinion on Arsenic in Food of the EFSA [2] was that the dietary exposure to iAs should be reduced. The two obvious approaches to reduce infant exposure to As are: (i) to use rice grain from areas or cultivars with low grain As, and (ii) to switch from rice to other grain crops, such as oat, barley, maize and wheat as cereal carbohydrate/protein source. The second approach will imply the bankrupt of all rice-based business and will be almost impossible to implement. However, other options also exist and deserve consideration. These options include (i) modifying the current agronomic practices of rice growing to reduce As availability to plants, (ii) making pre-treatments to rice before its use as a food ingredient, (iii) optimizing the conditions of unit operations involving contact of rice with water and the time and temperature of thermal treatments during rice-based infant food manufacture. Consequently in the next sections different practical approaches to reduce the content of As in rice-based products will be discussed. These approaches will be classified in two categories: (i) those that can be implemented during rice farming and (ii) those that can be implemented during rice processing at the facilities of food companies.

BEFORE AND DURING RICE FARMING

The As content in rice is related to environmental and genetic factors [32, 33], with soil As (concentration and species) having a dominant effect [10]. It has been demonstrated that different growing regions from one country can produce rice with differing As contents, for example the As content in rice from Cadiz and Seville regions of Spain is lower than in rice from other areas surveyed to date within Spain (Valencia, Delta del Ebro and Calasparra) [10, 24, 34]. Therefore, the first option to reduce As content in rice is to screen As levels in existing rice to identify varieties that have low As levels.

French rice grain is amongst the highest globally surveyed to date. This may be due to specific cultivars employed in the Camargue, or it may be due soil availability and/or management practice. Also, it seems that addition of organic matter, or organic rich soils, may lead to enhanced As methylation, which may be desirable as organic arsenic (oAs) species are less toxic than inorganic ones.

Studies by Norton et al. [33] have clearly shown that arsenic uptake, transport and accumulation in the edible rice grain are affected by cultivar. These authors studied the agronomic behavior of 76 cultivars from Bangladesh and reported stable genetic differences in As accumulation. Later Norton et al. [32] cultivated 13 rice cultivars with low arsenic accumulation under different climatic conditions at two sites in three countries, Bangladesh, China and India.

These authors concluded that breeding low grain As cultivars that will have consistently low grain tAs and iAs contents over multiple environments using traditional breeding approaches may be difficult, although four cultivars with interesting potential were identified and had low grain As across all field sites.

Finally, it must be mentioned that aerobic rice cultivation is now starting to be considered worldwide with the aim of increasing efficiency in the management of nitrogen fertilizers and mainly water, resource that for different reasons is getting more and more deficient worldwide.

Compared to continuous flooding, aerobic management lowers As availability to plant uptake and consequently As assimilation by rice plants and the content of As in the edible grain [35]. However, this practice opens new issues related to the adaptation capability of traditional European cultivars to the increased soil oxygen availability and, therefore, to their future quantitative and qualitative performances with respect to grain yield and quality. Soil redox potential strongly affects microbial activity in the rhizosphere, the biogeochemical cycles of nutrients, the morpho-physiological features of roots, the availability of toxic elements (e.g., Cd, As) and, in the end, to alter plants productivity and the quality of final products [36].

On the other hand, aerobic conditions can affect the availability of other toxic elements, such as Cd. Arao et al. [35] studied the effects of water management in rice paddy on As and Cd contents in Japanese rice, concluding that flooding increased As concentrations in rice grains, whereas aerobic treatment increased the concentration of Cd. Consequently, As and Cd have opposite behaviors in the soil-plant system.

When a paddy field is flooded and the soil is under reducing conditions, any Cd in the soil combines with sulfur to form CdS, which has a low solubility in water, while the dissolution of Fe-oxyhydroxides and the concurrent release of the co-precipated As will increase As availability to rice plants [14]. However, when the field is drained and the soil is under aerobic conditions, CdS is converted into $CdSO_4$, which is soluble in water, while As will be readily absorbed and/or co-precipitated onto hydrated iron oxides, mainly as arsenate. This means that the solubility of Cd and As change depending on the redox potential (Eh) of the soil. So, an equilibrium between soluble levels of As and Cd should be reached in paddy fields, otherwise rice grain will be either polluted with As or Cd, but polluted.

Summarizing, although it is desirable to reduce the iAs content of rice, alteration of agronomic practices should not detrimentally impact of micronutrients of nutritionally importance (.e.g., Fe and Zn), of vitamins, or of nutraceutic metabolites already scarce in rice grains [37] or increase the contents of other toxic elements such as Cd.

AFTER RICE HAS ENTERED THE FOOD INDUSTRY

Changes to the tAs content and to As species might take place during the manufacturing and/or preparation of food for human consumption. Various processes may cause a considerable increase or decrease in the As concentrations in food commodities and thus in the actual dietary exposure to As. For example, traditional washing and soaking of rice may significantly reduce the As levels. Changes in tAs content of the food can occur due (i) to losses (solubilization) to the cooking medium or preservation solution or (ii) to losses (volatilization) due to intense and long heat treatments. Additionally, toxic As species can be converted to other arsenicals or *vice versa* during food preparation.

Pre-Processing

Rice can be pre-treated before its incorporation into the manufacturing process of rice-based infant foods. Some of the external layers of the rice grain (Figure 8) contain significant contents of As and their removal will reduce the exposure to iAs of consumers of rice-based products.

Initially Signes et al. [38] compared the two rice dehusking processes (removal of the external hull or husk of the rice grain) currently in use in India, wet (soaking and boiling of rice and mechanical hulling, leading to parboiled rice) and dry (mechanical hulling, leading to atab rice). The dry method was recommended if As-free water was not available; however, soaking and light boiling resulted in lower As concentrations if non-polluted water was used. Therefore, the use of high volumes of water for washing and boiling the rice could be good ways of easily and significantly reducing the As content of rice, before starting the production of rice flour for rice-based infant foods.

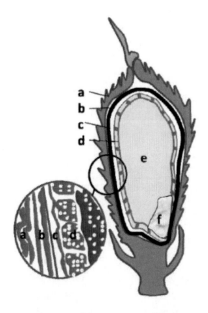

Figure 8. Morphology of the rice grain (a: hull; b: bran; c: polish; d: aleurone layer; e: starchy endosperm; f: embryo).

Simultaneously, Torres-Escribano et al. [39] reported a higher iAs concentration in brown rice, also known as whole-grain rice, compared to white rice (Figure 9), which might indicate that part of the As is attached to components of the bran. Sun et al. [30] also reported that iAs in rice bran and its products are an order of magnitude higher than in bulk grain, reaching concentrations of ~ 1000 μg/kg d.w. Besides, the layer of bran can be of different color, brown, reddish or even black. Norton et al. [33] found that Bangladeshi landraces with red bran had significantly more grain As than the cultivars with brown bran.

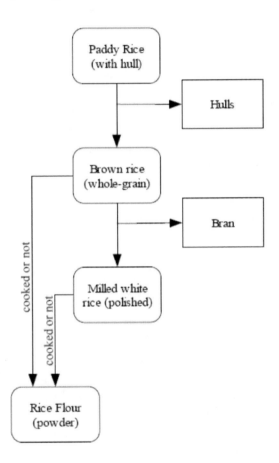

Figure 9. Pre-processing of paddy rice.

Summarizing, polishing brown rice to obtain white rice may lead to a substantial decrease in the contents of both tAs and iAs.

Later, brown or white rice can be cooked before entering the final manufacturing of rice-based products. The three most common methods of cooking rice in Asia are known as: a) traditional, b) intermediate and c) contemporary. In the *traditional method*, rice is washed until washings become clear, the washings are discarded and rice is boiled in excess water until cooked; finally, the remaining water is discarded. Rice cooked following the *intermediate method* is washed as above but is boiled with less water until no water is left to discard. Finally, in the *contemporary method* unwashed rice is boiled with low water volume until no water is left to discard [40]. Signes et al. [41] and Signes-Pastor et al. [7] simulated these three cooking methods in their facilities and recommended the use of the traditional method (using high volumes of water for the washing and cooking steps); this method significantly reduced the content of tAs from a maximum of 387 to 258 µg/kg. Similar conclusions were previously reached by *Sengupta* et al. [40], who cooked rice using low-As water (<3 µg As L^{-1}) using traditional and modern methods and found that the traditional method (wash until clear; cook with rice:water ratio of 1:6; discard excess water) removed up to 57% of the As from the initial rice. Approximately half of the As was lost in the wash water and the other hand in the discard water.

At the same time, Signes et al. [42] simulated cooking of rice with different levels of As (>50 µg/L) species in the cooking water and concluded that As concentration in cooked rice was always higher than that in the raw rice and ranged from 227 to 1642 µg/kg. Mondal and Polya [43] reported values of As in cooked rice of 170 µg/kg (during two household surveys in West Bengal, Nadia district), Smith et al. [44] of 350 µg/kg for a Bangladesh household survey, Bae et al. [45] of 270 µg/kg for a Bangladesh on site survey, Rahman et al. [46] of 320 µg/kg for a Bangladesh field survey and Roychowdhury et al. [47] of 370 µg/kg for a West Bengal household survey. Perhaps, the lower t-As concentrations reported in these later studies compared to the laboratory study by Signes et al. [42] were due to the use of cooking water with low-As contents or even free-As water.

In addition, Signes et al. [42] showed that the As speciation was not significantly affected by the cooking process, probably because the temperature reached during rice cooking, ~ 100°C, was lower than required for promoting species exchanges. Van Elteren and Slejokovek [48] studied the effect of high temperatures on aqueous standards of various species of As and concluded that temperatures above 150°C were required to found significant changes in As speciation. A later study carried out by Devesa et al. [49] agreed with this statement and concluded that these high temperatures can be attained in some cooking treatments in which the surface of the food is in direct contact with the heat source (grilling, frying or baking) and reaches temperatures close to 250°C; similar results were found by Hanaoka et al. [50] and Torres-Escribano et al. [39].

Finally, Raab et al. [51] systematically investigated tAs and iAs in different rice types basmati, long-grain, polished (white) and wholegrain (brown) that had undergone various types of cooking in non-contaminated water. The effects of rinse washing, low water volume (rice to water ratio 1:2.5) and high water volume (rice to water ratio 1:6) during cooking, as well as steaming, were investigated. Rinse washing was effective at removing about 10 % of the tAs and iAs from basmati rice, but was less effective for other rice types. While steaming reduced tAs and iAs rice content, it did not do so consistently across all rice types investigated. High volume water cooking effectively removed both tAs and iAs by 35 % and 45 %, respectively, in the long-grain and basmati rice, compared to uncooked (raw) rice. This study indicates that rinse washing and a high volume of non-contaminated cooking water are effective in reducing the As content of cooked rice, specifically the inorganic.

During Processing

The diagram on Figure 10 shows a general procedure for the manufacturing of a typical cereal-based baby food. As can be seen the only unit operation involving a heat treatment is "drum drying". The temperature reach in the surface of the drum during this drying process could be as high as that reported for grilling and therefore could significantly affect the content and speciation of As in the rice flour. However, the options to reduce the As content in cooked rice when entering the processing line are very limited.

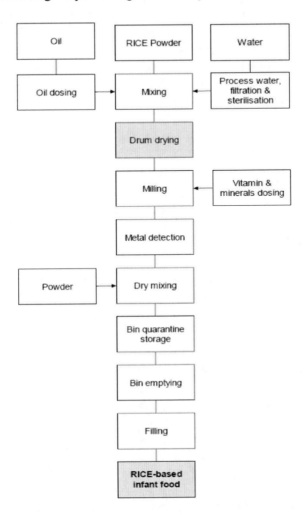

Figure 10. General manufacturing process of a rice based infant foods.

In summary, the effects of food processing on the concentrations of As in rice-based infant foods depends mainly on (i) proper selection of the raw ingredients (from low As accumulating cultivars and from geographical areas with low soil As contents) and (ii) proper pre-processing (removal of husk and/or bran layers, washing, rinsing and cooking with high volumes of water) of rice grain before entering the main manufacturing line.

ACKNOWLEDGMENTS

To all researchers involved in the preparation of the proposal "Inorganic Arsenic from Rice Products in Infant Diets, IARPID", which was submitted to the European Commission and finally was not funded. It is important that our work is available for the Scientific Community. Thanks to all of you.

REFERENCES

[1] J.O. Nriagu, "Arsenic in the Environment", John Wiley and Sons, Inc., New York (1994).

[2] EFSA Panel on Contaminants in the Food Chain, *EFSA J.* 7(10), 1351 (2009).

[3] W.H.O. (World Health Organization), "Environmental Health Criteria 18. Arsenic", WHO, Geneva, Switzerland (1981).

[4] G. Falcó, J.M. Llobet, A. Bocio and J.L. Domingo, *J. Agric. Food Chem.* 54, 6106 (2006).

[5] G. Perelló, R. Martí-Cid, J.L. Llobet and J.L. Domingo, *J. Agric. Food. Chem.* 56, 11262 (2008).

[6] F. Burló, I. Guijarro, A.A. Carbonell-Barrachina, D. Valero and F. Martinez-Sánchez, *J. Agric. Food Chem.* 47, 1247 (1999).

[7] A.J. Signes-Pastor, K. Mitra, M. Hobbes, F. Burló and A.A. Carbonell-Barrachina, *J. Agric. Food Chem.* 56, 9469 (2008).

[8] A.A. Carbonell-Barrachina, A.J. Signes-Pastor, L. Vázquez-Araújo, F. Burló, and B. Sengupta, *Mol. Nutr. Food Res.* 53, 531 (2009).

[9] P. Mitchell and D. Barr, *Environ. Geochem. Health.* 17, 57 (1995).

[10] P.N. Williams, A. Villada, C. Deacon, A. Raab, J. Figuerola, A.J. Green, J. Feldmann, and A.A. Meharg, *Environmental. Sci. Technol.* 41, 6854 (2007).

[11] P.H. Masscheleyn, R.D. Delaune and W.H. Patrick Jr., *Environ. Sci. Technol.* 25, 1414 (1991).

[12] P.H. Masscheleyn, R.D. DeLaune, and W.H. Patrick Jr., *J. Environ. Qual.* 20, 522 (1991).

[13] A.A. Carbonell-Barrachina, A. Rocamora, C. García-Gomis, F. Martínez-Sánchez and F. Burló, *Geoderma,* 122, 195 (2004).

[14] A. Signes-Pastor, F. Burló, K. Mitra, and A.A. Carbonell-Barrachina, *Geoderma,* 137, 504 (2007).

[15] A.R. Marin, P.H. Masscheleyn and W.H. Patrick Jr., *J. Plant. Nutr.* 16, 865 (1993).

[16] A.R. Marin, P.H. Masscheleyn and W.H. Patrick Jr., Plant. Soil. 152, 245 (1993).

[17] J.T. Gilmour and B.R. Wells, *J. Agron.* 72, 1066 (1980).

[18] A.R. Marin, P.H. Masscheleyn and W.H. Patrick Jr. *Plant. Soil.* 139, 175 (1992).

[19] A.A. Carbonell Barrachina, M.A. Aarabi, R.D. DeLaune, R,P. Gambrell and W.H. Patrick Jr. *Plant. Soil.* 198, 33 (1998).

[20] A.A. Carbonell-Barrachina, F. Burló, D. Valero, E. López, D. Martinez-Romero and F. Martinez-Sanchez, *J. Agri. Food Chem.* 47, 2288 (1999).

[21] Y. Lario, F. Burló, P. Aracil, D. Martínez-Romero, S. Castillo, D. Valero and A.A. Carbonell-Barrachina, *Food Addit. Contam.* 19, 417 (2002).

[22] A.A. Meharg, G. Sun, P.N. Williams, E. Adamako, C. Deacon, Y.G. Zhu, J. Feldmann and A. Raab, *Environ. Pollut.* 152, 746 (2008).

[23] A. Masotti, L. Da Sacco, G.F Bottazzo and E. Sturchio, *Environ. Pollut.* 157, 1771 (2009).

[24] F. Burló, A. Ramírez-Gandolfo, A.J. Signes-Pastor, P.I. Haris, and A.A. Carbonell-Barrachina, *J. Food Sci* in press (2011).

[25] A.J. Signes-Pastor, C. Deacon, R.O. Jenkins, P.I. Haris, A.A. Carbonell-Barrachina and

A.A. Meharg, *J. Environ. Monit.* 11, 1930 (2009).

[26] A.A. Carbonell-Barrachina, X. Wu, A. Ramírez-Gandolfo, G.J. Norton, F. Burló, C. Deacon, and A.A. Meharg, *Environ. Pollut.* 163, 77 (2012).

[27] A.A. Meharg, C. Deacon, R.C.J. Campbell, A.M. Carey, P.N. Williams, J. Feldmann and A. Raab, *J. Environ. Monitor.* 10, 428 (2008).

[28] C. Cascio, A. Raab, R.O. Jenkins, J. Feldmann, A.A. Meharg and P.I. Haris, *J. Environ. Monit.* 13, 257 (2011).

[29] UK FSA, 2009 at http://www.food.gov.uk/news/newsarchive/2009/may/arsenicinriceresearch

[30] G.X. Sun, P.N. Williams, Y. Zhu, C.M. Deacon, A. Carey, A. Raab, J. Feldmann and A.A. Meharg, *Environ. Int.* 35, 473 (2009).

[31] K. Ljung, B. Palm, M. Grandér and M. Vahter, *Food Chem.* 127, 943 (2011).

[32] G.J. Norton, G. Duan, T. Dasgupta, M.R. Islam, M. Lei, Y.G. Zhu, C.M. Deacon, A.C. Moran, S. Islam, F.J. Zhao, J.L. Stroud, S.P. McGrath, J. Feldmann, A.H. Price and A.A. Meharg, *Environ. Sci. Technol.* 43, 8381 (2009).

[33] G.J. Norton, M.R. Islam, C.M. Deacon, F.-J. Zhao, J.L. Stroud, S.P. Mcgrath, S. Islam, M. Jahiruddin, J. Feldmann, A.H. Price and A.A. Meharg, *Environ. Sci. Technol.* 43, 6070 (2009).

[34] A.A. Meharg, P.N. Williams, E. Adamako, Y.Y. Lawgali, C. Deacon, A. Villada, R.C.J. Cambell , G.-X. Sun, Y.G. Zhu, J. Feldmann, A. Raab, F.J. Zhao, R. Islam, S. Hossain and J. Yanai, *Environ. Sci. Technol.* 43, 1612 (2009).

[35] T. Arao, A. Kawasaki, K. Baba, S. Mori and S. Matsumoto, *Environ. Sci. Technol.* 43, 9361 (2009).

[36] W. Cheng, G. Zhang, G. Zhao, H. Yao and H. Xu, *Field. Crop. Res.* 80, 245-252 (2003).

[37] A. Oikawa, F. Matsuda, M. Kusano, Y. Okazaki and K. Saito, *Rice*, 1, 63-71 (2008).

[38] A. Signes, K. Mitra, F. Burló and A.A. Carbonell-Barrachina, *Eur. Food Res. Technol.* 226, 561 (2008b).

[39] S.Torres-Escribano, M. Leal, D. Velez and R. Montoro, *Environ. Sci. Technol.* 42, 3867-3872 (2008).

[40] M.K. Sengupta, M.A. Hossain, A. Mukherjee, S. Ahamed, B. Das, B. Nayak, A. Pal and D. Chakraborti, *Food Chem. Toxicol.* 44, 1823-1829 (2006).

[41] A. Signes, K. Mitra, F. Burló and A.A. Carbonell-Barrachina, A.A. *Eur. Food Res. Technol.* 226, 561-567 (2008).

[42] A. Signes, K. Mitra, F. Burló and A.A. Carbonell-Barrachina, *Food Addit. Cont.* 25, 41 (2008).

[43] D. Mondal and D.A. Polya, *App. Geochem.* 23, 2987-2998 (2008).

[44] A.H. Smith, R. Lee, D.T. Heitkemper, K.D. Cafferky, A. Haque and A.K. Henderson, *Sci. Total Environ.* 370, 294-301 (2006).

[45] M. Bae, C. Watanabe, T. Inaoka, M. Sekiyama, M. Sudo, M.H. Bokul and R. Ohtsuka, *Lancet.* 360, 1839-1840 (2002).

[46] M.A. Rahman, H. Hasegawa, M.M. Rahman and M.A.M Miah, *Sci. Total Environ.* 370, 51-60 (2006).

[47] T. Roychowdhury, H. Tokunaga and M. Ando, *Sci. Total Environ.* 308, 15 (2003).

[48] J.T. Van Elteren and Z. Slejkovec, *J. Chromatogr.* 789, 339-340 (1997).

[49] V. Devesa, D. Velez and R. Montoro, *Food Chem. Toxicol.* 46, 1-8 (2008).

[50] K. Hanaoka, W. Goessler, H. Ohno, K.J. Irgolic,and J. Kaise, *App. Organomet. Chem.* 15, 61-66 (2001).

[51] A. Raab, C. Baskaran, J. Feldmann and A.A. Meharg, *J. Environ. Monit.* 11, 41 (2009).

[52] A.A. Meharg, P.N. Williams, K.G. Scheckel, E. Lombi, J. Feldmann, A. Raab, Y. Zhu and R. Islam, *Environ. Sci. Technolol.* 42, 1051 (2008).

In: Arsenic
Editor: Andrea Masotti

ISBN: 978-1-62081-320-1
© 2013 Nova Science Publishers, Inc.

Chapter 8

ARSENIC IN DRINKING WATER AND PREGNANCY OUTCOMES: AN OVERVIEW OF THE HUNGARIAN FINDINGS (1985-2005)

Peter Rudnai, Mihály Csanády,
Mátyás Borsányi and Mihály Kádár
National Institute of Environmental Health,
Budapest, Hungary

ABSTRACT

In 1981 a country-wide survey revealed that in South-East Hungary more than 400,000 people were exposed to high levels (> 50 µg/L) of arsenic in the drinking water supplied by certain deep wells. This chapter presents detailed data of three retrospective ecological epidemiological studies conducted between 1985 and 2005 on pregnancy outcomes in the affected area. These studies used district nurses' yearly reports based on the records of pregnancy care units and revealed significant associations between high arsenic levels of the supplied drinking water and the frequency of adverse pregnancy outcomes (stillbirth, perinatal mortality and, especially, spontaneous abortion). As the arsenic level was gradually lowered, the frequency of these adverse birth outcomes also decreased. Significantly increased risk for spontaneous abortions could be observed only in settlements supplied with drinking water containing arsenic above 20 µg/L.

INTRODUCTION

In the southern part of Hungary deep well water has been used for drinking for a long time. In 1981 it was discovered that a considerable part of these deep wells contained arsenic of geological origin. A series of studies revealed that more than 400,000 people living in this area consumed drinking water containing arsenic higher than the health limit value of that time (50 µg/L): the values ranged up to 330 µg/L. The most affected area was Bekes county where 33 settlements were supplied with drinking water exceeding 50 µg/L arsenic level. A study was initiated to look for health consequences of the arsenic contamination by using available health statistics and registries. (Rudnai and Deák, 1988) Apart from an increased lung cancer mortality rate among the female population of Bekes county, the only apparent sign of the health impact was a statistically significant increase in the incidence of spontaneous abortions and stillbirths in settlements supplied with drinking water above 100 µg/L. The results of this epidemiological study significantly contributed to the decision so that an intervention programme was launched which included establishment of new water works, transportation of water from farther districts or various procedures of arsenic removal. As a result of the intervention arsenic concentration in most settlements was lowered to below 50 µg/L by 2000. However, the World Health Organization and the European Union established a new limit value of 10 µg/L arsenic in drinking water, which was also implemented by the Hungarian law in 2001. Management of a reduction of this size has an enormous financial burden on both the municipalities and the inhabitants, even if state and EU money are also involved. (Implementation of drinking water quality improvement programmes is still in progress in 2012).

Some non-health specialists even claimed that higher arsenic level in drinking water could be allowed in Hungary, because here much less arsenic is consumed with food than in countries surrounded by sea (e.g. Taiwan), from where the epidemiological findings were taken as a basis of the establishment of the new health limit value.

Between 1985 and 2005, three retrospective ecological epidemiological studies were conducted on pregnancy outcomes in the affected area using district nurses' yearly reports based on the records of pregnancy care units. These were the best available sources of health statistics in the area. These studies revealed significant associations between the arsenic level of the supplied drinking water and the frequency of adverse pregnancy outcomes. As the arsenic level was gradually lowered, the frequency of these adverse birth outcomes also decreased.

By the end of 1990-ies the arsenic level in drinking water was below 50 µg/L in most settlements of Hungary. After the new health limit value of 10 µg/L was introduced it was important to see what kind of health effects could be observed in the range between 10 and 50 µg/L arsenic levels. Therefore we conducted a retrospective ecological study on the association between the incidence of adverse pregnancy outcomes and the arsenic level of drinking water in all the 71 settlements of the most affected county (Bekes). Significant increase of risk for spontaneous abortions was found only above 20 µg/L arsenic level of the supplied drinking water and no similar association was observed related to the incidence of stillbirths.

DETAILS OF THE STUDIES

Study 1 (Partly Published in Börzsönyi et al., 1992)

Methods

In the first series of arsenic measurements carried out in 1982-83, a semi-quantitative (Gutzeit method) method was used in settlements where the expected concentration was in line with the aim to be able to detect one tenth (5 µg/L) of the limit value of that time. Where the arsenic concentration was found about 25 µg/L or higher, the concentration value was verified using photometric method with silverdithiocarbamate; hydride generation or electrothermal atomic absorption methods.

7 villages supplied with drinking water >100 µg/L arsenic level were selected as exposed settlements and 6 villages with drinking water <10 µg/L arsenic served as control ones. Town Bekes with its 22,000 inhabitants was considered an exposed settlement, too (Table 1.)

Pregnancy outcomes were evaluated on settlement level using district nurses' yearly reports since 1970, based on the records of pregnancy care units. The first ecological epidemiological analysis covered only the periods before any intervention in the water supply took place.

Results

The frequency and relative risks of spontaneous abortions and stillbirths in the studied settlements are shown in Table 2. There was significantly higher incidence of both pregnancy outcomes in the exposed settlements. There was no difference in the frequency between the exposed villages and the exposed town Bekes.

Study 2. (Published in Hugarian: Gulyas and Rudnai, 1997)

Before 1984 the arsenic content of the supplied drinking water in town Karcag, a town on the great plain of Hungary with 24,000 inhabitants was between 95 and 130µg/L, i.e. well over the hygienic limit value of 50 µg/L. Due to various interventions the arsenic content started to decrease from 1984 but a steady concentration below 50 µg/L could be reached only from 1990 on. In order to study the adverse effects of drinking water related arsenic exposure, some demographic data for years 1975-95 of Karcag were compared to those of Törökszentmiklós, a town with similar number of inhabitants and to the average values of County Jász-Nagykun-Szolnok. (Table 3.) Before 1984 the frequencies of spontaneous abortion, stillbirth, preterm birth, perinatal and infant mortality were all significantly higher in Karcag than in the reference populations.

From 1984 on, all the mentioned frequencies started to decrease and the difference between the data of Karcag and the whole county, especially in the case of stillbirth and perinatal mortality, became significantly smaller. Between 1990-95, when the arsenic level in the drinking water was steadily below the limit value, the frequencies of spontaneous abortion, stillbirth and perinatal mortality were lower than those in the control town although still slightly higher than the county values.

The relative risks of adverse pregnancy outcomes in the exposed town Karcag for the period between 1975-83 compared to the control town, Törökszentmiklós, and compared to the 1990-95 period of Karcag (after the remediation process) show that the high frequency of spontaneous abortion, stillbirth and perinatal mortality found in Karcag before 1984 may have been related to the high level of arsenic in the drinking water. (Table 4.) The statistical data also reflect the favourable trends, which may be explained by the effectiveness of interventions carried out since 1984 and especially since 1989-90.

Table 1. Characteristics of the study areas

EXPOSED SETTLEMENTS	Inhabitants (01.01. 1984)	As (ug/L) Before intervention	Year of intervention	As (ug/L) After intervention
BUCSA	2,580	100 – 261	1986	22 - 54
ECSEGFALVA	1,756	144 – 270	1991	22 - 48
KÖTEGYÁN	1,637	87 – 209	1990	2 - 22
TARHOS	1,243	92 – 210	1991	4 - 19
TELEKGERENDÁS	1,568	90 – 132	1991	7 - 20
FÜZESGYARMAT	6,337	64 – 112	1994	16 - 21
SZEGHALOM	10,527	60 – 190	1994	15 - 28
Together	25,648			
BÉKÉS (TOWN)	22,289	80 – 270	1987	1 - 45
CONTROL SETTLEMENTS	Inhabitants	(ug/L)		
MEZŐKOVÁCSHÁZA	7,123	Mean: 2.0 (0.0 - 10.0)		
VÉGEGYHÁZA	1,927	Mean: 2.0 (0.0 - 10.0)		
MEZŐHEGYES	7,385	Mean: 3.0 (0.0 - 17.0)		
DOMBEGYHÁZA	2,748	< 10.0		
HUNYA	917	< 10.0		
ÖRMÉNYKÚT	736	< 10.0		
Together	20,836			

Table 2. Relative risks of spontaneous abortions (SpAB) and stillbirths (SB) in settlements exposed to drinking water with arsenic level above 100 µg/L (1970-94)

	Number of live births	Incidence of SpAB (per 1000 live births)	Spontaneous abortions RR (95% C.I.)	Incidence of SB (per 1000 live births)	Stillbirths RR (95% C.I.)
Control villages	6519	42.9		3.23	
Exposed villages	9622	63.3	1.45 (1.26 – 1.66)	7.52	2.32 (1.43 - 3.81)
Town Békés	7786	65.5	1.49 (1.29 – 1.72)	7.71	2.38 (1.47 - 3.84)

Table 3. Demographic data of Karcag (exposed) and Törökszentmiklós (control) in relation to the quality of drinking water

DEMOGRAPHIC PARAMETERS	Karcag			Törökszentmiklós			County Jász-Nagykun-Szolnok		
	1975-83	1984-89	1990-95	1975-83	1984-89	1990-95	1975-83	1984-89	1990-95
Total number of live births	3 910	2 082	1 950	3 663	1 919	1 837	63 206	33 295	31 315
Total number of spontaneous abortions	245	123	90	159	120	99	No data available		
Number of spontaneous abortion per 1000 live births	62,7	59,1	46,2	43,4	62,5	53,9	No data available		
Total number of preterm births	523	241	218	330	174	156	6 601	3 307	2 749
Frequency of preterm births (%)	13,38	11,58	11,18	9,01	9,07	8,47	10,44	9,93	8,78
Total number of stillbirths	48	17	8	28	18	15	630	239	116

DEMOGRAPHIC PARAMETERS	Karcag			Törökszentmiklós			County Jász-Nagykun-Szolnok		
	1975-83	1984-89	1990-95		1975-83	1984-89	1990-95		1975-83
Number of stillbirths per 1000 live births	12,3	8,2	4,1	7,6	9,4	8,2	10,0	7,2	3,7
Total number of perinatal death cases	118	46	30	63	37	35	1 489	590	362
Perinatal mortality (per one thousand)	29,8	21,9	15,3	17,1	19,1	18,9	23,3	17,6	11,5
Total number of infant death cases	111	50	33	74	33	31	1 350	553	401
Infant mortality (per one thousand)	28,4	24,0	16,9	20,2	17,2	16,9	21,4	16,6	12,8

Table 4. Relative risks of adverse pregnancy outcomes of the high arsenic content of the drinking water in Karcag between 1975-83

	Compared to Törökszentmiklós (1975-83) RR (95% C.I.)	Compared to Karcag (1990-95) RR (95% C.I.)
Spontaneous abortions	1.42 (1.17 – 1.72)	1.36 (1.07 – 1.78)
Stillbirths	1.60 (1.01 – 2.54)	2.96 (1.40 – 6.25)
Preterm births	1.48 (1.30 – 1.69)	1.20 (1.03 – 1.39)
Perinatal mortality	1.75 (1.30 – 2.37)	1.96 (1.32 – 2.92)
Infant mortality	1.41 (1.05 – 1.88)	1.68 (1.14 – 2.46)

Study 3. (Published as an Abstract by Rudnai et al., 2006)

Methods

The official definitions of spontaneous abortion and stillbirth were changed from January 1st 1998 (Table 5.), so we could not continue to follow up further changes of these indices. Therefore we started to collect these data according to the new criteria and evaluated the situation in county Bekes.

Table 5. Criteria of spontaneous abortion and stillbirth in Hungary

	Spontaneous abortion		Stillbirth	
	up to 1997	from 1998	up to 1997	from 1998
Time spent in utero	< 28 weeks	< 24 weeks	>28 weeks	> 24 weeks
Body weight	< 1000 g	< 500 g	> 1000 g	> 500 g
Body length	< 35 cm	< 30 cm	> 35 cm	> 30 cm
Sign of life	-	-	No	No
When multiple birth	-	-	At least 1 live foetus	

Data on pregnancy outcomes between 1998 and 2002 were collected for all the 71 settlements of County Bekes, using the annual reports of the district nurses providing pregnancy care for newly registered pregnant women.

Data on the arsenic level of drinking water supplying the settlements were based on the measurements performed in 1999 as part of the National Environmental Health Action Programme. (No changes were recorded in the arsenic levels in these settlements between 1998 and 2002. The new legislation with the 10 µg/L health limit value appeared in 2001.) Poisson regression was used for analysis, using STATA 7.0 programme.

Results

In the 71 settlements during the 5-year period altogether 17,813 live births, 1,062 spontaneous abortions and 135 stillbirths were recorded. (Table 6)

Table 6. Demographic data of County Bekes between 1998-2002

Number of settlements	71
Number of live newborns	17,813
Number (and frequency) of spontaneous abortions	1,062 (5.96%)
Number (and frequency) of stillbirths	135 (7.58 per 1 thousand)

Rates of both spontaneous abortion and stillbirth changed with settlement size: stillbirth rate decreased and spontaneous abortion rate increased with increasing number of population. (Table 7)

Table 7. Pregnancy outcomes according to settlements size

Population	Settlements	Live births	Stillbirths	Stillbirth rate (per 1 thousand)	Spontaneous abortions	Spontaneous abortion rate (%)
1-1,000	11	353	3	8.50	13	3.68
1,001-5,000	38	4,073	41	10.07	207	5.08
5,001-10,000	13	3,927	29	7.38	236	6.01
>10,000	9	9,460	62	6.55	606	6.40

In 35 settlements the arsenic level of drinking water was below 10 µg/L. Taking this as a reference value, the incidence rate ratios (and their 95% confidence intervals) of spontaneous abortions (corrected for the number of inhabitants) were significantly increased above 20 µg/L concentrations. (Table 8.) Similar association between the arsenic concentrations and the incidence of stillbirths could not be observed. (Table 9).

Table 8. Risk of stillbirth (SB) above various concentrations of arsenic in the drinking water

As conc. cut-off point (µg/L)	Settlements above/below cut-off point	Live births above/below cut-off point	Number of SB above/below cut-off point	SB rates (per 1 thousand) above/below cut-off point	Crude Relative Risk RR (95% CI)	P value	Incidence Rate Ratio adjusted to population size RR (95% CI)	P value
10	36/35	11,335/6,478	83/52	7.3/8.0	0.91 (0.65-1.29)	0.60	1.04 (0.72-1.51)	0.823
15	27/44	9,655/8,158	69/66	7.1/8.1	0.88 (0.63-1.24)	0.47	1.00 (0.70-1.43)	0.999
20	13/58	3,509/14,304	30/105	8.5/7.3	1.16 (0.78-1.74)	0.46	1.27 (0.83-1.93)	0.271
25	9/62	2,892/14,921	24/111	8.3/7.4	1.11 (0.72-1.73)	0.63	1.21 (0.77-1.90)	0.404
30	7/64	1,374/16,439	9/126	6.6/7.7	0.86 (0.44-1.68)	0.66	0.86 (0.43-1.69)	0.656

Table 9. Risk of spontaneous abortions (SpAB) above various concentrations of arsenic in the drinking water

As conc. cut-off point (µg/L)	Settlements above/below cut-off point	Live births above/below cut-off point	Number of SpAB above/below cut-off point	SpAB rates (%) above/below cut-off point	Crude Relative Risk RR (95% CI)	P value	Incidence Rate Ratio adjusted to population size RR (95% CI)	P value
10	36/35	11,335/6,478	677/385	5.18/5.23	1.00 (0.89-1.13)	0.94	0.94 (0.83-1.08)	0.383
15	27/47	9,655/8,158	588/474	5.36/5.11	1.05 (0.93-1.18)	0.46	0.99 (0.87-1.12)	0.844
20	13/58	3,509/14,304	244/818	5.82/5.07	1.20 (1.05-1.38)	0.009	1.16 (1.001-1.34)	0.048
25	9/62	2,892/14,921	208/854	6.03/5.09	1.24 (1.07-1.44)	0.004	1.20 (1.03-1.42)	0.022
30	7/64	1,374/16,439	102/960	6.03/5.11	1.25 (1.03-1.53)	0.030	1.23 (1.01-1.52)	0.043

DISCUSSION

The arsenic level of drinking water in the study area has some special features. The first one is that neighbouring villages can have drinking water with very different arsenic level depending on the depth of the wells supplying the drinking water. This explains that in settlements supplied with drinking water of high or low arsenic content, the way of life and other environmental risk factors do not differ significantly from each other.

The second feature is that in several settlements more than one well is connected to the drinking water network. Therefore, depending on the seasonal variations of water demand, various mixing ratios of water from different sources may result in changing levels of arsenic in the water of the same settlement. Moreover, due to the ecological type of these studies exposure assessment was done on settlement level and the individual characteristics of exposure could not be considered. All these may, naturally, cause uncertainty in the exposure estimation. However, the improving statistical data on pregnancy outcomes with decreasing level of arsenic in drinking water support our findings on the associations between adverse pregnancy outcomes and arsenic level of drinking water.

Summing up the findings of our studies, we can conclude that, in spite of their ecological design, our studies provided evidence on the associations between the arsenic level of drinking water and some adverse pregnancy outcomes, first of all, spontaneous abortions.

REFERENCES

Börzsönyi M, Bereczky A., Rudnai P., Csanády M. and Horváth A.(1992) Epidemiological studies on human subjects exposed to arsenic in drinking water in Southeast Hungary. *Arch Toxicol* 66:77-78.

Gulyás E. and Rudnai P. (1997) Adverse effects of drinking water related arsenic exposure on some pregnancy outcomes in Karcag, Hungary. (in Hungarian) *Egészségtudomány* 41, 137-44.

Rudnai P. and Deák Zs. (1988) Epidemiological study on the health effects of drinking water with high arsenic content. (in Hungarian) In: A környezet arzén szennyezettségének településegészségügyi kérdései a dél-alföldi régióban. (ed by I. Dési) Szeged, pp. 63-68.

Rudnai P., Varró M.J., Borsányi M, Páldy A, von Hoff K, Sárkány E. and Szép H (2006) Arsenic in drinking water and pregnancy outcomes: an ecological study. *Epidemiology* 17(6 Suppl):S329-330.

In: Arsenic
Editor: Andrea Masotti

ISBN: 978-1-62081-320-1
© 2013 Nova Science Publishers, Inc.

Chapter 9

GEOPHAGY: IMPLICATIONS OF UNKNOWINGLY INGESTING ARSENIC – A CASE STUDY IN SELECTED AREAS IN GHANA

Emmanuel Arhin[*,1] *and Musah Saeed Zango*[2]
[1]University of Leicester, Department of Geology,
University Road, Leicester, UK
[2]University for Development Studies,
Department of Earth Science, Navrongo, Ghana

ABSTRACT

Geophagy is the practice of eating earthy or soil-like substances such as clay or chalk, supposedly to obtain essential nutrients from it or for cultural and spiritual purposes. The clays eaten by most geophagists in Ghana are molded into cylindrical or polygonal shapes and are products of rocks, reflecting the geogenic and geochemical variations of chemical elements in the parent rocks. This chapter examines clay-balls or tablets and hair samples from Navrongo, Kumasi and Anfoega markets that were analyzed for As, Pb, Ca, K, Mg, Zn, Fe, Cu, Mn and Se using Wet Digestion and ICP-MS methods respectively. The results showed insignificant concentrations of Pb and K and revealed relatively high concentrations of Ca and As. Ca has assay values ranging from 17.3 to 26.1 %; K has values ranging from 1.2 to 1.3 % whereas As has values ranging from 14.5 to 20.9 ppm. Ca, noted as an aid in fetal development during pregnancy, was present in all the samples analyzed while K, which is also used to curtail stomach aches and diarrhea, was relatively low compared with Ca. It was revealed for the first time that sun-baked clay balls and tablets, locally known as "Hyire" and consumed mostly by Ghanaian women, can be a source of arsenic exposure that has not been previously considered in risk assessment studies. It is concluded that the elemental constituents of the geophagic materials in Ghana be determined and intake quantities (kg) be established per day in order to make the practice of geophagy safer and more attractive.

Keywords: Pregnancy; arsenic; concentration; human health; Kumasi

---

[*] Corresponding author e-mail: ea163@le.ac.uk

INTRODUCTION

The deliberate eating of non-food substances is known as pica [1]. However, modern literature has shown that human health is intimately linked to diet and healthy eating. Some people consume non–food substances such as charcoal, uncooked rice and earthy materials [2] in order to cure some ailments or for cultural reasons. Geophagy is the most common type of the deliberate eating of earthy materials of non-nutritive substance such as clay. It is an ancient practice that still occurs in most parts of Ghana, especially in its rural areas. The precise reasons underlying the practice of geophagy in Ghana still remains unknown, although some geophagists and vendors of geophagic products suggest their consumption for nutritional [3, 4] and medicinal purposes [5]. Other geophagists, mostly from northern Ghana, also indulge in the practice for cultural and spiritual reasons. Typically, in Ghana, the practice is generally common among children and females, especially menstruating girls, lactating mothers and pregnant women [6-9]. Geophagic material consumption has often been recommended as a means to increase the intake of essential elements (especially Ca, Mg, Zn, Fe, Cu, Mn and Se). The practice of geophagy is not only observed in Ghana, but it is also common in Kenya, Namibia, Tanzania, Burkina Faso, Togo, and Cote I'voire as well.

The practice of geophagy in Ghana is of great concern as the levels of harmful elements in the clay balls or clay tablets have not been empirically determined. Some metals are essential to life; they can be crucial for the correct functioning of biological systems if maintained at required, permissible maximum daily intake levels (PMDIL) [10]. Therefore, there is a challenge associated with the lack of evidence of archival reports documenting the PMDIL of the different levels of toxic elements in the clay balls and with the failure to define the allowable consumption quantities per day. Trace elements or metals can become harmful to human health if their exposure levels exceed the permissible safe thresholds. Conversely, potentially harmful substances (PHS) such as trace elements (e.g. As, Pb, Sb, Cd, U etc.) whose prevalence and excesses in humans cause health problems are ubiquitous and appear present in the environment and in all living organisms [11]. This may contribute to the risk enhancement of PHS exposures if the orally ingested geophageous materials are not kept at acceptable limits. The acceptable limits of elements ingested into humans differ from location to location. For example, a study by Yan-Chu [12] in China used the US Environmental Protection Agency's quantified lifetime exposure of arsenic in drinking water at concentrations of 0.0017 mg/L, 0.00017 mg/L, and 0.000017 mg/L. It was realized that a lifetime skin cancer risk of 1 in 10,000, 1 in 100,000, and 1 in 1,000,000 were associated with the excesses of these limits respectively. The World Health Organization contends that levels of 0.01 mg/L poses a risk of 6 in 10,000 chances in a lifetime for skin cancer risk and argues that this level of risk is acceptable [13]. However, in Ghana, despite the high vulnerability of pregnant women and their unborn babies, the unknowing ingestion of toxic elements through geophagy and its potential adverse health effects has not been explored. There is no provisional tolerable daily intake levels of clay balls, and the concentrations of toxic elements, such as "As", Cd, Mn, Se and Pb, known to be associated with the Birimian system of Ghana have not been determined either.

The aim of this chapter is to raise the awareness of possible adverse health effects that may ensue from geophagy if studies do not embark on determining the *provisional maximum tolerable daily intake (PMTDI)* of the clay balls by establishing tolerable levels of the toxic elements in the clay-balls.

GEOPHAGY

Geophagy may be defined as deliberate and regular consumption by animals and humans of earthy materials such as soils, clays, and related mineral substances. It is a practice that is found in all continents, but is most common in the tropics [14]. Geophagy was known in the ancient world, being first mentioned by Aristotle [15], and was more fully described when tropical countries began to be extensively explored and developed in the nineteenth century. Geophagy has the widest distribution among the world's poorer or more tribally-orientated people [14]. Deliberate soil consumption is, therefore, still common in the tropics, yet geophagy largely remains an under-reported and misunderstood practice. The aetiology of soil eating is manifold, though the causal link between geophagy and the regulation of mineral imbalances still remains to be confirmed. Nevertheless, soils have the potential to be an important source of mineral nutrients, especially iron, to geophagists but the relative concentrations, either harmful or harmless, of the elements being ingested must be known.

Geophagy is a nearly universal, transcultural phenomenon. The practice is a complex, multi-causal behavior with roots in spiritual and religious beliefs, ritual oaths and ceremonies, medicinal practice, and perhaps nutritional need. Geophagists claim eating clays reduce abdominal pain caused by hookworm, ease the pangs of hunger, soothe heartburn and nausea, or simply tastes good. More so, the absorptive properties of clay minerals are well documented for healing skin and gastrointestinal ailments. However, the antibacterial properties of clays have received less scientific attention. Haydel et al. [16] also report the use of geological materials to heal skin infections. Clay treatments of bacterial diseases, including infections for which there are no effective antibiotics, such as Buruli ulcers and multi-drug resistant infections, have been evident since earliest recorded history [17].

Many worldwide published reports reveal that geophagy is the world's oldest medical practice, and clay is recorded as a pharmaceutical product throughout history [18]. The most common manifestation of geophagy as a medicinal product is exhibited during pregnancy because pregnancy exerts a large physiological cost in terms of calcium, iron and other nutrients that must be delivered to the growing fetus. However, because of poverty, the practitioners respond to their physiological needs by having the pregnant women eat clays to satisfy the physiological need for these nutrients. Modern society uses kaolin and smectite clays for treating gastrointestinal disorders (e.g., Tricilicate and Milk of Magnesia). It is evidenced from [19] and [20] that geophagists consume clays because the clay coats the gastrointestinal tract and absorbs dangerous toxins, also providing critical calcium for fetal development. Modern scientific research has also confirmed the healing powers of clay elsewhere in the world [21; 17]. Similarly, [22] has shown that the clays ingested serve as a medicine, food detoxifier, psychological comforter and supplier of mineral nutrients. While it is a cultural practice, geophagy also fills a physiological need for nutrients. In Africa, pregnant and lactating women are able to satisfy the very different nutritional needs of their bodies by eating clay. However, not everybody in Africa practices geophagy. For instance, in Ghana, pregnant women and lactating mothers are mostly the people that practice geophagy.

Children also ingest clay, but that is usually involuntary. It has been noted that during the first trimester of pregnancy, the ingested clays consumed adsorb dietary toxins that are potentially teratogenic to the embryo, while simultaneously suppressing the common symptoms of pregnancy sickness such as nausea [22]. In the second trimester, when pregnancy usually sickness subsides, clays serve as a source of mineral nutrition, and calcium supplementation which aids in the formation of the fetal skeleton and reduces the risk of pregnancy-induced hypertension [23]. There are reports that some geophagy practitioners sometimes get multiple benefits when they indulge in geophagy while seeking a particular outcome [18]. Among the clays most commonly used for medicinal purposes are kaolin and the smectite clays such as bentonite, montmorillonite, and Fuller's Earth. It is not of any surprise that geophagists mention the effectiveness of ingested clays in treating gastrodynia (stomach ache), dyspepsia (acid indigestion), nausea and diarrhea [22].

THE HIDDEN RISK OF GEOPHAGY

Clay contains a range of concentrations of potentially harmful substances, and some of these substances bioaccumulate more than others depending on their "binding" and environmental conditions conditions. In one account, [24] indicated that some slaves ingested clays as food, as well as for medical and cultural reasons, during the slave trade era. In another example, [25] indicated that some slaves indulged in excessive clay consumption not as a food source, but rather to become ill and thus avoid work. Furthermore, others ingested it to commit suicide. There are records where Negro cooks mixed arsenic with food to kill their superiors or masters. At times this was also done by plantation workers to commit suicide during the slave trade era. In all these accounts, the quantities of geophagic materials consumed and their relative concentrations of toxic elements were unknown. Earth materials can be variable in their trace element composition depending on their mineral composition, source, and diagenetic and geo-environmental history. The wobbly and dodgy side of this practice is that the earth materials consumed contained some toxic elements which probably caused the deaths. Apt knowledge of chemical content and concentrations of toxic elements must be known in order to predict potential health impacts. Geophagy, thus, has a double edged sword effect: it cures and ruins, but its usage could be beneficial if the concentrations of its chemical contents are known. In Ghana, there is very little evidence in published reports that addresses the balance between the benefits and pitfalls of geophagy. The chemical content of arsenic in geophageous materials and their speciation have not yet been investigated or clarified. Furthermore, much less is known about the quantities of clay balls consumed per day by person and whether there are other possible sources of toxic elements (e.g., through foods, water, diet, air, etc.) into the human system that needs to be studied.

The realization of the benefits associated with geophagy may be enhanced if the concentrations of elements in clay balls could be provided and their daily intake specified. Abrahams et al. [14] reported that sikor, equivalent to 3 – 4 tablets (Fig.1A), can be consumed per day by pregnant Bangledeshi women. This chapter seeks to highlight the need to determine the different elements in the clay balls and also establish a safe clay ball intake per day per person, especially to eliminate the potential hazards associated with clay consumption and the people in the three regions selected for the pilot geophagy studies in Ghana.

A. Sikor tablets from Bangladesh   B. 'Hyire' or clay-balls from Ghana

Figure 1. Typical geophagic materials: A) from Bangladesh and B) from Ghana. Quantities of oral intake of Sikor in Bangladesh and the contained elemental concentrations have been determined whereas elemental concentrations in Hyire from Ghana are yet to be determined.

LOCATION OF STUDY AREAS

Geophagy is widely practiced throughout Ghana. Three regions were selected as a pilot study to assess the geological and geochemical aspects of geophagy on human health. The selected regions are Upper Eastern, Ashanti and Volta (Figure 2). The locations of the study areas are geo-referenced to the precise sample points: Navrongo for Upper Eastern, Kumasi for Ashanti and Anfoega for Volta region respectively. Sampling points fall within UTM Zone 30 northern hemisphere.

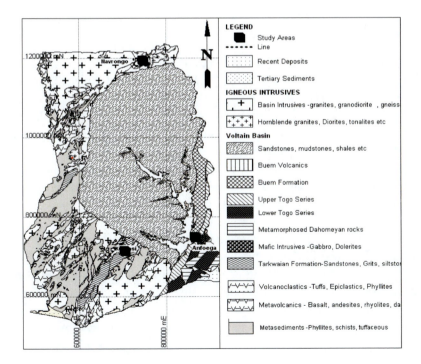

Figure 2. Geology and locations of study areas (modified after Kesse 1985).

MATERIALS AND METHODS

Field Survey

Two types of data were collected in the survey. The data comprised of clay field samples that have been brought to the market by the vendors and hair samples from some geophagy practitioners. Prior to the selection or sampling, the vendors were interviewed to determine the source of the geophagic materials, the reasons for selling them and the ailments the geophagic materials were thought to help, but the geophagists were not available to be interviewed to determine their diet for other sources of arsenic intake.

The next phase of the research intends to determine how much clay is consumed over a given period of time and also to investigate what other sources of arsenic these people are exposed to. The current research collected three samples from vendors selling traditional clay/herbal remedies at each of the selected sites (i.e. Navrongo, Kumasi and Anfoega respectively).

The cylindrically shaped clays, or "Hyire" as locally known, were disaggregated in a porcelain mortar with a pestle, and then sieved through a 2 mm sieve. The -2 mm sieved sample fraction was thoroughly homogenized, divided into two sub-samples, bagged and labeled. These labeled samples were sent to two different laboratories for to analyze Na, P, Ca, Mg, K and important nutrients, whereas As and Pb are generally toxic depending on their concentrations. All these elements are found in underlying rocks (e.g., metavolcanic and metasedimentary rocks, volcano-sedimentary units and granitic intrusives) in the study areas [26].

Clay Samples and Analysis

The samples were analyzed using a Wet Digestion Method. In this method, 0.5 to 2.0 g of the sieved samples were put into a 250 ml "Kjedahl" flask and weighed, after which the kjedahl digestion tablet was added.

The digestion tablet consisted of a mixture, of selenium metal, copper sulphate ($CuSO_4$) and lithium sulphate ($LiSO_4$) that had been made into a tablet. 20.0 ml of concentrated sulphuric acid (H_2SO_4) was then added to the digestion tablet. The digestion mixture was allowed to saturate for about 30 minutes at a temperature of 100°C and later increased to 300°C for two hours or until the mixture was permanently clear. Distilled water was then added to the cooled liquid. The digested sample was then transferred to a 100ml volumetric flask.

Aliquots of the digested samples were then taken to determine the requested major and toxic elements. Sodium and K were determined by using the Flame Photometer; Ca, Mg, As and Pb were determined by using Atomic Absorption Spectrometer (AAS); and phosphorus was determined by using the Phosphormolybdate Blue Complex reduction with ascorbic acid, and the blue color measured on a visible Spectro Photometer at a 600 nm wavelength. These analyses were performed by the crop and soil research institute laboratory in Kumasi, Ghana.

Hair Sampling and Analysis

Tests on hair and fingernails can measure exposure to high levels of arsenic over the past 6–12 months. These tests determine if one has been exposed to above-average levels of arsenic; however, they cannot predict whether the arsenic levels in the body will affect health. The World Health Organization recommends a limit of 0.01 mg/L (10 ppb) of arsenic in drinking water. This recommendation was established based on the limit of detection by available testing equipment at the time of publication of the WHO water quality guidelines [27].

Samples of scalp hair from 42 geophagic practitioners with non-dyed hair, aged between 18-60 years were obtained from Kumasi (n=21) and Navrongo (n=21). The third site was not included because of time and funds available for the study. However, in conforming to research ethics, all participating members consented to the study before hair samples were taken. External contaminations were eliminated by washing each hair sample in a sequence of acetone, bi-distilled water and acetone [28].

The hair samples were later air-dried at room temperature in a dust free area. They were then digested in aqua-regia at $95^{\circ}C$, for 2 hours, and dried inside a closed Teflon vessel for 30-45 minutes. As a safety precaution against contamination, the samples were kept in plastic bags away from dust. Solutions obtained from the hair samples were diluted and analyzed using a high resolution ICP-MS. In-house quality control samples and certified reference materials were ran together with the hair samples. Four duplicate samples were inserted into the 42 samples to determine the precision of the analysis. Two standard samples were also analyzed alongside the 42 samples collected during the studies to monitor the accuracy of the analysis.

RESULTS

Summary information on major elements (Na, Mg, K, Ca, and P) and potentially toxic elements (As and Pb) in sample clays are shown in Table 1. An examination of the data shows trace amounts of Pb in all analyzed geophagic samples. Arsenic has a concentration of about 21 ppm in the samples from Kumasi, 19 ppm at Anfoega and 15 ppm in Navrongo. Arsenic concentrations from Kumasi and Navrongo were compared with two major elements, K and Ca that are generally beneficial to the geophagists (within prescribed concentration ranges).

The consequent graphs produced from the selected major and trace elements are presented in Figure 3 and Figure 4. In addition, Figures 5 and 6 illustrate the bio-accumulation of As and Ca over a range of time periods, after analyzing hair samples in two of the study areas, Kumasi and Navrongo, respectively. Table 2 shows the arsenic concentrations in hair samples from Kumasi and Navrongo respectively. Arsenic concentrations in clay balls were variable from the three study areas. The highest concentration of about 21 ppm was in Kumasi while it was 14 ppm in Navrongo and about 16 ppm at Anfoega.

Table 1. Major and Toxic elements recorded from geophagy samples

SITE	MAJOR ELEMENTS					TOXIC ELEMENTS	
	Na (%)	Mg (%)	K (%)	Ca (%)	P (%)	Pb (ppm)	As (ppm)
Kumasi 1	0.33	0.29	1.19	1.33	0.09	Trace	20.89
Kumasi 2	0.31	0.28	1.26	26.00	0.10	Trace	16.00
Kumasi 3	0.33	0.18	1.24	19.00	0.11	Trace	14.45
Kumasi 4	0.35	0.29	1.37	17.39	0.98	Trace	20.84
Navrongo 1	0.33	0.31	1.18	17.29	1.00	Trace	20.94
Navrongo 2	0.30	0.31	1.24	26.02	0.10	Trace	16.32
Navrongo 3	0.35	0.21	1.24	19.01	0.11	Trace	14.46
Navrongo 4	0.40	0.29	1.23	17.39	0.43	Trace	14.69
Anfoega 1	0.32	0.30	1.17	17.30	0.90	Trace	20.92
Anfoega 2	0.30	0.32	1.24	26.01	0.29	Trace	16.30
Anfoega 3	0.34	0.20	1.23	17.32	0.33	Trace	14.46
Anfoega 4	0.36	0.40	1.34	26.71	0.39	Trace	16.60

**Table 2. Arsenic and Ca concentrations in hair samples
(approximated to the nearest whole number)**

Age/ Sample ID Elements	18-25 As (ppm)	18-25 Ca (%)	26-40 As (ppm)	26-40 Ca (%)	41-60 As (ppm)	41-60 Ca (%)
KS001	16	13	19	10	22	15
KS002	14	10	20	13	20	15
KS003	13	15	17	14	19	14
KS004	13	14	18	16	22	12
KS005	12	10	13	15	19	13
KS006	13	9	15	11	18	11
KS007	14	10	18	12	19	16
NV001	12	15	15	16	19	16
NV002	13	15	13	17	17	15
NV003	15	14	16	14	20	17
NV004	13	16	15	13	22	16
NV005	14	14	17	15	16	14
NV006	13	13	16	11	15	15
NV007	12	11	13	13	14	18

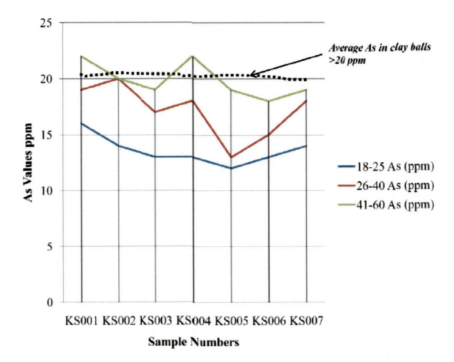

Figure 3. Arsenic in hair samples (Kumasi).

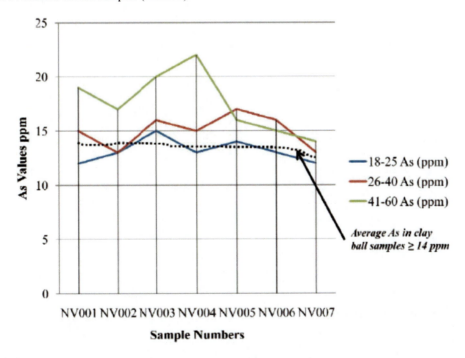

Figure 4. Arsenic in hair samples (Navrongo).

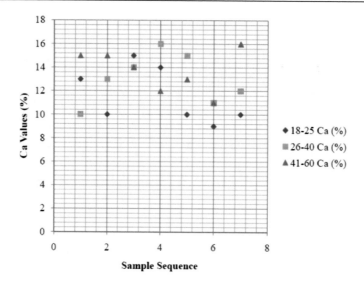

Figure 5. Calcium in hair samples (Kumasi).

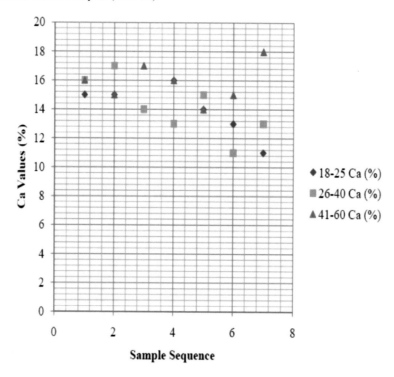

Figure 6. Calcium in hair samples (Navrongo).

BIOACCUMULATION

The Bioaccumulation Factors (BAFs) were calculated as the hair sample's total arsenic result (ppm) divided by the content of the geophagic clay ball sample's results (ppm). This is presented in Table 3.

Table 3. BAF for different age groups

Location/ages	Ave. As in Hair (ppm)			Ave. As in clay (ppm)	BAF value			Differences in As conc. (ppm)		
	18-25	26-40	41-60		18-25	26-40	41-60	18-25	26-40	41-60
Kumasi	13,57	17,14	19,85	20,92	0,64866156	0,819311663	0,948852772	-7,35	-3,78	-1,07
Navrongo	13,14	15	17,57	14,46	0,908713693	1,037344398	1,215076072	-1,32	0,54	3,11

DISCUSSION

Based anecdotal evidence, interviews conducted during the survey suggest that the consumption of clays can cure or ameliorate diarrhea, indigestion, stomach aches, etc., possibly because some clays contain anti-diarrheal properties. The clays and their therapeutic/cosmetic properties are determined by the prevalence of certain minerals/oxides in them which may have a relationship with the K in kaolin, montmorillonite and smectite clays. The concentrations of K in the samples analyzed at the study areas were just enough to serve as an anti-diarrheal ingredient; however, when ingested orally, the associated As (up to about 21ppm and may include intake from foods and water yet to be empirically ascertained) makes the practice very dangerous. This finding has relevance when establishing what quantities of clay balls should be taken by geophagists, mainly vulnerable pregnant women, lactating mothers and girls. Of these classes of geophagists, pregnant women are of particular concern in this context since As can transfer from the mother to the fetus, placing the health of the unborn baby at risk. Arsenic is a known, potentially hazardous substance. The World Health Organization's acceptable concentration of As in water is 10ppb and the As content in clay ranges between 4-10ppm or mg kg^{-1}.

Based on the analyzed clay balls from the three study areas, the arsenic concentrations exceeded the potentially maximum tolerance level (PMTDI) by 2.1 -fold. The current study was unable to establish the exact ingestion number of clay balls that yielded this value. It may be possible that the ingestion of 50g -100g/day of Hyire by pregnant women can produce high As levels and could present health problems to all practitioners; pregnant women may be at risk of harming their unborn baby.

From Tables 1 and 2, concentrations up to about 21 ppm of As from geophagic materials were exposed to the practitioners to be ingested as long as they remained in the practice. A possible safe dose for humans, considered as a dietary mineral, has been calculated to be 15-25 µg/day [29]. This amount could be absorbed from food and water without causing health problems. Toxicity differs between various arsenic compounds; for example, monomethyl arsenic acid and inorganic arsenide have a higher toxicity level than arsenic chlorine. Acute toxicity is generally higher for inorganic arsenic compounds than for organic arsenic compounds. Oral intake of more than 100mg is lethal. The lethal dose of arsenic trioxide is 10-180mg, and is 70-210mg for arsenide [29]. The mechanism of toxicity is binding and blocks sulphur enzymes. The arsenic content analyzed in the clay-balls is twice the concentration as compared to the permissible intake of arsenic to be orally ingested. Therefore, it is necessary to investigate arsenic speciation as well as the chemical contents of all elements in the clay balls. A report by the WHO [27] indicates the outcomes associated with health as a result of arsenic exposure depend on the dose, modality and duration of exposure, as well as the source of arsenic. This suggests a great calamity to practitioners in Ghana because of the lack of knowledge associated with the chemical contents of the clays desired to beaten to ameliorate their ailments, or to satisfy some traditional or cultural beliefs. Bioaccumulation is a result of longer exposure times at lower concentration levels through regular ingestion during menstrual periods, pregnancy and lactating periods. What is most critical in the voluntary ingestion of clay in Ghana is that there is no archival evidence that reports the efficacy of eating clays to cure nausea, stomach aches and diarrhoea, etc. However, the involvement of the lactating mothers, pregnant women and teenage girls is generally overwhelming. Part two of this research includes gathering empirical data in order to address the several health vulnerabilities such as skin rashes and lessions, circulatory and nervous disorders as well as dermal changes that appear to be common. During the current research it was realized that geophagists ingest clay balls to treat gastrodynia and dyspepsia (i.e., stomach aches and acid indigestion), as well as nausea and diarrhoea, which has been reported by the [22] study on geophagy.

Carl Ege [30] reports of the use of "Milk of Magnesia or Kaolin Pectate" to relieve stomach upset. The consumed clays in the study areas contain concentrations of K between 1.17 to 1.37% and may be suitable as a traditional medicine to cure abdominal colics and acid indigestions. Many of the practitioners confirmed the relief obtained after eating the clay balls, but could not specify the quantity per day. The scientific research to answer that question is underway. Also, [23] asserted that clays serve as a source of mineral nutrients and calcium supplements that build the fetal skeleton during the second trimester of pregnancy. This was a benefit to the practitioners in that the calcium concentrations recorded in the study areas, as shown in Tables 1 and 2, were significant, but the proportion that is bio-accessible needs to be determined for the purpose of bone and fetal development required by the practitioners. However, the presence of As, irrespective of the concentrations and the ignorance of the geophagists in Ghana about the presence of As, in the voluntarily eaten clay, is a matter of great concern. Apparently, it is not evident that the clays were ingested for the medicinal use of As as an agent of criminal intent; however, it is a serious error to consume something without knowing its chemical contents. Nonetheless, all the three regions studied had high arsenic concentrations in samples compared to the average level of 3-4 ppm in soils (Tables 1 and 2; Figs. 3 and 4).

In addition, as seen in Tables 1 and 2 and Figs. 3 and 4, the geophagists in Kumasi and Navrongo had arsenic concentrations in hair samples close to or greater than As concentrations recorded in clay balls. This suggests that the continuous and regular practice of geophagy may contribute to increased concentrations of the toxic element. Therefore, to reduce the impact of health hazards ensued from arsenic exposures via geophagy requires knowledge of the chemical compositions of the clay balls and what amounts of clay need to be ingested per day so as not to exceed the WHO's or country's standards over time. From Figs. 3 and 4 there is an increase in As in the different age brackets, with highest concentration occurring among the 41-60 year category; this is followed by the 26-40 year group and then the 18-25 year class, in both Kumasi and Navrongo respectively,. Conversely, there were insignificant changes in calcium concentrations as shown in Figs. 5 and 6. However, as portrayed in Table 3, arsenic bioaccumulates in the bodies of the practitioners, and the concentrations increase with duration of practice as well. Quantities ingested also contributed to the increase, but this research was not able to quantify the amounts as the practitioners were not able to estimate quantities ingested per each time. As presented in Figs. 3 and 4, arsenic in both regions show greater increases in hair samples than in clay samples. It may be possible that other As sources (e.g., from water, food etc.) were also added to the concentrations of arsenic in the hair. However, the BAF value in Kumasi was < 1.00 which appeared to be insignificant in terms of bioaccumulation. This does not imply there was no bioaccumulation, however. There may be other reasons such as Kumasi being a metropolitan city, the second biggest city in Ghana, and that accessibility to healthcare was not as widespread of a problem as it used to be a couple of decades ago. The number of women with rashes or lesions have reduced compared to rural Navrongo where many of the geophagists still use clay balls as medicine. The pregnant women and lactating mothers that form the core of the practice rather tend to go for medications at Health Centers instead of indulging in geophagic practices. These factors probably had an impact on the degree to which clay was ingested, iresulting in a lower bioaccumulation factor: BAF < 1.00 (Table 3). Meanwhile, the bioaccumulation situation in Navrongo was different: the BAF >1.00 for practitioners aged between 26-60 years of age (Figure 4).

The higher BAF in this area may be attributed to socio-cultural and traditional factors. Navrongo, contrary to Kumasi, falls within rural Ghana. Accessibility to healthcare is extremely difficult. Furthermore, the poverty level in this area is very high compared to Kumasi. Practitioners may prefer to indulge in geophagy that is more affordable and readily accessible rather than to seek medical care at Health Centers that are comparatively more expensive.

In addition, tradition and cultural adherences have taken a stronger hold of the people in the area. Likely, these are some of the causes of the 4% and 21% increases in As concentrations in age groups 26-40 years and 46-60 years respectively. On the whole, concentrations of As increased with age in the scalp hairs of the older practitioners (41-60 years). The latter increases could be associated with a changing diet, their drinking water, or the existence of a bioaccumulation effect over time [27].

CONCLUSION

The practice of geophagy as understood by the people that indulge in it in Ghana reiterate why it was embedded in some cultures. It was also reaffirmed to be one of the world's oldest medicines recorded as a pharmaceutical throughout history. It was their grandfathers' functional medicine; why should they abandon it, if their grandfathers could live long by using them? This confirms why the practice is still evident. It is still common among pregnant women who usually need about 800-1200 mg/day of Ca to provide suitable nutrients for fetal skeletal growth and development [22].

Invariably, the cost of geophagic materials is far more affordable than orthodox medicine. The study reveals, for the first time, that the sun-baked clay (locally known as "Hyire") consumed mostly by Ghanaian women can be a source of arsenic exposure that has not been previously considered in risk assessment studies. From an expert's perspective, the benefits of geophagy should break even with its nuisance by acknowledging that rock is an assemblage of minerals and those clays are weathered products of rocks. The minerals could either be harmful or harmless, and should not be consumed if the constituent elements and their biotic components are unknown.

It is being concluded that the elemental constituents of the geophagic materials in Ghana be determined and intake quantities (kg) be established per day in order to make the practice of geophagy safer and more attractive. Otherwise, geophagists may be slowly killing themselves as it was done during the slave trade era, where some slaves consumed soils in order to commit suicide. The authors recommend that those responsible for public health act to create awareness about the potential dangers of consuming sun-baked clays in populations where the practice is prevalent.

ACKNOWLEDGMENTS

The author acknowledges the assistance of Musah Salifu (Moba) for helping in the sample collection and the Medical Geology Association Chapter in Ghana for their encouragement during this research.

REFERENCES

[1]   Woywodt A, Kiss A: Geophagia: the history of earth-eating. *Journal of the Royal Society of Medicine 2002,* 95: 143-146.

[2]   Young SL, Wilson MJ, Miller D, Hillier S: Toward a Comprehensive Approach to the Collection and Analysis of Pica Substances, with Emphasis on Geophagic Materials. *Pros One* 2008, 3: 1-13.

[3]   Ghorbani H: Geophagia, a Soil – Environmental Related Disease. *International Meeting on Soil Fertility Land Management and Agroclimatology. Turkey* 2008, 957-967.

[4]   Yanai J, Noguishi J, Yamada H, Sugihara S, Kilasara M, Kosaki T: Function of geophagy as supplementation of micronutrients in Tanzania. *Soil Science and Plant Nutrition* 2009, 55: 215-223.

[5]   Ferrell RE: Medicinal clay and spiritual healing. *Clays and Clay Minerals* 2008, 56: 751-760.

[6]   Luoba AI, Geissler PW, Estambale B, Ouma JH, Magnussen P, Alusala D, Aya R, Mwaniki D, Friis H: Geophagy among pregnant and lactating women in Bondo District, western Kenya. *Transaction of the Royal Society of Tropical medicine and hygiene 2004.* 98: 734-741.

[7]   Vermeer DE: Geophagy among Ewe of Ghana. Ethnology 1971, 1056.

[8]   Thomson J: Anaemia in pregnant women in eastern Caprivi, Namibia. *South African Medical Journal* 1997, 87: 1544-1547.

[9]   Kawai K, Saathoff E, Antelman G, Msamanga G, Fawzi WW: Geophagy (soil eating) in relation to Anaemia and Helminth Infection among HIV-infected pregnant women in Tanzania.

[10]   Apostoli, P.: *J Chromatogr B,* Elements in environmental and occupational medicine; 778: 63-97, (2002)

[11]   Smedley, P.L. and Kinniburgh D.G: A review of the source, behavior and distribution of arsenic in natural waters. *Applied geochemistry,* 17: 517-568, (2002)

[12]   Yan-Chu, H., in *Arsenic in Environment. Part I: Cycling and Characterization* (ed. Nriagu, J. O.), John Wiley and Sons Inc., 1994, pp. 17–49.

[13]   Smedley P.L and Kinniburgh D.G.: Arsenic in groundwater and the environment. In (O. Selenus, B. Alloway, J.A Centeno, B.R. Finkelman, R. Fuge, U. Lindh and P.L. Smedley (Editors). Essentials of medical geology: Impacts of natural environments on public health. *Elsevier academic press,* London, (2005).

[14]   Abrahams P.W., and Parsons J.A.: Geophagy in the tropics: A literature review. *Geog. J.* 162: 63-72 (1996).

[15]　Mahaney, W. C., Milner, M. W., Mulyono, Hs, Hancock, R. G. V., Aufreiter, S., Reich, M., and Wink, M. 2000. Mineral and chemical analyzes of soils eaten by humans in Indonesia. *Int. J. Environ. Health Res.* 10:93–109.

[16]　Haydel, S.E, Christine M. Remenih, C.M and Lynda B. Williams, L.B. 2008. Broad-spectrum in vitro antibacterial activities of clay minerals against antibiotic-susceptible and antibiotic-resistant bacteria pathogens. *Journal of Antimicrobial Chemotherapy*; 61, 353–361.

[17]　Guarino A., Bisceglia M., Castellucci G., Iacono G, Casali L.G., Bruzzese E., Musetta A., Greco L., 2001. Smectite in the treatment of acute diarrhea: a nationwide randomized controlled study of the Italian Society of Pediatric Gastroenterology and Hepatology (SIGEP) in collaboration with primary care pediatricians. SIGEP Study Group for Smectite in Acute Diarrhea. *J Pediatr Gastroenterol Nutr.* 2001 Jan;32(1):71-5.

[18]　Selenus O., Alloway, B., Centeno J.A., Finkelman R.B., Fuge R., Lindh, U and Smedley P., 2005. Essentials of Medical Geology, *Impacts of the Natural Environment on Public Health,* Elsevier Academic Press, 812p.

[19]　Vermeer, D. (1971). "Geophagy Among the Ewe of Ghana". *Ethnology* 10 (1): 56–72.

[20]　Vermeer, D.E; Frate, D.A. (1975). "Geophagy in Mississippi County". Annals of the Association of American Geographers 65 (3): 414–416. doi:10.1111/j.1467-8306.1975.

[21]　Dupont, C. and Vernisse, B. 2009. Anti-Diarrheal Effects of Diosmectite in the Treatment of Acute Diarrhea in Children: A Review. *Pediatric Drugs*: 11(2): p. 89-99.

[22]　Abrahams P. W., 2005. Geophagy and Involuntary Ingestion of Soil. *In Essentials of Medical Geology, Impacts of the Natural Environment on Public Health,* P. 435-458. Elsevier academic press, Amsterdam

[23]　Wiley, A. S., and Katz, S. H. 1998. Geophagy in Pregnancy: A test of Hypothesis, *Curr. Anthropol.* 39, p. 532-545.

[24]　Hunter, J. M., 1973. Geophagy in Africa and the United States: A Culture-Nutrition Hypothesis, *Georgr. Rev.*, 63, p.170 – 195.

[25]　Haller, J. S., 1972. The Negro and the Southern Physician: A study of Medical and Racial Attitudes 1800-1860, *Med. Hist*, 16, p. 238-253.

[26]　Kesse, G. O., 1985. *The mineral and Rock Resources of Ghana. A.*A. Balkema, Rotterdam, Netherlands, 610 p.

[27]　WHO (World Health Organization) 2001. WHO Fact Sheet No. 210. Bulletin of the World Health Organization, 78(9): p 1096. Prepared for World Water Day 2001. Reviewed by staff and experts from the Programme for Promotion of Chemical Safety (PCS), and the Water, Sanitation and Health unit (WSH), World Health Organization (WHO), Geneva.

[28]　Ryabukkin YS., 1978. Activation analysis of hair as an indicator of contamination of man by environmental trace element pollutants. Vienna: IAEA.

[29]　Elsom, M.H., 1992. *Staying healthy with nutrition: the complete guide to diet and nutritional medicine.* Celestial Arts Publishing. California.

[30]　Carl Ege 2002: What are minerals used for. *Survey notes,* 34, (2).

In: Arsenic
Editor: Andrea Masotti

ISBN: 978-1-62081-320-1
© 2013 Nova Science Publishers, Inc.

Chapter 10

CENTRAL NERVOUS SYSTEM TARGETS OF CHRONIC ARSENIC EXPOSURE

Sergio Zarazúa[1]*, María E. Jiménez-Capdeville[2], Reinhard Schliebs[3] and Rosalva Ríos[2]

[1]Department of Toxicology, Chemistry Faculty,
University of San Luis Potosi, San Luis Potosi, S.L.P., Mexico
[2]Department of Biochemistry, Medicine Faculty,
University of San Luis Potosi, San Luis Potosi, S.L.P., Mexico
[3]Paul Flechsig Institute for Brain Research, Medical Faculty,
University of Leipzig, Leipzig, Germany

ABSTRACT

The worldwide presence of arsenic and its impact on human health, affecting millions of exposed individuals through polluted drinking water and soils, has motivated intensive research during the last two decades. The central nervous system is one of the main arsenic targets, and although it has been well documented since the last century that arsenic poisoning causes serious damage ranging from peripheral neuropathy to death, low chronic exposure is today the most relevant concern. Neurotoxic effects include cognitive dysfunction, alterations in verbal comprehension, decrease in attention and long-term memory in children and adolescents. In adults, alterations in learning, concentration and recent memory formation have been reported. With the help of animal models, the underlying basis of these alterations have been investigated from morphological, neurochemical and behavioral perspectives. This chapter focuses on disturbances associated with the process of arsenic methylation that have been observed in essential nervous system constituents such as myelin and neurotransmitters, and the connection with hepatic transformation of arsenic in its methylated species for excretion. The experiments referred to here demonstrate that chronic exposure to low arsenic levels decrease performance in behavioral tasks and modify the methylation of DNA regions encoding for proteins related with neuronal plasticity, leading potentially to epigenetic long lasting changes. Behavioral effects of arsenic are observed in the absence of any toxic manifestation and could be related to interferences with signaling pathways,

* E-mail address: sergio.zarazua@uaslp.mx.

enzyme and DNA changes which may trigger degenerative processes later in life. Indeed, cytoskeletal modifications, altered processing of amyloid precursor protein (APP) and demyelination are hallmarks of low level arsenic exposure. In light of current hypothesis of arsenic neurotoxic mechanisms, the recent discoveries are gathered in this review to provide links between investigation lines on arsenic toxicity and future research directions.

INTRODUCTION

Neurotoxicity is a prime concern of environmental arsenic exposure. The two main reasons to consider arsenic neurotoxic effect as health threatening are its widely recognized carcinogenic potential and cardiovascular effects. Firstly, infantile exposure compromises the development of cognitive abilities of children, which results in a severe and difficult to reverse handicap for the entire life of exposed individuals. Second, even temporal exposure during development may trigger changes in the nervous system contributing to neurodegenerative diseases later in life, which today constitute one of the main health challenges, due to the increase in life expectancy of the population. This chapter will focus on the consequences of arsenic exposure at levels below that displaying carcinogenic effects, primarily those associated with effects in the nervous system.

ARSENIC METABOLISM AND NEUROTOXICITY

Cellular arsenic metabolism is situated in the confluence of a number of pathways that provide exchange and recycle key metabolites for divergent functions such as lipid transport, amino acid synthesis, metabolism of membrane components, oxidative stress defenses, neurotransmitter synthesis and the crucial methyl group trafficking, that are essential for modification of a vast number of biomolecules (Figure 1). Arsenic thus shares with other neurotoxicants the characteristic of eliciting toxic effects not only through direct interference of normal cellular processes, but principally by disturbing the balance of basic metabolic pathways, and as a result of the organic species generated, which are blamed responsible for other cellular damaging effects [1]. At cellular level, arsenic toxic effects include a wide range of alterations such as activation of several cell signaling pathways (i.e. p38 MAPK, ERK, JNK among others [2, 3]; generation of oxygen reactive species (ROS) and the consequent alterations in membrane lipids through lipid peroxidation [4,5], depletion in glutathione levels [6] and chromatin modifications [7]. Its main systemic effects are considered to be the induction of several types of cancer such as lung, skin, liver, bladder, cardiovascular and neural damage.

Of importance in analyzing the mechanisms of arsenic toxicity in the brain, two aspects must be considered. Firstly, species differences in arsenic biotransformation make it difficult to extrapolate results obtained in rodents even within them - rat, mice, hamster, guinea pig, etc-, and consequently to extrapolate to human metabolism what has been learned through animal models. Second, most of the studies demonstrating the damage induced by methylated metabolites have been performed in vitro, and very few in cells from the nervous system [8, 9]. Therefore, we still do not have enough evidence supporting the extent to which arsenic is metabolized in brain cells and how far that metabolism and organic species can be blamed for arsenic neurotoxic effects.

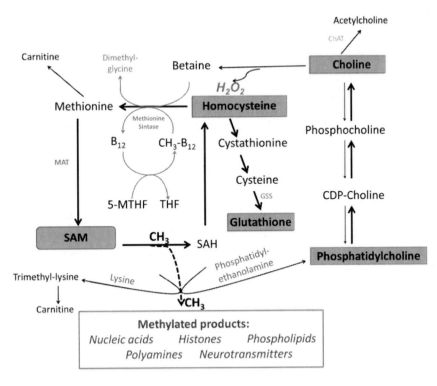

Figure 1. One-carbon metabolism in normal circumstances. After transfering its methyl group S-adenosyl methionine (SAM) becomes homocysteine, which can be recycled to methionine through the folate cycle and restore SAM levels. Glutathione (GSH) arises from homocysteine through the transsulfuration pathway. Choline can be oxided to betaine, but it releases hydrogen peroxide as co-product when providing methyl groups for methionine regeneration.

However, some similarities are consistent among animal models that suggest common metabolic steps among mammals which explains the presence and actions of arsenic in the brain. Arsenic metabolism consists of methylation of each arsenic atom that enters the cell, which is a process carried out by methyltransferases in an S-adenosylmethionine-dependent pathway that includes also the consumption of 2 to 3 molecules of glutathione and other thiols that serve as reducing agents [1]. The main methylation site is the liver, and from there the metabolites are systemically distributed, although other organs are also able to methylate arsenic, deriving in 3 to 5 methylated species, namely monomethylarsonous acid (MMA3), monomethylarsenic acid (MMA5), dimethylarsinous acid (DMA3), dimethylarsinic acid (DMA5) and trimethylarsine oxide (TMA) From them, those that are trivalent have a short half-life, but they are actually present in the liver of different rodent species after arsenate exposure. The major end product in most tissues, including the brain, is DMA5 since it becomes rarely trimethylated. It is not completely clear whether arsenic is indeed methylated in the brain. The presence of the enzymes required for arsenic transformation in brain tissue [10], and the detection of MMA5 and DMA5 in mouse brain slices following incubation with arsenite [11], strongly suggests that arsenic methylation can take place in the brain. However, in vivo studies performed in mice and rats are inconclusive. After a single oral administration of arsenate (As5) to mice the detectable arsenic in the brain is in form of DMA5 , even when the other species of the methylation pathway are present in liver, namely arsenate, arsenite, MMA5 and DMA5 [12]. On the other hand, chronic exposure of rats to arsenic results in depletion of S-adenosylmethionine in the liver, as confirmed by several authors, but not in the brain [13], suggesting that the methylated metabolites detected in the brain arise mainly from liver metabolism. Possibly, the methylation process in nervous tissue occurs either rarely or very fast, which may result in concentrations of intermediate products that are below the detection level of the analytical methods usually employed under in vivo conditions.

After methyl donation, S-adenosylmethionine (SAM) is converted to S-adenosylhomocysteine (SAH) which can be used for glutathione synthesis or regenerated to SAM by receiving a methyl group from methionine catalyzed by the methionine adenosyltransferase (MAT). Methionine synthesis from homocysteine requires the methyl-donation from 5-methylen tetrahydrofolate (5MTHF) using vitamin B12 as cofactor [14]. During a condition of high methyl demand, other molecules such as betaine and choline participate as important sources of methyl groups due to folate exhaustion. It has been demonstrated [15, 16] that dietary intake of folate and other components of one –carbon metabolism such as choline and betaine by arsenic-exposed individuals increases the methylation rate of arsenic. Moreover, the high prevalence of hyperhomocysteinemia among exposed individuals is significantly associated with a poor intake of methyl donors [17, 18] (Figure 2).

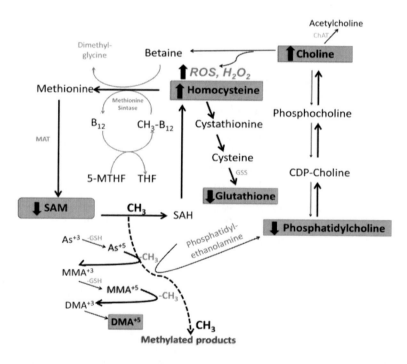

Figure 2. Arsenic demand of methyl groups for its own biotransformation disbalances one carbon metabolism. Decreased levels of SAM and glutathione in liver as immediate consequences activate a series of compensatory changes. Free choline increases in order to provide methyl groups both through decreased synthesis of phosphatidyl choline and even perhaps an increased breakdown of this phospholipid or other membrane lipids containing choline. Increased ROS due both to GSH depletion and increased formation by choline oxidation can lead to oxidative stress.

Arsenic and its methylated metabolites can easily cross the placenta and blood brain barrier, and thus may intefere with influencing neurochemical processes during the early development of the nervous system, such as the timely and adequate supply of key molecules [19]. Methyl groups are required by numerous acceptor molecules such as nucleic acids, histones and other proteins, phospholipids, polyamines, nitrogen bases, neurotransmitters, creatine, and carnitine, among other metabolites [20, 21]. Arsenic exposure during pregnancy requires a higher demand of methyl groups for its own metabolization, and thus may lead to a methyl group imbalance with consequences on the developmental brain, which will be discussed in more detail in the next section.

EARLY EXPOSURE AND EPIGENETIC EFFECTS

A number of fundamental reports that appeared during the first decade of the XXI century clearly established that environmental arsenic exposure has remarkable consequences on brain performance of exposed children.

Mainly through drinking water but also due to the presence of contaminated soils, exposure to arsenic of people in Mexico, China, USA and Bangladesh is significantly associated with deficits in cognitive performance in healthy children even when the data are corrected for a series of factors such as socioeconomic, nutritional and educational condition as well as exposure to other neurotoxicants.

The sequence of epidemiological studies performed in this decade starts with a report from 2001 of Mexican children living close to a copper smelter and exposed to arsenic and several heavy metals [22].

Even when these 80 children had a significant lead exposure indicated by mean blood lead levels close to 10 µg/dL (ranging between 4 and 26 µg/dL), deficits on verbal IQ, long term memory and linguistic abstraction were significantly associated with arsenic exposure. The study by Tsai and collaborators in Taiwan [23], followed by two studies by Wasserman and collaborators [24, 25] in more than 300 Bangladeshi children found intellectual deficits when the chronic arsenic exposure in drinking water ranged between .118 and .188 ± .225 ppm. Again, in a study of 31 children exposed to contaminated soil containing mainly arsenic, manganese and cadmium, a significant correlation between arsenic exposure levels and decrease of general intelligence scores, verbal IQ and memory was reported in 2006 [26]. Finally, in 2007 appeared two studies that refer to deficits of cognitive performance significantly associated with urine arsenic levels, in two different and geographically distant populations of Mexican children [27, 28].

By these investigations, an association between arsenic exposure and cognitive deficits has been firmly established by means of different protocols, in different countries and in individuals either exposed to other contaminants and/or affected by poverty and poor nutritional status (Table 1).

Now, researchers are turning to explore other functions such as emotional behavior [29] and motor skills [30] initiating another level of identification of arsenic targets in functions of the central nervous system.

**Table 1. Studies performed during the last decade demonstrating cognitive deficits
in arsenic exposed children**

Reference	Location	Type of exposure	Mean arsenic levels (mg/L)	Mean urinary arsenic (µg/L)	Co-exposure identified
Calderon 2001	Mexico	Hazardous waste		36% had levels above 50	Lead Cadmium
Tsai 2003	Taiwan	Drinking water	0.131-0.411		
Wasserman 2004	Bangladesh	Drinking water	0.120	347 ± 352	Manganese
Wasserman 2006	Bangladesh	Drinking water	0.118	110 ± 132	Manganese
Wright 2006	EUA	Hazardous waste		(Only hair levels were reported)	Manganese Cadmium
Rosado 2007	Mexico	Hazardous Waste		58 ± 33	Lead
Rocha-Amador 2007	Mexico	Drinking water	0.006-0.194		Fluoride

The maximum contaminant level for drinking water recommended by the World Health Organization
(WHO) is 0.01 mg/L; the action level established by the Centers for Disease Control and
Prevention(CDC) for total urinary arsenic 50µg/L.

Animal models of chronic exposure to environmentally relevant arsenic concentrations in
drinking water have recently appeared with the objective of elucidating the mechanisms
underlying cognitive deficits,. From the findings summarized in Table 2, subtle behavioral
and neurochemical changes can be detected at low to moderate- levels of exposure in some
models. One of the models that attempts to get as close as possible to environmental
exposure, consist of rats exposed to inorganic arsenic (3 ppm) starting from gestation
throughout the period of lactation, until four months of postnatal age [31] . Moderate damage
corresponding to partial lack of myelin has been observed by electron-microscopy, and it was
present in all arsenic treated animals (Figure 3). Also, severe damage to the myelin sheath
was found in 33% and 66% of animals exposed to arsenic for up to ages of two and three
months, respectively.

All these observations were accompanied by a decrease in dimethyl-arginine levels in rat
striatum measured by HPLC. Furthermore, not only did the dimethylarginines necessary for
myelin assembly show an important reduction associated with arsenic exposure but also
immunoreactivity to myelin basic protein itself [32]. These changes are accompanied by an
imbalance between SAM and lipid compounds in the liver of exposed animals [13].These
data demonstrate that myelin composition is a target of arsenic by interference with arginine
methylation, and they suggest that disturbances in signal transmission through de-myelinated
fibers are an important component of arsenic neurotoxicity.

In addition to altering neurodevelopmental programs through metabolic interference, it
has been reported that arsenic exposure modifies the expression of certain genes [33]. Now
we know that these changes involve epigenetic mechanism such DNA methylation, histone
methylation and acetylation and PARP inhibition [34]. These modifications are of crucial
importance for the central nervous system since not only during development but also in fully

differentiated neurons, epigenetic mechanisms are involved in the continuous formation, reinforcing and elimination of synapses that constitute neuronal plasticity.

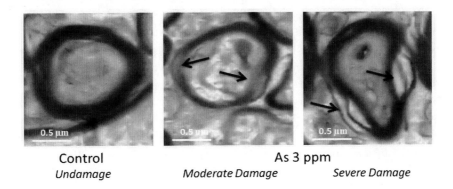

Control
Undamage

As 3 ppm

Moderate Damage   *Severe Damage*

Figure 3. Photomicrographs of fibers tracts from corpus striatum of the rat. A. Control animal, normal morphology of fiber tracts with homogeneous myelin sheath. B and C. Notice that the group exposed to 3 ppm of arsenic presents damage from moderate to severe, showing empty spaces on myelin sheath of fibers tracts (arrows). Scale bar equals 5 μm.

Table 2. Recent animal models of chronic arsenic exposure at levels ≤ 5 mg/L

Reference	Exposure/ species	Arsenic levels (mg/L)	Key findings	Comment
Ríos 2009 [31]	Drinking water/ rats	3	Damaged fiber tracts and axons, decreased immunoreactivity to nitric oxide synthase in striatum	Only slight changes in cell bodies, no apparent neuronal death
Wang 2009 [53]	Drinking water/ mice	1 and 4	Down- regulation of Ca/calmodulin-dependent protein kinase in cerebellum and deficits on spatial learning	
Bardullas 2009 [54]	Drinking water/ mice	0.05, 0.5 and 5	Increased locomotor activity. Changes of dopamine and its metabolites in cerebral regions. Changes of mRNA expression for antioxidant and tyrosine hydroxylase genes	
Luo 2009 [55]	Drinking water/ mice	2.7		No changes of NMDAR expression, spatial learning
Hong 2009 [56]	Drinking water/ mice	1 and 4	Decreased expression and activity of mitochondrial respiratory complexes, increase of oxidative stress markers	
Zarazúa 2010 [32]	Drinking water/ rats	3	Alteration of striatal myelinated tracts and decrease of dimethylarginines	
Rodríguez 2010 [57]	Drinking water/ rats	0.5 5	Changes of mRNA expression for antioxidant and dopamine receptor genes	No behavioral/brain monoamine changes at these exposure levels
Ríos 2011 [13]	Drinking water/ rats	3	Altered components of one-carbon metabolism in liver and decreased myelin basic protein in striatal fiber tracts	

Morphological, behavioral and neurochemical studies performed under chronic-low level arsenic exposure through drinking water in rodents. Reports appeared during the last 3 years.

In a recent publication, Martinez and collaborators [35] analyzed the pattern of DNA methylation in hippocampus and cortex of arsenic exposed rats during pre- and postnatal development, as well as the methylation of promoter regions of the genes RELN and PP1, which are involved in synaptic plasticity. DNA methylation was significantly higher in both regions of exposed animals compared to controls at one month of postnatal age, and there was a significant increase in the non-methylated form of PP1 gene promoter, at two and three months of age, either in cortex or hippocampus. Moreover, the arsenic exposure was associated with progressive deficits of fear memory.

These findings open the possibility to further analyze the epigenetic regulation of neuronal plasticity as a key target of arsenic. Moreover, since developmental consequences of altered gene expression may not become evident until the aging process advances, fetal and early postnatal arsenic exposure may elicit silent modifications that contribute to neurodegeneration later in life.

LONG TERM CONSEQUENCES OF ARSENIC EXPOSURE: POSSIBLE ROLE IN NEURODEGENERATIVE DISEASES

In contrast to several epidemiological studies searching for cognitive alterations associated with arsenic exposure in children, only one report has recently been published addressing this issue in adults with a mean age of 62 years [36]. Arsenic exposure through well water was estimated in a community of U.S. participants who showed low scores in cognitive testing (mini-mental state examination). Although this study did not include a correction for other contaminants present in the same community, or for differences in educational level, a significant association between exposure and cognitive deficits has been reported, while at very low arsenic well water levels (.003 to .016 ppm), but for several years of exposure. By contrast, Dakeishi and coworkers [37] reviewed the neurobehavioral sequelae in adults who were highly exposed to arsenic-contaminated dried milk as babies, but only for few days (estimated: 540 – 610 µg/Kg/day), during the middle 1950's in Morinaga, Japan. They report a series of neurological alterations including epilepsy, minimal brain damage and mental retardation (Intelligence Quotient less than 85 %). In post mortem brain studies cerebral edema, petechial hemorrhages in cerebellum and degeneration of myelin sheath in optic nerves were observed.

These studies suggest that arsenic exposure can be considered as important as other environmental factors such as nutrition and stress, which are playing a role in adult onset of neurodegenerative diseases. In fact a recent publication has highlighted the possibility that arsenic exposure may lead to the development of Alzheimer's disease (AD) because this metalloid is able to activate apoptosis-related signaling pathways such as JNK3 and p38MAP, and thus to contribute to neuronal death [38] Also, Gong and O'Bryant [39] proposed the hypothesis of arsenic exposure being a major factor in development and progression of AD, based on previous reports of arsenic exposure-mediated increase in tau hyperphosphorylation [40] and overexpression of the gene encoding amyloid precursor protein (APP) [41], including the induction of inflammatory responses similar as observed in AD patients [9, 42].

Alzheimer's disease (AD) is characterized by β-amyloid (Aβ) deposition and tau-pathology. Aβ peptides are formed through proteolytic processing of the amyloid precursor protein (APP) by subsequent catalytic actions of the β-and γ-secretase (Figure 4A - C). Abnormal APP processing has been assumed to play a causative role in AD development resulting in an imbalance of production and clearance of Aβ (amyloid cascade hypothesis of AD). Aβ accumulates in brain parenchyme as axonal/synaptotoxic oligomeres and as fibrillar aggregates (senile plaques) initiating a number of further pathologic events such as neuronal cell death and synaptic dysfunction, inflammation with activation of microglia and reactive astrocytes and chronic release of pro-inflammatory cytokines, as well as tau-pathology, and finally dementia (for review, see [45] and [46])

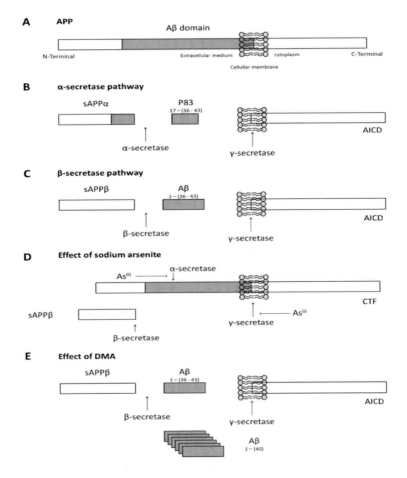

Figure 4. Processing of APP and suggested arsenic mechanisms. (A). Amyloid precursor protein (APP) is a transmembrane protein with a N-terminal tail in extracellular medium. It possesses an amyloid β (Aβ) domain that includes 28 residues outside of the cellular membrane and a small fragment (12 to 14 residues) in the transmembrane domain. APP can be processed in two pathways: α-secretase (B) and β-secretase pathways (C). In α-secretase pathway, Aβ domain is cleaved by α-secretase, releasing the soluble APP fragment (sAPPα) and the remaining carboxy terminal fragment (CTF) is then cleaved by the intramembranal complex γ-secretase, releasing p83 peptide. In β-secretase pathway, cleavage occurs just before Aβ domain, releasing soluble APP fragment (sAPPβ). The remaining CTF is also cleaved by γ-secretase complex, releasing Amyloid β peptide (Aβ). The APP intracellular domains (AICD) that remains are further metabolized in cytoplasm. (D) Incubation with sodium arsenite increases sAPPβ levels released into the medium and decreases the release of Aβ, suggesting the inhibition of α- and γ- secretases. (E) Incubation with DMA resulted in increase of sAPPα, sAPPβ and Aβ (1-40) released to the medium.

There are few publications that demonstrate a direct relation between arsenic exposure and markers of Alzheimer's disease. Studies performed in neuroblastoma SK-N-SH cell cultures show an association between exposure to different arsenic compounds and cytoskeletal gene expression, including tau protein and neurofilaments, and the possibility of alterations of cytoskeletal composition [44]. Furthermore, it is known that sodium arsenite is able to induce hyperphosphorylation of tau protein in Chinese hamster ovary T40 cells expressing human tau protein [40].

Exposure of Hela and pluripotent embryonal carcinoma NT-2 cells transiently transfected with Alzheimer's APP by 100 μM arsenite resulted in an increase in APP expression ([41]. Furthermore, we recently reported that arsenite and DMA alter APP processing in primary neuronal cultures derived from transgenic mouse Tg2576 overexpressing the Swedish mutation of human APP, by quantifying the amounts of sAPPβ and Aβ released into the culture medium as indicators of APP processing.

Sodium arsenite (5μM) treatment of primary neuronal cultures for 12h resulted in increased sAPPβ levels released into the culture medium, while the amounts of Aβ(1-40) and Aβ(1-42) were significant lower as compared to the control group, suggesting an inhibition of α- and γ-secretase activities induced by inorganic arsenic (Figure 3D) [47]. Notch-participating γ-secretase activity has also been found decreased in human keratinocyte cultures exposed to sodium arsenite [48]. However, DMA treated neuronal cultures result in an increase of total sAPP levels, indicating an increase in sAPPα and sAPPβ production, accompanied by an increase in Aβ(1-40) secretion, favoring the amyloidogenic path of APP processing (Figure 3E).

The mechanisms for arsenic-induced alterations in APP processing still remain to be clarified but it is known that arsenic exposure induces increase in heat shock proteins (HSP) [49] which are also related with increase of APP levels in Alzheimer's disease [50, 51]. In summary, the involvement of arsenic as one factor contributing to the development of AD represents a working hypothesis that still needs to be addressed using appropriate in vivo and in vitro approaches.

CONCLUSION

Taking together, the diverse cellular targets of arsenic as well as its recognized interference with chemical neurotransmission [52] suggest that the presence of arsenic and its methylated metabolites in the brain has profound consequences on brain maturation and function which should be considered for people exposed to arsenic due to contaminated environment.

ACKNOWLEDGMENTS

The authors like to thank Professor Michael G. Stewart, The Open University, Milton Keynes, U.K., for his support in revising the English manuscript.

REFERENCES

[1]   M. Vahter, G. Concha, Pharmacol. *Toxicol.* 89, 1 (2001).

[2]   U. Namgung, Z. Xia, *J. Neurosci.* 20, 6442 (2000).

[3]   U. Namgung, Z. Xia, *Toxicol. Appl. Pharmacol.* 174, 130 (2001).

[4]   S. Zarazua, F. Perez-Severiano, J.M. Delgado, L.M. Martinez, D. Ortiz-Perez, M.E. Jimenez-Capdeville, *Neurochem. Res.* 31, 1069 (2006).

[5]   S. Ahmed, S. Mahabbat-e Khoda, R.S. Rekha, R.M. Gardner, S.S. Ameer, S. Moore, E.C. Ekström, M. Vahter, R. Raqib, *Environ. Health Perspect.* 119, 258 (2011).

[6]   K. Jomova, Z. Jenisova, M. Feszterova, S. Baros, J. Liska, D. Hudecova, C.J. Rhodes, M. Valko, *J. Appl. Toxicol.* 31, 95 (2011).

[7]   J.F. Reichard, M. Schnekenburger, A. Puga, *Biochem. Biophys. Res. Commun.* 352, 188 (2007).

[8]   M. Styblo, L.M. Del Razo, L. Vega, D.R. Germolec, E.L. LeCluyse, G.A. Hamilton, W. Reed, C. Wang, W.R. Cullen, D.J. Thomas, *Arch. Toxicol.* 74, 289 (2000).

[9]   L. Vega, M. Styblo, R. Patterson, W. Cullen, C. Wang, D. Germolec, *Toxicol. Appl. Pharmacol.* 172, 225 (2001).

[10]   A. Sampayo-Reyes, R.A. Zakharyan, S.M. Healy, H.V. Aposhian, *Chem. Res. Toxicol.* 13, 1181 (2000).

[11]   V.M. Rodriguez, L.M. Del Razo, J.H. Limon-Pacheco, M. Giordano, L.C. Sanchez-Pena, E. Uribe-Querol, G. Gutiérrez-Ospina, M.E. Gonsebatt, *Toxicol. Sci.* 84, 157 (2005).

[12]   A. Juarez-Reyes, M.E. Jimenez-Capdeville, J.M. Delgado, D. Ortiz-Perez, Arch. *Toxicol.* 83, 557 (2009).

[13]   Ríos R., Delgado JM, Zarazua S., Ceballos F., Ortiz MD, Santoyo M., Jiménez-Capdeville ME. S-Adenosylmethionine levesl and histological damage in the brain of Wistar rats expossed to hypomethylating agents. 40th Meetting of the Society for Neurosciences (2010).

[14]   G. DE LA HABA, G.L. Cantoni, J. Biol. Chem. 234, 603 (1959).

[15]   M.V. Gamble, X. Liu, V. Slavkovich, J.R. Pilsner, V. Ilievski, P. Factor-Litvak, D. Levy, S. Alam, M. Islam, F. Parvez, H. Ahsan, J.H. Graziano, *Am. J. Clin. Nutr.* 86, 1202 (2007).

[16]   J.E. Heck, M.V. Gamble, Y. Chen, J.H. Graziano, V. Slavkovich, F. Parvez, J.A. Baron, G.R. Howe, H. Ahsan, *Am. J. Clin. Nutr.* 85, 1367 (2007).

[17]   M. Hall, M. Gamble, V. Slavkovich, X. Liu, D. Levy, Z. Cheng, A. van Geen, M. Yunus, M. Rahman, J.R. Pilsner, J. Graziano, *Environ. Health Perspect.* 115, 1503 (2007).

[18]   J.R. Pilsner, X. Liu, H. Ahsan, V. Ilievski, V. Slavkovich, D. Levy, P. Factor-Litvak, J.H. Graziano, M.V. Gamble, *Environ. Health Perspect.* 117, 254 (2009).

[19]   R.M. Gardner, B. Nermell, M. Kippler, M. Grander, L. Li, E.C. Ekström, A. Rahman, B. Lönnerdal, A.M. Hoque, M. Vahter, *Reprod. Toxicol.* 31, 210 (2011).

[20]   P.K. Chiang, R.K. Gordon, J. Tal, G.C. Zeng, B.P. Doctor, K. Pardhasaradhi, P.P. McCann, *FASEB J.* 10, 471 (1996).

[21]   J.M. Mato, F.J. Corrales, S.C. Lu, M.A. Avila, *FASEB J.* 16, 15 (2002).

[22]   J. Calderon, M.E. Navarro, M.E. Jimenez-Capdeville, M.A. Santos -Diaz, A. Golden, I. Rodriguez-Leyva, V. Borja-Aburto, F. Díaz-Barriga, *Environ. Res.* 85, 69 (2001).

[23]   S.Y. Tsai, H.Y. Chou, H.W. The, C.M. Chen, C.J. Chen, *Neurotoxicology,* 24, 747 (2003).

[24]   G.A. Wasserman, X. Liu, F. Parvez, H. Ahsan, P. Factor-Litvak, A. van Geen, V. Slavkovich, N.J. Lolacono, Z. Cheng, I. Hussain, H. Momotaj, J.H. Graziano, *Environ. Health Perspect.* 112, 1329 (2004).

[25]   GA Wasserman, X Liu, F Parvez, H Ahsan, P Factor-Litvak, J Kline, A van Geen, V Slavkovich, NJ Loiacono, D Levy, Z Cheng, JH Graziano, *Environ. Health Perspect.* 115, 285 (2007).

[26]   R.O. Wright, C. Amarasiriwardena, A.D. Woolf, R. Jim, D.C. Bellinger. *Neurotoxicology,* 27, 210 (2006).

[27]   J.L. Rosado, D. Ronquillo, K. Kordas, O. Rojas, J. Alatorre, P. Lopez, G. Garcia-Vargas, M. C. Caamaño, ME Cebrián , RJ Stoltzfus , Environ. Health Perspect. 115, 1371 (2007).

[28]   D. Rocha-Amador, M.E. Navarro, L. Carrizales, R. Morales, J. Calderon, *Cad. Saude Publica,* 23, S579 (2007).

[29]   A. Roy, K. Kordas, P. Lopez, J.L. Rosado, M.E. Cebrian, G.G. Vargas, D. Ronquillo, R.J. *Stoltzfus, Environ. Res.* 111, 670 (2011).

[30]   F. Parvez, G.A. Wasserman, P. Factor-Litvak, X. Liu, V. Slavkovich, A.B. Siddique, R. Sultana, R. Sultana, T. Islam, D. Levy, J.L. Mey, A. van Geen, K. Khan, J. Kline, H. Ahsan, J.H. Graziano, *Environ. Health Perspect.* In Press (2011).

[31]   R. Rios, S. Zarazua, M.E. Santoyo, J. Sepulveda-Saavedra, V. Romero-Diaz, V. Jimenez, F. Pérez-Severiano, G. Vidal-Cantú, J.M. Delgado, M.E. Jiménez-Capdeville, *Toxicol.* 261, 68 (2009).

[32]   S. Zarazua, R. Rios, J.M. Delgado, M.E. Santoyo, D. Ortiz-Perez, M.E. Jimenez-Capdeville, *Neurotoxicology,* 31, 94 (2010).

[33]   M. Salgado-Bustamante, M.D. Ortiz-Perez, E. Calderon-Aranda, L. Estrada-Capetillo, P. Nino-Moreno, R. Gonzalez-Amaro, D. Portales-Pérez, *Sci. Total Environ.* 408, 760 (2010).

[34]   J.F. Reichard, A. Puga. *Epigenomics* 2, 87 (2010).

[35]   L. Martinez, V. Jimenez, C. Garcia-Sepulveda, F. Ceballos, J.M. Delgado, P. Nino-Moreno, L. Doniz, V. Saavedra-Alanís, C.G. Castillo, M.E. Santoyo, R. González-Amaro, M.E. Jiménez-Capdeville, *Neurochem. Int.* 58, 574 (2011).

[36]   S.E. O'Bryant, M. Edwards, C.V. Menon, G.Gong, R. Bardeb, *Int. J. Environ. Res. Public Health,* 8, 861 (2011).

[37]   M. Dakeishi, K. Murata, P. Grandjean, *Environ. Health.* 5, 31 (2006).

[38]   S. Gharibzadeh, *J. Neuropsychiatry Clin. Neurosci,* 20, 4 (2008).

[39]   G. Gong, S.E. O'Bryant, *Alzheimer Dis. Assoc. Disord.* (2010).

[40]   B.I. Giasson, D.M. Sampathu, C.A. Wilson, V. Vogelsberg-Ragaglia, W.E. Mushynski, V.M. Lee, *Biochemistry,* 41, 15376 (2002).

[41]   N.N. Dewji, C. Do, R.M. Bayney, *Brain Res. Mol. Brain Res.* 33, 245 (1995).

[42]   R.C. Fry, P. Navasumrit, C. Valiathan, J.P. Svensson, B.J. Hogan , M. Luo, S. Bhattacharya, K. Kandjanapa, S. Soontararuks, S. Nookabkaew, C. Mahidol, M. Ruchirawat, L.D. Samson, *PLoS Genet.* 3, 207 (2007).

[43]   K. Heinitz, M. Beck, R. Schliebs, J.R. Perez-Polo, J. Neurochem. 98, 1930 (2006).

[44] A. Vahidnia, R.J. van der Straaten, F. Romijn, J. van Pelt, G.B. van der Voet, F.A. de Wolff, *Toxicol. In Vitro,* 21, 1104 (2007)

[45] R. Schliebs, T. Arendt, *Behav. Brain Res.* 221, 555 (2011).

[46] Y.W. Zhang, R. Thompson, H. Zhang, X. Huaxi, *Mol. Brain,* 4, 3 (2011).

[47] S. Zarazua, S. Bürger, J.M. Delgado, M.E. Jimenez, R. Schliebs, *Int. J. Devl. Neuroscience,* 29, 389 (2011).

[48] T.V. Reznikova, M.A. Phillips, R.H. Rice, J. Invest. Dermatol. 129, 155 (2009)

[49] L.M. Del Razo, B. Quintanilla-Vega, E. Brambila-Colombres, E.S. Calderón-Aranda, M. Manno, A. Albores, *Toxicol. Appl. Pharmacol.* 177, 132 (2001).

[50] J.E. Hamos, B. Oblas, D. Pulaski-Salo, W.J. Welch, D.G. Bole, D.A. Drachman, *Neurology,* 41, 345 (1991).

[51] N. Perez, J. Sugar, S. Charya, G. Johnson, C. Merril, L. Bierer, D. Perl, V. Haroutunian, W. Wallace, *Brain Res. Mol. Brain Res.* 11, 249 (1991).

[52] V.M. Rodriguez, M.E. Jimenez-Capdeville, M. Giordano, *Toxicol. Lett.* 145, 1 (2003).

[53] Y. Wang, S. Li, F. Piao, Y. Hong, P. Liu, Y. Zhao, *Neurotoxicol. Teratol.* 31, 318 (2009).

[54] U. Bardullas, J.H. Limón-Pacheco, M. Giordano, L. Carrizales, M.S. Mendoza-Trejo, V.M. Rodríguez, *Toxicol. Appl. Pharmacol.* 239, 169 (2009).

[55] J.H. Luo, Z.Q. Qiu, W.Q. Shu, Y.Y. Zhang, L. Zhang, J.A. Chen, *Toxicol. Lett.* 184, 121 (2009).

[56] Y. Hong, F. Piao, Y. Zhao, S. Li, Y. Wang, P. Liu, *Neurotoxicology,* 30, 538 (2009).

[57] V.M. Rodríguez, J.H. Limón-Pacheco, L. Carrizales, M.S. Mendoza-Trejo, M. Giordano, *Neurotoxicol. Teratol.* 32, 640 (2010).

In: Arsenic
Editor: Andrea Masotti

ISBN: 978-1-62081-320-1
© 2013 Nova Science Publishers, Inc.

Chapter 11

DRINKING WATER SANITATION, ARSENIC REMOVAL SYSTEM, AND METAL INTERACTION: A TRIANGULAR RELATIONSHIP ON HUMAN HEALTH

Victor Raj Mohan Chandrasekaran and Ming-Yie Liu[*]
Department of Environmental and Occupational Health,
College of Medicine, National Cheng Kung University,
Tainan, Taiwan

ABSTRACT

Contaminated drinking water is a frequent cause of diseases such as cholera, typhoid, viral hepatitis A, and dysentery. Water may be contaminated with naturally occurring inorganic elements such as arsenic, radon, or fluoride. Many metals are essential to living organisms, but some of them are highly toxic or become toxic at high concentrations. The risk for developing a human disease derived from environmental exposure is not based solely on the environmental exposure, but is modified by mitigating conditions, such as genetic or other environmental factors. Arsenic is a well-documented toxic element in drinking water, and many arsenic-associated health complications are reported worldwide. Removing arsenic from drinking water is widely dependent on iron-based techniques. Although inorganic arsenic has long been known to be toxic to humans, little is known about its metabolite toxicity and the interaction between arsenic and its metabolites with other metals such as iron. Further, the influence of water quality on the interaction between the metalloid arsenic and metals is poorly understood. Therefore, this chapter deals with the possible effect on human health of arsenic interacting with metals, systems to remove arsenic from water, and water quality.

INTRODUCTION

In areas where the drinking water supply contains unsafe levels of arsenic, the immediate concern is finding a safe source of drinking water. There are two main options: finding a new

---

[*] Email: myliu@mail.ncku.edu.tw.

safe source, and removing arsenic from the contaminated source. In either case, the drinking water supplied must be free from harmful levels of arsenic, but also from bacteriological contamination, and from other chemical contaminants. Metals are unique among pollutant toxicants in that they all occur naturally and, in many cases, are ubiquitous within the human environment.

In addition, all life has evolved in the presence of metals, and organisms have been forced to deal with these potentially toxic but omnipresent elements [1]. Unlike organic waste, metals do not degrade in the body, which facilitates accumulation to toxic levels [2]. In recent years, evidence has been presented that exposure to some metals may be more dangerous than expected [3-5].

Therefore, metal toxicity is one of the challenges for the ecosystem as well as for environmental safety.

WATER CONTAMINATION AND HUMAN HEALTH

Waterborne diseases arise from the contamination of water, either by pathogenic viruses, bacteria, or protozoa, or by chemical substances. These agents are directly transmitted to people when the water is used for drinking, preparing food, recreation, or other domestic purposes.

An outbreak of waterborne disease is usually defined as an event meeting two criteria: (a) at least two people have experienced similar illness after exposure to water; and (b) epidemiological evidence implicates water as the probable source of the illness. The occurrence of outbreaks of waterborne diseases is not limited to developing countries; affluent countries are also affected [6-9].

DRINKING WATER CONTAMINATION

The risk of outbreaks of waterborne diseases increases where standards of water, sanitation, and personal hygiene are low. Worldwide, the proportion of people with access to safe drinking water and basic sanitation rose from 78% in 1990 to 83% in 2004. Despite this progress, however, an estimated 425 million children under 18 years old still have no access to a safe water supply. In 2004, it was estimated that diarrhea due to unsafe water and a lack of basic sanitation annually contributes to the death of 1.5 million children under 5 years old [10].

In the European Region, the annual burden of diarrheal disease attributable to poor water quality, sanitation, and hygiene in children 0–14 years old is estimated at 13,548 deaths (5.3% of all deaths) and 31.5 disability-adjusted life years per 10,000 children [11]. Contaminated drinking water is a frequent cause of diseases such as cholera, typhoid, viral hepatitis A, and dysentery. Water may be contaminated with naturally occurring inorganic elements such as arsenic, radon, or fluoride. Human activity may also cause water to become contaminated with substances such as lead, nitrates, and pesticides [12].

METAL CONTAMINATION AND HUMAN HEALTH

A decisive step in the development of human technology and culture was the discovery of metals below the surface of our planet, and their excavation, extraction, and use as tools to fulfill human needs. Because of anthropogenic activities, freshwater systems worldwide are confronted with thousands of compounds. A major contribution to chemical contamination originates from wastewater discharges that affect surface water quality because of their incompletely removed organic contaminants [13, 14]. These days, the metal wastes are distributed over the soils and waters of the Earth's surface and have detrimental effects on life in the environment and on human health. Unlike organic waste, metals and their compounds are not degraded by living organisms and may accumulate up to harmful levels [2]. Metals accumulate in the soft tissue and become toxic when they are not metabolized by the body [15-17]. Metals may enter the human body through food, water, air, or absorption through the skin when they come in contact with humans in agriculture and in manufacturing, pharmaceutical, industrial, or residential settings [15].

ARSENIC AND HUMAN HEALTH

Arsenic, a metalloid, has a long history of use in human civilization. Today, arsenic compounds are still widely used in industry and agriculture. However, arsenic has been identified as a human carcinogen by the International Agency for Research on Cancer [18]. Arsenic exposure results in both chronic and acute toxicity in humans. Chronic arsenic poisoning is a global health problem affecting millions of people, especially in India, Bangladesh, and China [19-21]. The main cause of widespread chronic arsenicosis is the consumption of underground drinking water that is naturally contaminated by arsenic. Arsenic contamination of drinking water may also result from mining and other industrial processes. Acute arsenic poisoning is relatively less common, but it has been documented after accidental ingestion of insecticides or pesticides, and attempted suicides or murders with arsenicals [22, 23].

IRON AND HUMAN HEALTH

Iron is an essential trace element for the growth, development, and long-term survival of most organisms. Iron absorption and distribution are homeostatically regulated to reduce the risk of deficiency and overload. These mechanisms interact, in part, with the mechanisms of oxidative stress that make iron available to pathogens [24]. Cases of acute iron toxicity are relatively rare and mostly related to hepatotoxicity [25]. From a compilation of biochemical, animal, and human data, links have been proposed between increased levels of iron in the body and an increased risk of a variety of diseases, including vascular disease, cancer, and certain neurological conditions [26, 27]. Iron-mediated formation of reactive oxygen species leading to lipid peroxidation appears to result from an exaggeration of the normal function of iron, which is to transport oxygen to tissue. Iron-induced free radical damage to cells appears to be important for the development of cancer [28].

ARSENIC REMOVAL FROM DRINKING WATER

Arsenic intoxication occurs because people drink well water or eat foods containing inorganic arsenicals [29]. Environmental exposure to arsenic will apparently continue to be common. Arsenic contamination of drinking water is reported in many parts of the world. Metal salts have been used to remove arsenic since at least 1934. The most commonly used metal salts, aluminum salts such as alum, and ferric salts such as ferric chloride and ferric sulfate [30], provide excellent arsenic removal. Laboratories report over 99% removal under optimal conditions, and residual arsenic concentrations of less than 1 µg/l [31]. Zero-valent iron filings can be used either in situ or ex situ to reduce arsenate and produce ferrous iron. However, treated water is very high in ferrous iron; therefore, it must undergo iron removal treatment before it is distributed and drunk [32]. Although the iron concentration is less after treatment (0.2 mg/l), the removal process has its limitations, which results in highly unacceptable iron concentrations [33]. In addition, long-term exposure to low concentrations of arsenic and iron can be fatal [34, 35].

Arsenic is removed from drinking water using various technologies: anion exchange, activated alumina, reverse osmosis, modified coagulation, filtration, etc. [36]. Because many arsenic removal plants in arsenic contaminated areas use filtration technologies that contain iron, precautions must be taken to reduce the risk of ionized iron's interaction with arsenic. A recent study [37], which tested higher doses of arsenic (10 mg/kg) and iron (50 mg/kg) to determine the maximum level of health complications, reported an important public health issue: the arsenic removal system may cause liver damage. However, (i) the lowest-dose interaction and health effects of arsenic and iron, and (ii) the role of the metabolites of arsenic and their interaction with iron in contaminated water, require additional investigation.

DIMETHYLARSINIC ACID (DMA) AND WATER SAFETY

Arsenic is associated with an increased risk of cancer in the skin, lungs, liver, kidneys, and urinary bladder in humans [38]. In the liver, and possibly in other organs, inorganic arsenic is metabolized to methylated compounds (DMA and monomethylarsinic acid [MMA]). Although methylation is believed to decrease the biological activity of arsenic [39], this has also been questioned [40]. Data from free radical, biochemical, and carcinogenic studies of DMA [41-45] suggest that the methylation of arsenic may instead be a pathway of toxification.

Concern about human exposure to DMA has focused on its use as a herbicide and on its residual presence in food products. However, there are other significant sources of exposure to this agent [46].

Inorganic arsenic ingested as a contaminant of drinking water and some foods [47-49] is biologically methylated, yielding MMA and DMA. Another significant source is seafood that contains a high concentration of DMA [50, 51]. The toxicity of inorganic arsenicals, arsenite, and arsenate has been extensively studied; however, the toxic manifestations of methylated arsenic metabolites have not been well characterized.

INTERACTION OF DMA, IRON, AND LIPOPOLYSACCHARIDE (LPS)

The risk for developing a human disease derived from environmental exposure is not based solely on environmental exposure, but is also modified by mitigating conditions, such as genetic or other environmental factors [35]. DMA releases iron from ferritin in vitro, which initiates the generation of hydrogen peroxide [43]. Iron may be important in arsenic-associated toxicity [37, 52]. In addition, sub-hepatotoxic arsenic exposure increases LPS-induced liver damage [35]. Many arsenic removal plants in arsenic-contaminated areas use filtration technologies that contain iron. DMA and DMA-plus-iron treatments do not affect the mouse hepatic system. In contrast, co-treatment with DMA, iron, and LPS significantly increases hepatic injury in rats. The effects of sub-chronic exposure of LPS on DMA-plus-iron show hepatic toxicity in rats (Fig. 1).

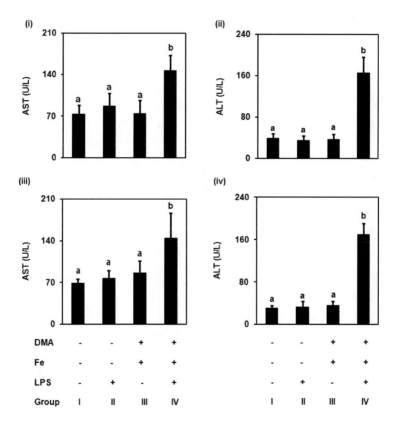

Figure 1. Effect on rat liver of co-treatment with DMA, iron, and LPS. Two sets of twenty-four rats (Study I and Study II) were divided into four groups. Study I: Group I, received normal drinking water. Group II, received LPS [0.01 mg/kg; intraperitoneally (i.p.)]. Group III received DMA and iron (1000 and 300 ppm, respectively) in drinking water. Group IV first received DMA and iron and then a single injection of LPS (0.01 mg/kg; i.p.). In Groups II and IV, LPS was given just 12 h before they were killed with Groups I and III on day 7, after which their serum aspartate transaminase [AST (i)] and alanine transaminase [ALT (ii)] levels were measured. Study II, the same conditions as in Study I, except that Study II continued for 14 days. After 14 days, the rats were killed and their serum AST (iii) and ALT (iv) levels were measured. Data are means ± standard deviation. Different letters (a, b) indicate a significant difference between groups (P < 0.05; one-way ANOVA and then Tukey's multiple comparison tests).

INTERACTION BETWEEN DMA-PLUS-IRON AND LPS
ON HEPATIC INJURY

Waterborne diseases arise from the contamination of water, either by pathogenic bacteria or protozoa or by chemical substances. These agents are directly transmitted to people when the water is used for drinking, preparing food, recreation, or other domestic purposes. Although a high dose of LPS, the outer membrane of bacteria, induces liver damage [35], a single ostensibly non-effective dose of LPS given to rats after they have been sub-chronically exposed to DMA-plus-iron for 7 or 14 days also induces significant hepatic damage. This indicates that the interaction between DMA-plus-iron and LPS causes hepatic injury in rats.

CONCLUSION

Drinking water, after food, is a secondary source of the inorganic arsenic found in the human system. Long-term exposure to arsenic causes a wide range of health effects. The World Health Organization established new guidelines for removing arsenic from drinking water. Further, various arsenic removal technologies include conventional adsorbents; a variety of different chemicals have also been tested. It is possible that bacteria-contaminated water results in the interaction between LPS and the iron-based precipitation systems that remove arsenic from drinking water, and that this interaction causes hepatic injury. We raise an important ecological and environmental public health issue: the arsenic removal system may cause liver damage by replacing the arsenic it has removed with almost equally dangerous arsenic metabolites.

REFERENCES

[1] R. Kutlubay, E. O. Oguz, B. Can, M. C. Guven, Z. Sinik and O. L. Tuncay, *Int. J. Toxicol.* 26, 297 (2007).
[2] D. Beyersmann and A. Hartwig, *Arch. Toxicol.* 82, 512 (2008).
[3] J. G. Hengstler, V. U. Bolm-Audor, A. Faldum, K. Janssen, M. Reifenrath, W. Götte, D. Jung, O. Mayer-Popken, et al., *Carcinogenesis* 24, 63 (2003).
[4] H. M. Bolt and J. G. Hengstler, *Arch. Toxicol.* 82, 1 (2008).
[5] D. A. Cristol, R. L. Brasso, A. M. Condon, R. E. Fovargue, S. L. Friedman, K. K. Hallinger, A. P. Monroe and A.E. White, *Science* 320, 335 (2008).
[6] S. E. Hrudey and E. J. Hrudey, *Safe Drinking Water. Lessons from Recent Outbreaks in Affluent Nations.* International Water Association Publishing, London (2004).
[7] M. F. Craun, G. F. Craun, R. L. Calderon and M. J. Beach, *J. Water Health* 4, 30 (2006).
[8] M. F. Blasi, M. Carere, M. G. Pompa, E. Rizzuto and E. Funari, *J. Water Health* 6, 432 (2008).
[9] P. Beaudeau, H. de Valk, V. Vaillant, C. Mannschott, C. Tillier, D. Mouly and M. Ledrans, *J. Water Health* 6, 503 (2008).

[10] World Health Organization, *Water, sanitation and hygiene links to health: facts and figures.* Geneva, 2004. At http://www.who.int/water_sanitation_health/ factsfigures2005.pdf (updated November 2004) or http://www.who.int/water_ sanitation_health/publications/factsfigures04/en/ (updated November 2004) (both accessed 17 February 2012).

[11] World Health Organization, *Outbreaks of waterborne diseases,* (Fact Sheet 1.1 CODE: RPG1_WatSan_E1) 2007. At http://www.euro.who.int/__data/assets/pdf_file/0006/ 97359/1.1.pdf (accessed 17 February 2012).

[12] World Health Organization, *Guidelines for drinking water quality, 3rd ed. Vol. 1. Recommendations.* Geneva, 2004. At http://www.who.int/water_sanitation_health/dwq/ GDWQ2004web.pdf (accessed 17 February 2012).

[13] S. A. Snyder, D. L. Villeneuve, E. M. Snyder and J. P. Giesy, *Environ. Sci. Technol.* 35, 3625 (2001).

[14] D. W. Kolpin, M. Skopec, M. T. Meyer, E. T. Furlong and S. D. Zaugg, *Sci. Total Environ.* 328, 130 (2004).

[15] D. Beyersmann, "Physiochemical aspects of the interference of detrimental metal ions with normal metal metabolism". In: G. Berthon (ed.), *Handbook on Metal-ligand Interactions in Biological Fluids,* Marcel Dekker, New York (1995) pp. 813-826.

[16] E. Nieboer, G. G. Fletcher and Y. Thomassen, *J. Environ. Monit.* 1, 14 (1999).

[17] A. Hartwig, *Antioxid. Redox. Signal* 3, 634 (2001).

[18] IARC, *IARC Monogr. Eval. Carcniog. Risk Hum.* 84, 477 (2004).

[19] D. N. Mazumder, J. Das Gupta, A. Santra, A, Pal, A, Ghose and S. Sarkar, *JAMA* 96, 7 (1998).

[20] H. Ahsan, M. Perrin, A. Rahman, F. Parvez, M. Stute, Y. Zheng, A. H. Milton, P. Brandt-Rauf, A. van Geen and J. Graziano, *J. Occup. Environ. Med.* 42, 1201 (2000).

[21] G. Sun, *Toxicol. Appl. Pharmacol.* 198, 271 (2004).

[22] T. Lech and F. Trela, *Forensic Sci. Int.* 151, 277 (2005).

[23] R. N. Ratnaike, *Postgrad Med. J.* 79, 396 (2003).

[24] G. J. Anderson, *Am. J. Hematol.* 82, 1131 (2007).

[25] M. Tenenbein, *J. Toxicol. Clin. Toxicol.* 39, 726 (2001).

[26] D. Berg, M. Gerlach, M. B. Youdim, K. L. Double, L. Zecca, P. Riederer and G. Becker, *J. Neurochem.* 79, 236 (2001).

[27] C. W. Siah, D. Trinder and J. K. Olynyk, *Clin. Chim. Acta* 358, 36 (2005).

[28] H. Ullen, K. Augustsson, C. Gustavsson and G. Steineck, *Cancer Lett.* 114, 216 (1997).

[29] National Research Council, *Arsenic in Drinking Water,* National Academy Press, Washington, DC (1999).

[30] P. Ratna Kumar, S. Chaudhari, K. C. Khilar and S. P. Mahajan, *Chemosphere* 55, 1251 (2004).

[31] R. C. Cheng, S. Liang, H. C. Wang and M. D. Beuhler, *J. Am. Water Works Assoc.* 86, 90 (1994).

[32] J. A. Lackovic, N. P. Nikolaidis and G. Dobbs, *Environ. Eng. Sci.* 17, 39 (2000).

[33] A. Ramaswami, S. Tawachsupa and M. Isleyen, *Water Res.* 35, 4479 (2001).

[34] G. Papanikolaou and K. Pantopoulos, *Toxicol. Appl. Pharmacol.* 202, 211 (2005).

[35] G. E. Arteel, L. Guo, T. Schlierf, J. I. Beier, J. P. Kaiser, T. S. Chen, M. Liu, D. J. Conklin, H. L. Miller, C. V. Montfort and J. C. States, *Toxicol. Appl. Pharmacol.* 226, 139 (2008).

[36] U.S. Environmental Protection Agency, *National primary drinking water regulations; arsenic and clarifications to compliance and new source contaminants monitoring,* Federal Register, Jan. 22, 2001 (Vol. 66, No. 14), Rules and Regulations, (Washington, DC) pp. 6975-7066. At http://www.uas.alaska.edu/attac/dlfiles/arsenic_finalrule.pdf (accessed 17 February 2012).

[37] V. R. Chandrasekaran, I. Mutahiyan, P. C. Huang and M. Y. Liu, *Water Res.* 44, 5823 (2010).

[38] M. N. Bates, A. H. Smith and R. C. Hopenhayn, *Am. J. Epidemiol.* 135, 462 (1992).

[39] R. J. Lewis, *Sax's Dangerous Properties of Industrial Materials,* Van Nostrand Reinhold, New York (1992).

[40] T. Murai, H. Iwata, T. Otoshi, G. Endo, S. Horiguchi and S. Fukushima, *Toxicol. Lett.* 66, 53 (1993).

[41] S. Ahmad, K. T. Kitchin and W. R. Cullen, *Arch. Biochem. Biophys.* 382, 195 (2000).

[42] B. K. Mandal, Y. Ogra and K. T. Suzuki, *Chem. Res. Toxicol.* 14, 371 (2001).

[43] S. Ahmad, K. T. Kitchin and W. R. Cullen, *Toxicol. Lett.* 133, 47 (2002).

[44] S. Nesnow, B. C. Roop, G. Lambert, M. Kadiisska, R. P. Mason, W. R. Cullen and M. J. Mass, *Chem. Res. Toxicol.* 15, 1627 (2002).

[45] E. I. Salmi, H. Wanibuchi, K. Morimura, M. Wei, M. Mitsuhashi, K. Yoshida, G. Endo and S. Fukushima, Carcinogenesis 24, 335 (2003).

[46] B. M. Adair, T. Moore, S. D. Conklin, J. T. Creed, D. C. Wolf and D. J. Thomas, *Toxicol. Appl. Pharmacol.* 222, 235 (2007).

[47] E. M. Kenyon and M. F. Hughes, *Toxicology* 160, 227 (2001).

[48] A. H. Ackerman, P. A. Creed, A. N. Parks, M. W. Fricke, C. A. Schwegel, J. T. Creed, D. T. Heitkemper and N. P. Velal, *Environ. Sci. Technol.* 39, 5241 (2005).

[49] P. N. Williams, A. H. Price, A. Raab, S. A. Hossain, J. Feldmann and A. A. Meharg, *Environ. Sci. Technol.* 39, 5531 (2005).

[50] K. A. Francesconi, R. Tanggaar, C. J. McKenzie and W. Goessler, *Clin. Chem.* 48, 92 (2002).

[51] R. Raml, W. Goessler, P. Traar, T. Ochi and K. A. Francesconi, *Chem. Res. Toxicol.* 18, 1444 (2005).

[52] M. Modi and S. J. Flora, *Cell Biol. Toxicol.* 23, 429 (2007).

Reviewed by Dr. Chih-Ching Chang, M.D., Ph.D., Department of Environmental and Occupational Health, College of Medicine, National Cheng Kung University, 138 Sheng-Li road, Tainan, Taiwan. E-mail: chang3@mail.ncku.edu.tw

SECTION III:
ARSENIC REMOVAL TECHNIQUES AND TOOLS

In: Arsenic
Editor: Andrea Masotti

ISBN: 978-1-62081-320-1
© 2013 Nova Science Publishers, Inc.

Chapter 12

BIOLOGICAL FILTRATION TECHNOLOGY FOR REMOVAL OF ARSENIC FROM GROUNDWATER

Danladi Mahuta Sahabi[1] and Minoru Takeda[2]*

[1]Department of Biochemistry,
Usmanu Danfodiyo University, Sokoto, Nigeria
[2]Division of Materials Science and Chemical Engineering,
Graduate School of Engineering, Yokohama National University,
Yokohama, Japan

ABSTRACT

Arsenic (As) is an environmental toxicant to which millions of people are exposed worldwide. The impact of arsenic on human health has led its drinking water maximum concentration limit to be drastically reduced from 50 to 10 µg L^{-1}. Biological filtration of arsenic in conjunction with iron or/and manganese removal has lately been advanced as an innovative alternative process for attaining this limit of 10 µg L^{-1} or even lower. The method comprises of the steps of biological oxidation of iron and manganese forming amorphous iron and manganese oxides, which coat the surface of the filter medium. Arsenic species are subsequently removed from the groundwater by a combination of biological and physico-chemicals sorption processes, including oxidation and adsorption by biogenic iron and manganese oxides. The bacteria in the system are directly involved in the oxidation of As(III) and indirectly, through the generation of reactive manganese oxides and adsorptive iron oxides, which remove both As(III) and As(V) from the water. The biological approach is more economical and eco-friendly as no chemicals are added and the sorbent materials are continuously produced in situ. Furthermore, it is a combined process for simultaneous removal of iron, manganese and arsenic from groundwater. This chapter reviews the history, development and mechanism of action of biological filtration technology for integrated treatment of iron, manganese and arsenic from groundwater.

---

* E-mail address: dmsahabi@yahoo.com.

INTRODUCTION

Biological filtration is an emerging technology for removal of arsenic in conjunction with biological iron or/and manganese oxidation and precipitation from groundwater. This highly innovative technology is gaining acceptance in many parts of the world because of its numerous advantages over the conventional physical-chemical methods, which include rapid filtration rates, low operation and maintenance costs, and the possibility of using single stage filtration. Furthermore, no chemicals are added as the sorbent materials are continuously produced in situ. Nowadays biological processes to remove Fe and Mn are well established and their integration with arsenic removal is under intense investigation in Europe, United States, Canada, and Latin America [1–9]. Similarly, in Japan, China and many other parts of Asia, a number of biological water treatment plants are designed to exploit this biocatalyzation potential of microorganisms for simultaneous removal of iron, manganese, arsenic and other contaminants from groundwater (Figure 1)

As biological filtration becomes a more popular groundwater treatment technology, there is the need for better understanding of key mechanisms and range of operational characteristics of the system. This chapter is aimed at furnishing more information on these important issues. The chapter begins with the concepts of biological filtration of iron and manganese, followed by the integration of the biological oxidation of iron or/and manganese with arsenic removal from groundwater. The possibility of exploiting the biogenic metals oxides (from established biofilters) in innovative wastewater treatment facilities for removal of other toxic heavy metals such as lead and cadmium is also discussed.

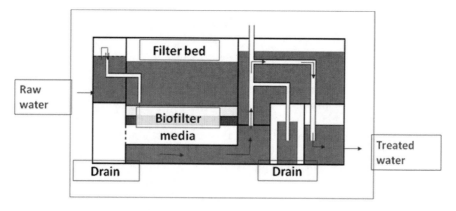

Figure 1. An outline of biological filtration process.

PROBLEMS OF IRON AND MANGANESE IN GROUNDWATER

Groundwater is an important source of municipal drinking water for many small and medium-sized communities across the world. Many favour groundwater over surface water because of its excellent and consistent quality, and because, generally, it requires little or no treatment before consumption. Unfortunately, many groundwater supplies are contaminated by varying levels of iron and manganese in concentrations that exceed the drinking water guidelines [9]. When iron and manganese are present in water supplies even at low concentrations, they can be linked to a number of problems including discoloration, turbidity, taste and odor problems, formation and accumulation of iron or manganese oxides slime in pipes. Both metals promote the growth of certain types of chlorine tolerant microorganisms in water distribution systems. This biota can provide protected sites for noxious organisms, and consequently, vastly increase the costs of cleaning and sterilizing systems that contain organisms dangerous to human health. Although there is little evidence that the consumption of water with natural concentrations of iron and manganese have adverse effects on public health, and they are in fact essential elements for human diet, they do remain problematic from an aesthetic, technical and economic point of view [2]. The WHO recommended limits of iron and manganese for drinking water are set at 0.3 and 0.05 mg g^{-1} respectively.

CONVENTIONAL TREATMENT

Iron and manganese usually exist in groundwater as soluble Fe^{2+} and Mn^{2+}, their removal is, therefore, based on their oxidation into Fe^{3+} and Mn^{4+} states, which are characterized by low solubility and therefore easily removed by sedimentation and filtration. Conventional iron and manganese removal plants rely on physical-chemical reactions using manganese greensands, intense aeration or chemical oxidation (with O_3, $KMnO_4$ or ClO_2). These processes provide treatment but they can have operating problems and they do not always provide an effluent that meets the water quality objectives [2]. The major drawbacks of the various conventional treatments for removal of iron and manganese from drinking water were outlined by Mouchet [3]. According to his findings, only 50 to 60 percent of the conventional treatment plants in US produced water that met standards for iron and manganese levels. Other problems associated with the conventional methods include iron complexation, negative effect of chlorination and flocculation of oxidized iron or manganese.

THE DISCOVERY OF BIOLOGICAL TECHNIQUES

The use of aeration plus filtration line, as employed in most of the conventional treatment plants, led to the accidental discovery of the biological phenomenon [3]. Iron was the first element for which biological removal techniques were discovered and developed because water rich in iron is more common and also because iron is more readily removed biologically. It was noticed that in some physical-chemical iron removal plants, in spite of raw water quality that was not well suited for conventional iron removal, satisfactory iron removal occurred anyway. Large amounts of bacteria, either stalked species such as *Gallionella ferruginea*, or filamentous ones such as *Leptothrix orhracea* were found present in the backwash sludge and those bacteria were in fact responsible for removing iron. In addition it was noticed that when iron or manganese was complexed with substances like humic acids, polyphosphates, silica etc., which normally interfered with the ability of physical-chemical treatment plants to work, the bacteria could still remove the metals. It was realized that these bacteria could be used for iron removal in water treatment plants [2, 3].

The discovery of iron-oxidizing bacteria (IOB) has led to a new generation of biological treatment plants in which a number of naturally occurring microorganisms mediate the oxidation of iron and manganese, and other chemical species such as ammonium. Biological oxidation of iron and manganese is faster, eco-friendly and more effective than the physical-chemical methods [3–5]. Sogaard et al. [10] reported almost a 1000 times faster rate of biological iron oxidation/precipitation compared to abiotic oxidation as reported by other researchers. Katsoyiannis and Zouboulis [8] also found that the biological oxidation kinetics of iron and manganese was several orders of magnitude greater than the respective abiotic oxidation.

MATURATION

The major distinguishing feature of biological phenomenon from the physical-chemical one is the requirement of the former for maturation – a period for the formation of oxides rich-biofilm on the filter media. Once a biological iron or manganese removal plant is constructed, the system must be given time to mature. The coating of the media surface with microbial community and catalytic oxides layer is a spontaneous process that depends on the chemical and biological characteristics of the groundwater [11]. For iron removal plants, the maturing period is quite short requiring anywhere from a day to about one week. For manganese removal plants, the time can be considerably longer, anywhere from 2 weeks to 3 months. At the end of the maturation period, the metal concentration in the effluent falls to near detection limits. In well aerated systems, complete removal of iron sometimes occurs before the formation of the microbial layer, indicating that substantial iron oxidation is achieved through aeration [12]. Many works were reported on how to reduce the filter maturation time, especially for manganese. Hopes and Bott [11] observed that by initially operating the biofilter as a re-circulating batch culture of manganese-oxidizing bacteria such as *Leptothrix discophora*, with an initial Mn concentration of about 2.5 mg L^{-1}, the filter maturation time could be reduce from months to days. A special inoculation start-up procedure was also reported [12], using microbial biomass of backwash sludge from long-time established biological plant for removal of iron, manganese and ammonium. Further investigation [5] demonstrated similar Mn removal is achievable by inoculating biological filters with either a pure culture of *L. discophora* SP-6 (which was emended to *L. cholodnii* SP-6 [13]) or an indigenous biofilm.

The biologically produced oxide coated media is termed as "mature" or "aged" biofilter media (Figure 2). Such biofilter media are able to remove dissolved iron, manganese and other contaminants such as arsenic from water by a combination of biological and physico-chemical sorption processes, without the addition of chemical oxidants [8]. The mechanism of Mn^{2+} removal by aged biofilter media is thought to involve adsorption of Mn^{2+} onto the preformed oxides layers, followed by catalytic and/or bacteriological oxidation [13, 14]. But, the actual contribution of microorganisms and that of the catalytic oxides layers to the manganese oxidation activity of mature biofilter media is difficult to elucidate.

Results of our previous work showed that the accumulated oxides layer on mature biofilter media indeed plays a significant role in Mn oxidation [15]. Indeed as earlier posited by Berbenni et al. [2] that biological activity may be essential for the start-up of Mn^{2+} oxidation, which then proceeds mainly by autocatalytic reaction.

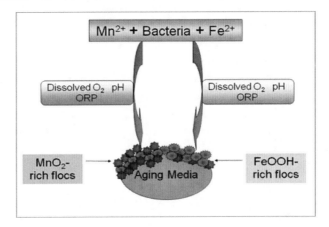

Figure 2. Schematic representation of reactions at the filter bed leading to deposition of biogenic iron and manganese oxides on the surface of aged biofilter medium.

DO, pH, Eh and Temperature Requirements

The environmental conditions required for the growth of different groups of iron- and manganese-oxidizing bacteria are different. The most important operational parameters in the operation of biological filters are pH, which is recommended to be between 6–8, and redox potential (Eh), which is recommended to be strongly oxidizing, i.e. >300 mV. The bacteria involved in biological Fe and Mn removal need different pH, redox potential (Eh), and dissolve oxygen (DO) concentrations for each metal. Iron-oxidizing bacteria may be completely aerobic or microaerophilic, depending on the pH, whereas manganese-oxidizing bacteria require a fully aerobic environment (DO > 4 mg L^{-1}) to precipitate Mn. Given the correct pH, and Eh, the bacteria are normally able to oxidize iron and manganese at temperatures ranging from 5 to 50°C [2–3].

Because of the significantly different conditions required for iron and manganese removal, a two-stage process is often required when both metals are present. Higher concentrations of iron are especially problematic. Precipitation of large amounts of Fe clogs the upper portion of the filter bed and necessitates frequent backwashing. Because an iron-oxidizing community develops much faster than a manganese-oxidizing one (20 hours versus 1-2 months), this hinders the development of a manganese-oxidizing community [14]. In general, patented systems used for the removal of these two metals include initial aeration followed by rapid filtration for Fe removal; and secondary aeration, pH adjustment and secondary rapid filtration for Mn removal [1]. These systems need devices to control DO, pH and Eh in order to limit abiotic Fe oxidation. In several regions of northern Europe, slow sand filter processes including one or two pre-treatment steps are also applied [1, 2, 16]. However, under optimal operating conditions, some biological plants can achieve successful removal of both iron and manganese in single-stage filtration with medium velocity. It was demonstrated that biological Mn removal is possible over a broader "field of activity" (e.g. Mn removal occurred at a pH level as low as 6.5) than has previously been reported [5]. Therefore, the ability of this treatment technology to work over a broader range of environmental conditions allows for more communities to consider biological treatment as an option to remove Fe and Mn from their drinking water.

IRON- AND MANGANESE-OXIDIZING COMMUNITIES

The iron- and manganese-oxidizing microbial community is the core of the technology for the biological removal of iron and manganese from ground water [17]. Consequently, the functions of these water treatment facilities rely on the type and metabolism of the microorganisms contained within the biofilm. The commonly called 'iron bacteria', *Gallionella*, *Sphaerotilus*, *Leptothrix*, and *Clonothrix* groups have long been well-known for their ability to deposit iron hydroxides or manganese oxides on their cell surface [18]. The environmental conditions required for the growth of different groups of these bacteria are different.

The stalked bacteria, e.g., the genus *Gallionella*, which are chemolithotrophic and microaerophilic thrive on Fe(II) and oxygen [18]. The energy released from Fe(II) oxidation allows the bacteria to reduce and assimilate carbon from CO_2. The ideal environmental conditions for the growth of *Gallionella* strains are a pH of 5.4 and DO of 6.5 mg L^{-1} [19]. The filamentous sheathed bacteria, the genera *Sphaerotilus* and *Leptothrix*, are obligate heterotrophic aerobes and require organic carbon for their growth. They grow favorably at a temperature of 15–40°C and a pH of 6.5–8.1 [20]. Another group of iron bacteria are the unicellular bacteria, e.g. the members of the genera *Siderocapsa*, *Siderocystis*, etc., which are also heterotrophic but more difficult to recognize by microscopic observation than the previous ones [3].

With the exception of the genera, *Gallionella* and *Sphearotilus*, which oxidize only iron, many of these bacteria, such as *Pseudomonas manganoxidans* and members of the genera *Leptothrix*, *Crenothrix*, *Hyphomicrobium*, *Siderocapsa*, *Siderocystis* and *Metallagenium* have the dual capacity of oxidizing both iron and manganese indifferently, although under different environmental conditions [3].

A variety of iron- and manganese-oxidizing bacterial species may be naturally present in groundwater, one type or another is able to thrive under a given environmental conditions and colonized the filter media. The amounts of these bacteria in groundwater are usually small, as compared to their corresponding concentration on biofilter surfaces. It has been reported that Mn-oxidizing bacteria amounted to 1–2% of the raw water in an Austrian pilot plant, but this increased to 35% in an aged biofilter medium [21].

Similarly, of the enumerated bacteria in the raw water of a biological plant in Belgium, only 2.25% demonstrated Mn^{2+} oxidizing ability, but the share of Mn-oxidizing bacteria on the sand filter amounted to 25–33% [14].

In addition to the iron- and manganese-oxidizing microorganisms, the activity of nitrifying bacteria in the biofilm may be critical, at least for initiation of Mn oxidation. This is often associated with the long time requirements (usually, 2–4 months) for the start–up of new biofilters, especially if the water contains relatively high concentration of ammonia [2, 21, 22].

Ammonium is removed by a two-stage nitrification process, which transforms it into the less toxic nitrate ions. In the first stage, ammonium is oxidized to nitrite by bacteria such as the genus *Nitrosomonas* and the second stage uses bacteria such as the genus *Nitrobacter* to oxidize the nitrites into nitrates. High concentrations of iron and ammonia may affect manganese oxidation negatively.

MOLECULAR BIOLOGY OF IRON - AND MANGANESE-OXIDIZING BIOFILMS

The development of culture-independent techniques has further revealed the presence of microorganisms in biological water treatment plants that were not hitherto detectable by classical cultivation techniques, thus changing the previous concepts on the biofilms structures and functions. For instance, such investigations based on modern molecular biological methods showed that, instead of the genera *Nitrosomonas* and *Nitrobacter*, species like the genus *Nitrospira* are more important nitrifying bacteria [23–25]. An interesting finding on microbial community of iron-bacterial biofilm was reported by Casiot et al. [26]. They showed that, contrary to expectation, the major population of bacteria in an iron-removal plant was not a member of the well-known iron-oxidizing taxon, the genus *Leptothrix* or *Gallionella*, but the dominant population was a newly identified strain that was found to simultaneously mediate arsenic oxidation with that of iron. The microbial diversity in Joyo biological treatment plant was earlier investigated in our laboratory, using a combination of culture-independent techniques (J. Takezaki, 2006, unpublished Master's thesis, Yokohama National University, Japan).

Briefly, the genomic DNA was extracted and the 16S rDNA were amplified by PCR using the universal primer 8F and 1510R. Sixty clones of the 16S rDNA library were classified, based on PCR-RFLP analysis, into 32 operational taxonomic units (OTUs). Phylogenetic analysis based on the nucleotide sequence of the V3 region in the 16S rDNA of each OTU was carried out. The results showed that several of the OTUs exhibited high similarities to the sequences of iron-oxidizing bacterium (*Gallionella ferruginea*), ammonium-oxidizing bacteria (*Nitrosomonas* sp. *and Nitrospira* sp.), nitrite-oxidizing bacterium (*Nitrospira* sp.), iron-reducing bacterium (*Geothrix fermentans*), and denitrifying bacteria (*Thiobacillus denitrificans, Azoarcus denitrificans*) or manganese-oxidizing bacterium (*Hyphomicrobium* sp.) respectively. Interestingly, *Leptothrix* sp. or its related species, which are considered to be the dominant manganese-oxidizing bacteria in water treatment plants were not detected at this stage. However, subsequent analyses using PSP-6 primer set [27] specific for *Leptothrix* revealed the presence of *Leptothrix mobilis* and its related bacterium, *Caldimonas manganoxidans*. The later oxidizes manganese but do not produce sheath. These results implied that *Leptothrix* sp. is a minority in the microbial consortium of Joyo biological treatment plant.

Another study of this kind was repeated in the same laboratory but with different biofilter media collected from a pilot plant in "M" city, Japan (R. Sasada, 2008, unpublished Master's thesis, Yokohama National University, Japan). The pilot plant achieves simultaneous removal of Fe(II), Mn(II) and As(III). The results were somewhat similar to those of Joyo biological treatment plants in terms of the 16S rDNA clone library.

Interestingly, however, ammonia-oxidizing bacteria and As(III)-oxidizing bacteria were not detected even though both ions are removed from the groundwater of M city. The absence of As(III)-oxidizing bacteria as suggested by these results may means that arsenic removal in M city is achieved through physic-chemical oxidation and adsorption by the biogenic iron and manganese oxides. It is obvious from the foregoing discussion that much is yet to be known about the nature and activity of the complex microbial communities present in groundwater biofilms.

BIOCHEMISTRY OF IRON AND MANGANESE OXIDATION

In anaerobic groundwater, iron and manganese may remain in their soluble forms as Fe^{2+} and Mn^{2+}. The complex metabolic activities of iron-oxidizing bacteria are not fully understood. The mechanism of bacterial iron oxidation is thought to be either a primary intracellular enzymatic oxidation, for autotrophic bacteria such as *Gallionella* sp., or secondary extracellular oxidation, caused by the catalytic action of the excreted extracellular polymer, namely the filaments comprising the stalks produced by *Gallionella* sp., or the sheaths from *Sphaerotilus* and *Leptothrix* as well as the extracellular polymers excreted by various *Siderocapsaceae* [3]. Energetically, the reaction can be described by the equation below [28]:

$$Fe^{2+} + 0.25O_2 + H^+ \rightarrow Fe^{3+} + \frac{1}{2}H_2O \qquad (1)$$

The $\Delta G^{o/}$ of the process yields only about 29 kJ per mol of Fe. From this equation, the oxidation of one mole of Fe(II) produces less energy compared to the oxidation of one mole of glucose. This means that a large amount of Fe(II) has to be oxidized to support lithotrophic growth of iron-oxidizing bacteria [29].

One major problem with biological iron oxidation is that, at neutral pH and under aerated conditions, there is rapid chemical oxidation of Fe^{2+} [30]. The kinetics that describe chemical Fe oxidation are

$$\frac{-d[Fe(II)]}{dt} = k[Fe(II)][OH^-]^2 p_{O2} \qquad (2)$$

where $k = (8 \pm 2.5) \times 10^{13}$ min^{-1} atm^{-1} mol^{-1} L^{-1} [31]. As is evident from this equation, pH has a strong influence on the reaction rate, which is why under very acidic conditions (pH < 4), Fe^{2+} is quite stable in the presence of air. These conditions are needed for the growth of acidophilic Fe oxidizers like *Thiobacillus ferrooxidans*. Some bacteria such as *Gallionella ferruginea* overcome Fe^{2+} stability problem by growing at a very low O_2 concentration at circumneutral pH, where the half-life of Fe^{2+} may be much longer [30].

Another problem faced by these organisms is that, at a pH of 6-7, the products of Fe^{2+} oxidation are insoluble ferric hydroxides, e.g., $Fe(OH)_3$ or FeOOH. To avoid mineralizing the cytoplasm of the cell, iron oxidation must occur at the exterior of the cell surface. This requires that cells possess a chemical mechanism for transporting electrons across the periplasm to cytoplasmic membrane, where a chemiosmotic potential is established. It is thus common to find the iron oxidase, as well as soluble electron transport components at the exterior of cytoplasmic membrane of some iron-oxidizing bacteria such as *Thiobacillus ferrooxidans* [30].

Unlike Fe^{2+}, Mn^{2+} is quite stable under fully aerobic conditions at neutral pH, and only at pH of 9 or greater does chemical oxidation of Mn occurs and it proceeds extremely slowly at lower pH values like those found in groundwater [2-3]. These properties have made it easier to demonstrate biological Mn oxidation. Mn^{2+} is thought to be oxidized enzymatically or indirectly through the rise in pH resulting from bacterial growth [30]. Bacterial Mn^{2+}

oxidation to the manganic form, Mn^{4+}, is a two-electron transfer, which can proceed directly or via one-electron steps through an unstable intermediate, Mn^{3+} [32]. In terms of energetics, the two-electron transfer to oxygen can be represented by the following equation:

$$Mn^{2+} + 0.5O_2 + H_2O \rightarrow MnO_2 + 2H^+ \qquad (3)$$

The standard free energy at pH 7 ($\Delta G^{o\prime}$) of this reaction is -70.9 kJ mol^{-1} [33]. Compared to the free energy for ATP formation (-30 kJ mol^{-1}), this would suggest that Mn oxidation could quite easily support lithotrophic energy metabolism. However, it remains to be conclusively demonstrated that any prokaryote can grow lithotrophically using Mn [30]

The products of biological Mn oxidation are deposited in the form of black precipitates coating the cells or as pustules on the sheaths of the filamentous bacteria. The pre-formed MnO_x is catalytic in nature and can facilitate Mn oxidation abiotically as described by the autocatalytic rate equation below [34]:

$$\frac{-d[Mn(II)]}{dt} = k_o[Mn(II)] + k_1[Mn(II)][MnO_x] \qquad (4)$$

In this expression, MnO_x is used to denote a general empirical formula for the oxidation product, so that the values of x of 1, 1.5 and 2.0 would correspond to Mn(II), Mn(III) and Mn(IV) oxides respectively, without concern for the degree of hydration.

Whatever the metabolic pathway for the iron and manganese oxidation reactions, the biological process is catalytic in nature and causes a rapid oxidation. The red insoluble precipitates formed are all slightly hydrated iron and manganese oxides that, beneficially, are more compact forms than the precipitates formed when using physical-chemical processes. This feature partially explains the greater iron retention capacity between backwashes of biological filters when compared to physical chemical treatment filters [2].

ARSENIC CONTAMINATION OF GROUNDWATER

Arsenic (As) exposure, often through drinking water sources, can contribute to a large spectrum of diseases, collectively called arsenicosis, which in turn, increases the risk of cancer. Contamination of groundwater with arsenic has emerged in the last three decades, as one of the major environmental problems affecting millions of people worldwide. Higher than expected concentration of As was observed in the surface water and shallow zones of ground water of many countries including Bangladesh, India, Pakistan, Argentina, Mexico, Mongolia, Germany, Thailand, China, Chile, USA, Canada, Vietnam, Hungary, Romania, Finland, Greece, and Italy [22-25]. In some regions of Bangladesh, arsenic concentration in groundwater is as high as 1000 mg L^{-1} [26].

In natural water arsenic is mainly present in inorganic forms as arsenate [As(V)] and arsenite [As(III)] species. The distribution of these two species in water depends mainly on redox potential and pH conditions. Under oxidizing conditions such as those prevailing in surface water, arsenates ($H_2AsO_4^-$, $HAsO_4^{2-}$ with pKa = 2.19, pKab = 6.94 respectively) are the major species; whereas, arsenites (H_3AsO_3, pKa = 9.22) predominate under reduced conditions and circumneutral pH values such as found in groundwater samples [35]. As(III) is more toxic to biological systems and more mobile in soil and aquatic environments than As(V). Arsenate resembles phosphate structurally and can compete with phosphate in several biochemical reactions, whereas, As(III) may inhibits certain enzymes and proteins by reacting with their thiol groups [36–38].

CONVENTIONAL ARSENIC REMOVAL TECHNIQUES

Because of increased concern over possible health risks associated with chronic ingestion of low levels of arsenic in drinking water, the WHO guideline value for arsenic in drinking water was reduced in 1993, from 50 to 10 $\mu g\ L^{-1}$ [39]. By 1998, the European Commission has directed that all drinking water supply systems in member' countries should comply with this new standard [40], and this was followed by USEPA' directive in 2003 [41]. To meet the challenge of reducing arsenic concentration to the MCL value of 10 $\mu g\ L^{-1}$ and its treatment cost, several water treatment processes have been reported for removal of arsenic species from drinking water [42, 43].

Some of these water treatment technologies, such as membrane processes, ion-exchange and electrodialysis are able to reach these legal requirements; they are, however, expensive for small water production units. A number of less expensive techniques for arsenic removal from water such as auto-attenuation, precipitation-coagulation, and adsorption on various coated filter media including, iron or manganese oxi-hydroxides coated alumina, iron-oxide-coated polymeric materials [44–46], have been promoted for application in small rural communities of developing economies like Bangladesh.

Among the emerging water treatment technologies, adsorption of arsenic onto thin layer of iron or manganese oxides coated on porous support media appears to be the method of choice and is termed as adsorptive filtration [44]. This method is based on the oxidative powers of manganese oxides and, to a lesser extent, iron oxides for As(III) [47], and their varying adsorption capacities for As(III) and As(V) species under different environmental conditions [48].

In the last two decades, coated media technology has received enormous attention, which led to the emergence of a variety of chemically synthesized iron or/and manganese oxides based media for arsenic removal from water [49]. Although these media can effectively remove arsenic from water, they have their limitations. Pre-oxidation of As(III) to As(V) or regeneration of the oxidative sites may be required, and a breakthrough point must be monitored due to exhaustation of the adsorbent [49]. The addition of chemical oxidants for regeneration of the oxidant or sorbent materials not only increases the operational cost, but may also generate arsenic-laden hazardous by-products.

BIOLOGICAL ARSENIC REMOVAL TECHNIQUES

The term 'biological removal' has been used disproportionately in the literature to refer to the use of different bio-based materials for sorption of arsenic species or its transformation. The concept of biological removal of arsenic could be classified under the three (3) main subthemes, namely, biosorption, biofiltration using living microbes and biological adsorptive filtration.

In biosorption, dead and metabolically inactive biomass is use as adsorbent. The economics of the process are also improved by using waste biomass such as agricultural by-products and seaweeds, instead of purposely produced biomass. Like adsorption of other heavy metals, the sorption mechanism is complex and may include ion-exchange, complexation, coordination and/or electrostatic interactions between the metal ions and different functional groups of the biomass [50].

Several modifications of the biomass have been reported to increase the biosorption efficiency [51]. Although numerous papers on the arsenic–biomass interactions are available in the literature, still large uncertainties exist, and no information yet on even a pilot plant application of biosorption for arsenic removal.

The mechanism of arsenic removal by biofiltration using living microbes differs from that of biosorption technique. In this case, the biotransformation of arsenic species (for example, conversion of As(III) to As(V)) may be performed by extracellular enzymes and the oxidized species may be adsorbed or precipitated/co precipitated onto the biofilm formed on the solid support medium.

In some cases, it may involve metabolism-dependent intracellular uptake, whereby arsenic is transported across the cell membrane into the cell, where methylation or oxidation–reduction of arsenic species may take place. It has been reported that in most of the cases the removal efficiency of this technique is controlled by plasmid genes. Hence, there is a high possibility for improving the efficiency by genetic modification of the microbes [52]. However, the evaluation of arsenic removal efficiency by living microbes remains a laboratory exercise and the reliability of the process is yet to be tested in the field.

BIOLOGICAL ADSORPTIVE FILTRATION

The third and probably the most established biological removal techniques is the so-called biological adsorptive filtration technology. This highly innovative technology combines the advantages of the well-understood coated media technology with the fact that the oxides coatings on the media are developed naturally (Figure 2) by the activity of iron- and manganese-oxidizing microorganisms native in the groundwater [53, 54]. The method comprises of the steps of biological oxidation of iron and manganese forming amorphous iron and manganese oxides, which coat the surface of the filter medium [3, 55]. Arsenic species are subsequently removed from the groundwater by a combination of biological and physicochemical sorption processes, including oxidation and adsorption by biogenic iron and manganese oxides [53, 56].

The bacteria in the system could be directly involved in the oxidation of As(III) and/or indirectly, through the generation of reactive manganese oxides and adsorptive iron oxides, which remove both As(III) and As(V) from the water [57, 58]. The biological approach is more economical and eco-friendly as no chemicals are added and the sorbent materials are continuously produced in situ. Furthermore, it is a combined process for simultaneous removal of iron, manganese and arsenic from groundwater [56, 57].

FACTORS AFFECTING ARSENIC REMOVAL BY BIOLOGICAL ADSORPTIVE FILTRATION

As mentioned earlier, removal of arsenic by biological adsorptive filtration is based on establishment of biological iron and manganese oxidation, therefore, all the factors discussed above for biological iron and manganese removal can be also applied. In this section, the additional factors that govern the interaction between arsenic and the composite materials in the filter bed would be discussed. These include the natural concentrations of iron and manganese in the ground water, composition of the biofilter media, pH and presence of other interfering ions in the groundwater.

LEVELS OF IRON AND MANGANESE IN GROUNDWATER

Iron, manganese and arsenic concentrations in groundwater and surface water are often related because arsenic can adsorb on, or coprecipitate with iron and manganese and adsorb onto clay mineral surfaces under oxidizing conditions. Arsenic becomes mobile when reducing conditions are sufficient to dissolve iron and manganese but not enough to produce sulfide [59]. Anderson and Bruland [60] noted that in surface waters, greater concentrations of arsenic, iron and manganese occurred in the absence of dissolved oxygen. Within oxygenated zones (groundwater or surface water), As(V) is stable, while under anoxic conditions, As(III) is stable [61]. Since groundwater quality in natural systems is a result of many environmental factors, it is difficult to predict a concentration pattern of these elements in natural water. However, water rich in iron is more common and the relative concentration of iron is often higher than that of manganese.

Mechanism of arsenic removal by biological adsorptive filtration strongly depends on the nature of the composite materials at the filter bed, which in turn depend, among other things, on the relative concentrations of iron and manganese in the groundwater. It has been reported that [62], for the natural composite materials with Mn/Fe ratio greater than 0.1, As(III) oxidation would be the dominant pathway, and at ratio less than 0.02, As(III) adsorption would be the dominant pathway for the removal of arsenite from solution. Previous reports have shown that manganese oxides are the primary oxidants for As(III), while iron oxides adsorbs the produced As(V) more strongly than As(III). One of the consequences of absence or very low levels of manganese in the groundwater is the need for frequent backwashing to avoid leaching of the loosely bound As(III) from the iron oxides-rich flocs at the filter bed - a common problem observed in pilot plants for arsenic removal in conjunction with biological iron oxidation. It remains to be elucidated whether artificial addition of Mn^{2+} to the groundwater will increase the oxidation efficiency of the system for As(III).

COMPOSITION OF THE BIOFILTER MEDIA

Biofilter media is the core of the technology for biological removal of iron, manganese and arsenic from groundwater [63]. Consequently, the functions of these water treatment facilities rely heavily on the nature and composition of the biogenic surface coatings on these filter media. The biogenic surface coating on the biofilter media is both chemically heterogeneous and dynamic. In our previous study on the nature of the biogenic surface coatings of aged biofilter media, we characterized two different (age-wise) biofilter media from a biological iron and manganese water treatment plant [15]. The results showed that, the surface coatings are composed of variable layers of iron and manganese oxides. Microbial biomass constitutes about 3.5 and 1.4% of the dry weight of the surface coatings on the 3 and 15 years old biofilter media respectively. A four-fold increase in iron and manganese oxides (combined) was observed between the 3- and 15-years filter media and a Mn/Fe ratio of 1.47 and 1.54 respectively, showing that manganese oxides are selectively accumulated over iron oxides despite the higher concentrations of Fe in the raw water (Figure 3).

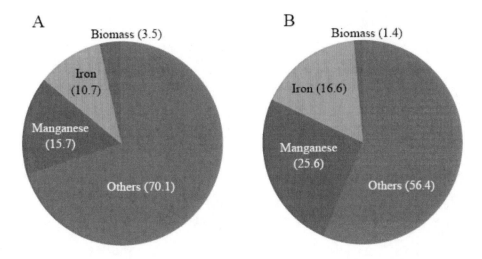

Figure 3. Percent compositions of biomass, iron and manganese in surface coatings of 3-year biofilter media (A) and 15-year biofilter media (B).

Consequently, the surface area of the aged biofilter media increased from 1.01 to 5.48 m^2 g^{-1}, but the point of zero charge (pH$_{PZC}$ - the pH at which the net charge on the surface of the filter medium is zero) decreased from pH 5.3 to 4.3 in the two media respectively. These changes in the surface characteristics of the biofilter media have remarkable effects on arsenic removal efficiency of the media. Obviously, the increased Mn/Fe ratio will increase the As(III) oxidation efficiency of the biofilter media, while the decreased pH$_{PZC}$ will negatively impact on the adsorption capacity of the media for both As(III) and As(V) species at the circumneutral pH of the groundwater.

The oxidation and adsorption efficiency of Mn oxides-rich surface coatings on aged biofilter medium was demonstrated with a simple batch experiment (unpublished data) using the 3-year anthracite biofilter medium at 10 g L^{-1} biofilter loading and an initial As(III) concentration of 1000 mg L^{-1} (at pH 7). As shown in Figure 4, the aged biofilter media can efficiently oxidaze As(III) to As(V), and complete oxidation of 1000 ppb As(III) was achieved within 5 hours. However, the adsorption of the As(V) was not as efficient as its production, and an equilibrium was reached within the first 3 hours of the reaction. The explanation for these behaviors of the oxide coated biofilter media can be deduced from the relative amounts of manganese and iron oxides on the surface coatings of the medium [15]; manganese oxides were about 3.5 times the concentration of iron oxides on the surface coating of the media. While the manganese oxides can sufficiently oxidize all the As(III) to As(V), the amount of iron oxides was not proportionate enough to remove the oxidized As(V) by adsorption.

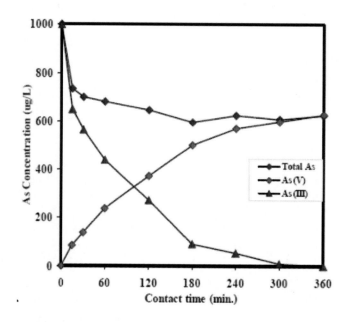

Figure 4. Kinetics of As(III) oxidation and adsorption by the 3-year biofilter media; biofilter media loading, 10 g L^{-1} and an initial As(III) concentration of 1000 mg L^{-1}; pH 7.

EFFECTS OF pH ON ARSENIC OXIDATION AND ADSORPTION

Inorganic arsenic in groundwater occurs in two oxidation states, As(III) (arsenite) and As(V) (arsenate). The two species show different pH adsorption behaviors onto mineral oxides surfaces, partly, due to differences in their acid-base characteristics. For instance, at acidic pH, As(V) generally sorbs more favorably on iron oxides than As(III), whereas at basic pH, the trend is reversed [64]. The acid–base behaviors of aqueous As(III) and As(V) within the pH range of 3–10 were modeled using MINEQL+ software program. Arsenious acid (H_3AsO_3) has a pKa value of 9.2, so in the pH range of 3–10, there is relatively little of its conjugate base $H_2AsO_3^-$ in the solution (Figure 5). The sum of concentrations of H_3AsO_3 and $H_2AsO_3^-$ is denoted As(III). Arsenic acid (H_3AsO_4) has pKa values of 2.7, 6.8, and 11.5, which are similar to those of phosphoric acid. In the pH range of 3–10 the concentrations of $H_2AsO_4^-$ and $HAsO_4^{2-}$ are much greater than those of H_3AsO_4 and AsO_4^{3-} (Figure 6). The sum of concentrations of H_3AsO_4, $H_2AsO_4^-$, $HAsO_4^{2-}$, and AsO_4^{3-} is denoted As(V) [65].

The effects of pH on arsenic adsorption/oxidation were evaluated using the 3-year anthracite biofilter medium [66]. Figure 7 presents the kinetics of arsenic adsorption/oxidation by the biofilter medium at pH values 4, 7 and 8.5. It is noted from the results that, on one hand, the depletion of As(III) from the solution was faster and less affected by pH-variations, but, on the other hand, the release of As(V) into the solution was much slower and strongly influenced by pH of the solution. The half-life of the initial As(III) of 18 μmol L^{-1} in the solutions were 0.96, 1.42 and 1.20 h at pH 4, 7 and 8.5 respectively. Because of the high Mn/Fe ratio, the rates of As(III) oxidation by the biofilter medium were almost comparable to those of pure manganese oxides [67]. In contrast to the pure manganese oxides, the release of As(V) was much slower than the depletion of As(III), indicating the contribution of iron oxides both in the initial adsorption of As(III) as well as in the re-adsorption of the produced As(V) [68]. The Mn oxides-rich biofilter media can easily oxidize As(III) to As(V) and the rate of As(III) oxidation was less affected by the pH-variations; while, the retention capacity of the media for the produced or added As(V) depends strongly on the pH of the solution. The high rates of As(III) oxidation were attributed to the higher Mn/Fe ratio on the surface of the media, whereas, the effects of pH on the adsorption of arsenates could be explained on the basis of the pH_{PZC} of the medium.

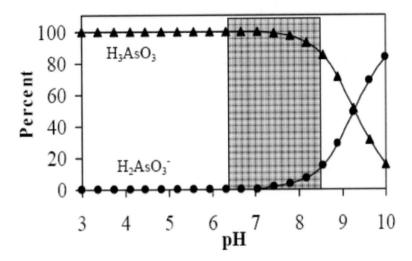

Figure 5. Concentrations of the As(III) species, H_3AsO_3 and $H_2AsO_3^-$, at pH values 3 to 10, obtained by MINEQL+ software program. The shaded area is the pH range of groundwater.

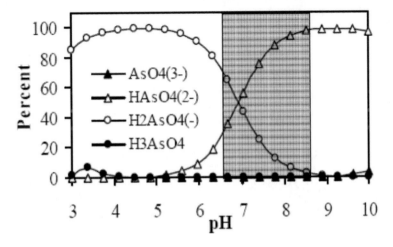

Figure 6. Concentrations of the As(V) species, AsO_4^{3-}, $HAsO_4^{2-}$, $H_2AsO_4^-$ and H_3AsO_4, at pH values 3 to 10, obtained by MINEQL+ software program. The shaded area is the pH range of groundwater.

Figure 8 depicts the adsorption profiles of arsenite and arsenate onto the 3-year biofilter medium at different pH values. Within the environmental reasonable arsenic concentration range of 20 μmol/L and pH values 4–9, As(III) and As(V) exhibited somewhat similar pattern of adsorption on the biofilter medium, though the relative uptake of As(III) was slightly higher than that of As(V) This trend was due to the oxidation of the As(III) to As(V) during the equilibration time followed by the adsorption of the produced As(V) onto the biofilter medium.

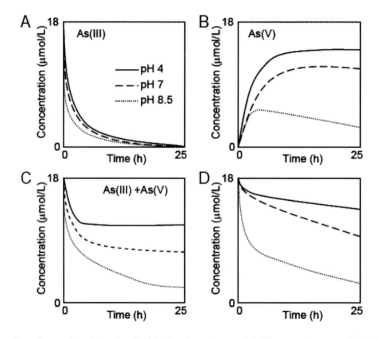

Figure 7. Kinetics of arsenic adsorption/oxidation by a 3-year biofilter medium at pH 4, 7 and 8.5. Depletion of As(III) (A); Release of As(V) (B); Removal of As(III) + As(V) (C); Removal of added As(V) (D). (A) - (C) were obtained by the same run supplemented with As(III) initial concentration of 18 μmol L^{-1}, while (D) was obtained from another run supplemented with As(V) initial concentration of 18 μmol L^{-1}; T = 25°C; biofilter loading = 10 g L^{-1}.

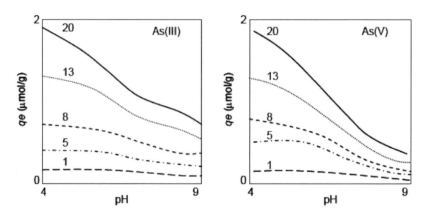

Figure 8. Illustrations of amounts of As(III) and As(V) adsorbed per gram (qe) of a 3-year biofilter medium as a function of the final pH for five different initial concentrations. Concentrations in μmol L-1 As(III) = As(V): 1; 5; 8; 13; 20; biofilter loading = 10 g L-1; T = 25oC.

In both cases, adsorptions were highest at pH 4 and decrease with increasing pH and initial concentrations. At pH 4, the amounts of As(III) removed from the solution were 100 and 97% of the initial concentrations of 1 and 20 µmol L^{-1} respectively. At pH 9, however, the amount removed decreased to 47 and 30% of the same initial arsenite concentrations. The effect of pH was more pronounced in the case of As(V), in which the percentage removal were 100 and 94% at pH 4, but, only 19 and 18% of the respective initial arsenate concentrations of 1 and 20 µmol L^{-1} were removed at pH 9. These results can be explained on the basis of pH$_{ZPC}$ of the biofilter medum, which was found at pH 5. At pH below the pH$_{ZPC}$, the surface of the biofilter medium is positively charged, thus, adsorption of negatively charged arsenate ions is enhanced by non-specific adsorptions (electrostatic attractions). Conversely, at pH above the pH$_{ZPC}$, electrostatic repulsion comes into play and increases with increasing pH. Under these conditions, adsorption of anions on the negatively charged biofilter surface is driven mainly by specific adsorptions [69] and may depend on the initial concentrations of arsenate in the solution.

The adsorption data were well fitted to the Freundlich and Sips isotherm equations. Freundlich isotherm is expressed below [70]:

$$q = k_F . C_e^{\,1/n_F} \tag{5}$$

where, q (µmol g^{-1}) is the amount adsorbed per unit mass of the adsorbent, C_e is the concentration at equilibrium (µmol L^{-1}), and k_F and n_F are Freundlich constants relating to adsorption capacity and intensity respectively.

Sips isotherm is expressed as follows [70]:

$$q = \frac{q_m b C_e^{\,1/n}}{1 + b C_e^{\,1/n}} \tag{6}$$

where, q and q_m (µmol g^{-1}) are the amount adsorbed per unit mass of the adsorbent and the maximum adsorption capacity respectively, C_e is the concentration at equilibrium (µmol L^{-1}), b is the equilibrium constant relating to binding strength, and n is the parameter of heterogeneity.

The equilibrium data (Figure 9) were well fitted to the Freundlich and Sips isotherm representing the expectation of multilayer adsorption on heterogeneous surface. Both isotherms assumed that the stronger binding sites are occupied first and that the binding strength decreases with increasing degree of site occupation [70, 71]. The maximum adsorption capacity, q_m (µmol g^{-1}), of the biofilter medium for both As(III) and As(V) decreases with increasing pH of the solution, whereas the equilibrium constant, b, relating to the binding energy, increases as the pH increases, suggesting the contribution of weaker binding sites at lower pH conditions. Similarly, both the Freundlich exponent ($1/n_F$) and the Sips heterogeneity parameter ($1/n$) for As(III) and As(V) alike, were less than unity up to pH 7, indicating that the adsorption is more favorable in acidic pH. This was also reflected in the curvature of the isotherms (Figure 9), which changes from H-curve at pH 4 to L-curve at pH 5, 6 and 7 and finally to C-curve at pH 8 and 9 [72]. The H-curve (high affinity) isotherm usually results from extremely strong adsorption at very low concentrations giving rise to a large initial slope. The L-curve is characterized by an initial region, which is concave to the concentration axis. Type L suggests that there is no strong competition between the adsorbate and the solvent to occupy the adsorption sites. C-curve isotherm shows an initial slope that remains independent of adsorbate concentration until the maximum possible adsorption. C type isotherm suggests a proportionate increase in the amount of adsorbing sites as the amount of arsenate adsorbed increases [73]. The decreased affinity of the biofilter medium for arsenate with increasing pH may be due to increasing competition for binding sites with OH⁻ or other pre-adsorbed ligand ions such as phosphates [73]. This possibility was tested further by plotting the amount of phosphate released by the biofilter medium against arsenate adsorbed at pH 4, 7 and 9. The results (data not shown) indicated that, the release of phosphate was a strong function of pH and a lesser function of adsorbed arsenate. It is obvious that arsenic adsorptions onto aged biofilter media is a complex process controlled by the interrelationship between the properties of the biofilter media surface, the pH, the arsenic speciation, initial arsenic concentration and pre-adsorbed competing ions.

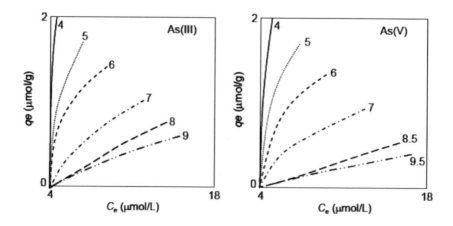

Figure 9. Adsorption isotherms of As(III) and As(V) on a 3-year biofilter medium at different pH values. Initial concentrations of As(III) = As(V) = 0.67 – 20 μmol L^{-1}; T = 25°C; biofilter loading = 10 g L^{-1}.

PROPOSED MODEL FOR AS(III) CONCERTED OXIDATION-ADSORPTION MECHANISM

The mechanism of As(III) heterogeneous oxidation as earlier reported for synthetic birnessite [47] was also observed with the biogenic manganese oxides on the surface coatings. Oxidation of As(III) is coupled with the reductive dissolution of the MnO_2 surface (Figure 10), leading to the release of Mn(II) and As(V) into the solution at low pH. At pH 6 and above, however, the released Mn(II) was readorbed and probably, reoxidized autocatalytically. A conceptual model for As(III) concerted oxidation-adsorption mechanism was proposed based on the assumptions that surface manganese oxides are directly involved in the oxidation of As(III) to As(V).

The adsorptive and oxidative sites on the biofilter medium can be represented by the generic terms SOH and SMnOOH respectively.

Where, SOH represents different surface sites of varying affinity for arsenic, including, iron oxides, organic matter and solvated cations such as Ca^{2+}; and SMnOOH represents manganese-oxides-rich binding sites bearing a surface manganese oxide molecule. The adsorption process of As(III) onto the biofilter medium surface can be represented by the following simplified non-stoichiometric equations [74]:

$$S - OH + H_3AsO_3 \xleftarrow{\text{weak oxidant}} S - OAsO_2H_2 + H_2O \tag{7}$$

$$S - O - Mn - OH + H_3AsO_3 \xleftarrow{\text{strong oxidant}} S - O - MnOAsO_2H_2 + H_2O \tag{8}$$

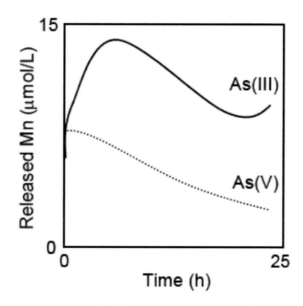

Figure 10. Illustration of progress in the release of Mn^{2+} into the solution by a 3-year biofilter medium in the presence of As(III) or As(V) at pH 4. Initial concentration of As(III) = As(V) = 18 μmol L^{-1}; T = 25°C; biofilter loading = 10 g L^{-1}. Mn^{2+} was not detected in the solutions at pH 7 and 8.5, probably because of complete readsorption of the released Mn^{2+} onto the biofilter medium surface at pH above 6.

The binding of As(III) onto the oxidative sites triggers fast redox reactions involving electrons transfer from the surface manganese oxides to As(III). Previous studies [74,75] have shown that the oxidation of As(III) to As(V) by manganese oxides occurs via a multi-steps process mechanism, which are schematically represented in Figure 11:

The release of reduced Mn^{2+} exposes new binding sites for the next round of adsorption. The newly exposed binding sites are now available for another molecule of As(III) or the produced As(V). At pH above 4, the adsorption of the reduced Mn^{2+} on the biofilter media surface was quantitative [74, 76], making the surface more positive, which enhances the uptake of negatively charged arsenate ions. It is conceivable that adsorption of As(III) onto the biofilter media surface causes oxidation, which, in turn, leads to increased adsorption; thus, the concerted adsorption-oxidation mechanism.

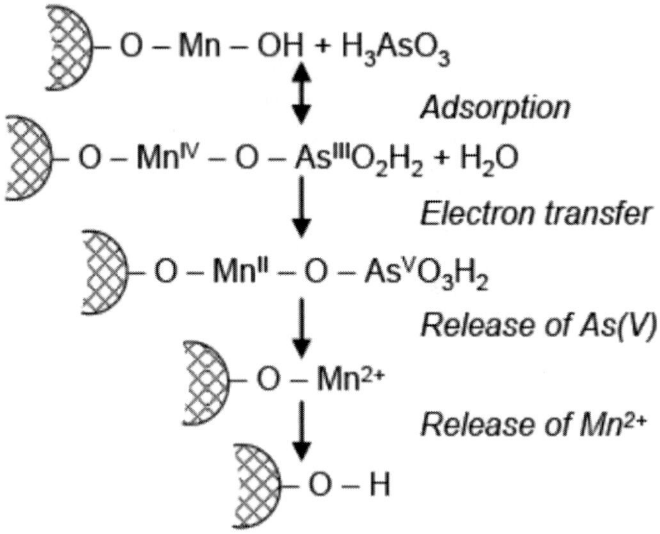

Figure 11. Schematic multi-steps redox reaction between As(III) and manganese oxides on the surface of aged biofilter media (– O–H represents a newly exposed binding sites on the media surface).

In the case of the produced or added As(V), the pH dependent adsorption reactions on the adsorptive and oxidative sites can be represented by the following simplified non-stoichiometric equations [57]:

$$S - OH + H_3AsO_4 \xrightarrow{pH\,dependent} S - OAsO_3H_2 + H_2O \qquad (9)$$

$$S - O - Mn - OH + H_3AsO_4 \xrightarrow{pH\,dependent} S - O - MnOAsO_3H_2 + H_2O \qquad (10)$$

The implication of these findings to biological arsenic removal is that the partitioning of the arsenate between the iron-oxides-rich flocs and the manganese-oxides-rich biofilter media, at any given time, would be determined by the pH of the microenvironment at the filter bed.

EFFECTS OF OTHER INTERFERING IONS

The iron and manganese oxides in the heterogeneous composite materials at the filter bed and the surface coating of biological filter media may be selectively associated with various amounts of metal cations and specifically adsorbed ligand ions [15]. Surface coatings on biofilter media of a biological water treatment plant typically consist of Mn, Fe and Ca, with trace amounts of Si, P and S. Some of these ligands may competitively interfere with arsenic adsorption onto the biofilter binding sites. Katsoyiannis et al. [58] observed that phosphates present at concentrations of about 600 µg L^{-1} had an adverse effect on As(III) removal (competitive adsorption) and reduced the overall removal efficiency from 80 to 30%.

Phosphates, however, did not affect the oxidation of As(III) [58]. The effects of other associating ions on arsenic removal efficiency of biofilter media await further research.

BIOLOGICAL FILTRATION AND ADVANCED WASTEWATER TREATMENT

In the preceding paragraphs, mineralogy of the surface coatings of aged biofilter media have shown that the coating materials are composed of irregularly distributed alternating layers of iron and manganese oxides. In addition to the organic matter, the oxides are also associated with various amounts of co-precipitated and specifically adsorbed elements and ligand ions such as Ca, Na, S and P etc. The potentials of the biogenic surface coatings for heterogeneous oxidation of As(III) to As(V) were also demonstrated. Biogenic iron and manganese oxides and their associated biomass are known to have diverse adsorption capabilities for a variety of toxic heavy metals. It is therefore conceivable that, such potentials could be harnessed for in situ remediation of toxic heavy metals in groundwater, or the biogenic oxides could be exploited for innovative waste water treatment to remove toxic metals.

Previous researches using fresh water natural surface coatings have shown that, the relative contributions of Fe and Mn oxides and organic matter to adsorption of Pb and Cd are remarkably different. Adsorption of Cd was dominated by Fe oxides with lesser roles attributed to Mn oxides and organic matter, while adsorption of Pb was dominated by Mn oxides with lesser roles attributed to Fe oxides and organic matter [76, 77]. Desorption experiments and speciation studies based on selective extraction techniques both indicated that, the affinity of NSCs for Pb is much higher than that of Cd; in order words, the retention of Cd by NSCs is weaker than that of Pb [77]. However, biogenic surface coatings of aged biofilter media may have higher metal oxides contents and lower organic matter than the weeks' old NSCs. Moreover, the oxides are associated with various co-precipitated and specifically adsorbed metal cations and ligands characteristics of the groundwater preliminary study, we combined macro adsorption experiments with electron probe analysis to evaluate the adsorption properties of the biogenic surface coatings on aged biofilter media for removal of cationic lead and cadmium as well as arsenate anion from contaminated water [78]. Results of both batch (Figure 12) and column (Figure 13) adsorption data indicated that the 3-year biofilter medium has stronger affinity for cationic heavy metals as compared to arsenate anion [79]. This information is useful in the removal of these toxic elements from groundwater in conjunction with biological iron and manganese oxidation. As shown above (Figure 3), in biological filtration of Fe and Mn from groundwater, Mn oxides are selectively accumulated on the biofilter media surface, a situation that leads to the presence of Mn oxides-rich biofilter media and Fe oxides rich flocs in the filter bed. Therefore, the adsorption and retention of any given metal contaminant should be driven by its varying affinity for these composite materials, at the prevailing water chemistry in the filter bed. The low affinity of such Mn oxides-rich biofilter media for As(V) suggests that, most of the As(V) originally in the water or produced from As(III) oxidation may be possibly bound to the Fe oxides-rich flocs rather than the Mn oxides-rich biofilter media. When the former are removed during backwashing and the As-laden sludge incinerated, the arsenate can be easily taken care of as hazardous land fill.

Heavy metal contamination of groundwater may results from both natural sources and anthropogenic activity. The results of this study suggest that Pb removal would be more efficient in conjunction with biological Mn^{2+} oxidation. The decreased affinity of the biofilter medium for cadmium, particularly in the column experiment (Figure 13), was attributed to the high Mn/Fe ratio of the biofilter medium and competitive effects of pre-adsorbed Ca ions, and this should be taken into consideration in the use of biological Fe and Mn oxidation process for Cd(II) contaminated groundwater.

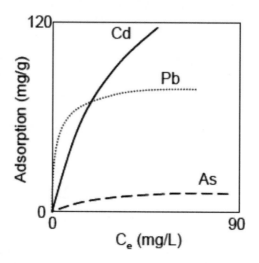

Figure 12. Adsorption isotherm: Initial concentration range of As(V) = Pb(II) = Cd(II) = 5 -100 mg L^{-1}; T = 30 °C; pH 5.5; biofilter loading = 10 mg L^{-1}.

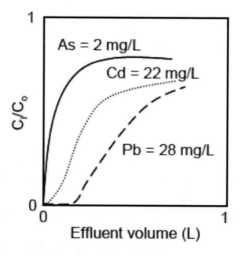

Figure 13. Breakthrough curves for As(V), Pb(II) and Cd(II) adsorption onto a 3-year biofilter medium: Flow rate = 0.5 mL min^{-1}; influent concentration as indicated in the figure; T = 30°C; pH 5.5; biofilter loading = 5 g L^{-1}.

CONCLUSION

Biological filtration emerged as a cost-effective and eco-friendly technology for remediation of metal contaminants from groundwater. As discussed in this chapter, the biological and physicochemical requirements of biological filtration technology for arsenic removal are quite liberal and not much different from those of the well-known biological iron and manganese removal systems. Full understanding of these requirements would allow for more communities across the world to exploit the biocatalyzation potential of microorganisms in drinking water treatments. The chapter also highlighted the prospect of this innovative technology in waste water treatment for removal of other toxic metals.

REFERENCES

[1]   V.A. Pacini, A.M. Ingallinella, and G. Sanguinetti, *Water Res.* 39, 4463 (2005).
[2]   B. Gage, P. Williams, *Proceedings of the Ontario Water Works Association Conference.* Ontario, Canada (2001).
[3]   P. Mouchet, *J. AWWA.* 84, 158 (1992).
[4]   P. Mouchet, *Proceedings of the Water Quality Technology Conference, AWWA.* New Orleans, 2287 (1995).
[5]   M.S. Burger, S.S. Mercer, G.D. Shupe and G.A. Gagnon, *Water Res.* 42, 4733 (2008).
[6]   I.A. Katsoyiannis and A.I. Zouboulis, *Water Res.* 38, 1922 (2004).
[7]   T. Hennebel, B. De Gusseme, N. Boon and W. Verstraete, *Trends in Biotechnol.* 27, 90 (2008).
[8]   I.A. Katsoyiannis, A.I. Zouboulis and M. Jekel, *Ind. Eng. Chem. Res.* 43, 486 (2004).
[9]   P. Williams. Preprint, 2009 at http://www.esemag.com/0302 /biological.html.
[10]   E.G. Sogaard, R. Medenwaldt and J.A. Peskir, *Water Res.* 34, 2675 (2000).
[11]   C.K. Hopes and T.R. Bott, *Water Res.* 38, 1853 (2004).
[12]   T. Stembal, M. Markic, F. Briski and L. Sipos, *J.Water Supply: Res and Technol - AQUA.* 53, 509 (2004).
[13]   S. Spring, P. Kampfer, W. Ludnig and K.H. Schleifer, *Syst. Appl. Microbiol.* 19, 634 (1996).
[14]   D. Vandenabeele, D. de Beer, R. Germonpre and W. Verstraete, *Microb. Ecol.* 24, 19 (1992).
[15]   D.M. Sahabi, M. Takeda, I. Suzuki and J. Koizumi, *J. Biosci. Bioeng.* 107, 151(2008).
[16]   T. Hatva, *Water Sci. Technol.* 20, 141 (1988).
[17]   H. Yang, D. Li, J. Zhang, R. Hao and B. Li, *J. Environ. Sci. Health A. Tox. Hazard. Subst. Environ. Eng.* 39, 1447 (2004).
[18]   W.C. Ghiorse, *Ann. Rev. Microb.* 38, 515 (1984).
[19]   E.G. Sogaard, R. Aruna, J.A. Peskir and C.B. Koch, *Appl. Geochem.* 16, 1129 (2001).
[20]   W.L. van Veen, E.G. Mulder, M.H. Deinema, *Microb. Rev.* 42, 329 (1978).
[21]   H. Frischhertz F. Zibuschka, H. Jung and W. Zerobin, *Water Supply.* 3, 125 (1985).
[22]   J. Vandenabeele, F. Houwen, D. Van de Sande, R. Germonpré and W. Verstraete, *FEMS Microbial. Ecol.* 29, 83 (1995).
[23]   D.M. Ward, R. Weller and M.M. Bateson, *Nature.* 345, 63 (1990).

[24] G.E. Muyzer, E.C. Waal and A.G. Uitterlinden, *Appl. Environ. Microbial.* 59, 695 (1993).

[25] R.I. Amann, W. Ludwig and K.H. Schleifer, *Microbial. Rev.* 59, 143 (1995).

[26] C. Casiot, V. Pedron, O. Bruneel, R. Duran, J.C. Personne, G. Grapin, C. Drakides and F. Elbaz-Poulichet, *Chemosphere.* 64, 492 (2006).

[27] P.L. Siering and W.C. Ghiprse, *Appl. Environ. Microbiol.*63, 644 (1997).

[28] H.L. Ehrlich, W.J. Ingledew and J.C. Salerno, iron- and manganese-oxidizing bacteria. In: J.M. Shively and L.L. Barton (ed.), Academic press Inc., San Diago, Califonia. 147 (1991).

[29] I.A Katsoyiannis and A.I. Zouboulis, *Water Qual. Res. J. Can.* 41, 117 (2006).

[30] D. Emerson, Microbial oxidation of Fe(II) and Mn(II) at circumneutral pH. In: D.R. Lovley (ed.), *Environmental microbe-metal interactions.* ASM press, Washington, D.C. 31 (2000).

[31] S.F. Tyrrel and P. Howsam, *Biofouling* 8, 65 (1994).

[32] S.M. Webb, G.J. Dick, J.R. Bargar and B.M. Tebo, *Microbialogy.* 102, 5558 (2005).

[33] B.M. Tebo, W.C. Ghiorse, L.G. van waasbergen, P.L. Siering and R. Caspi, *Rev. Mineral.* 35, 225 (1997).

[34] M.A. Kessick, *Environ. Sci. Technol.* 9, 157 (1975).

[35] M. Edwards, *J. AWWA.* 86 (9), 64 (1994).

[36] J.F. Ferguson and J. Davis, *Water Res.* 6, 1259 (1972).

[37] W.R. Cullen, K.J. Reimer, *Chem. Rev.* 89, 713 (1989).

[38] F.W. Pontius, K.G. Brown and C.J. Chen, *J. AWWA.* 86 (9), 52 (1994).

[39] WHO, Arsenic Compounds Environmental Health Criteria, 2nd ed., World Health Organisation, Geneva, 224 (1996).

[40] European Council Directive 98/83/EC, Off. J., 330 32 (1998).

[41] U.S. EPA, National Primary Drinking Water Regulation, Federal Register, 66, 6976, (2001).

[42] E.O. Kartinen and C.J. Martin, *Desalination.* 103, 78 (1995).

[43] M.A. KHAN and Y. HO, *Asian J. Chemistry.* 23, 1889 (2011).

[44] I.A. Katsoyianni and A.I. Zouboulis, Water Res. 36, 5141(2002).

[45] I.A. Katsoyiannis and A.I. Zouboulis, *Rev. Environ. Health.* 21, 25 (2006)

[46] O.X. Leupin, S.J. Hug and A.B.M. Badruzzaman, *Environ. Sci. Technol.* 39, 8032 (2005).

[47] D.W. Oscarson, P.M. Huang, C. Defosse, A. Herbillon, *Nature.* 291,51 (1981).

[48] M.L. Pierce and C.B. Moore, *Water Res.* 16, 1247 (1982).

[49] D. Mohana and C.U. Pittman Jr., *J. Hazard. Mater.* 142, 1 (2007).

[50] K. Chojnacka, A. Chojnacki and H. Go´recka, *Chemosphere.* 59 75 (2005).

[51] B. Volesky and Z.R. Holan, *Biotechnol. Prog.* 11, 235 (1995).

[52] P. Mondal, C.B. Majumder.and B. Mohanty, *J. Hazard. Mater.* B137, 464 (2006).

[53] G.F. Lehimas, J.I. Chapman and F.P. Bourgine, www.saur.co.uk/poster /html 10/2009.

[54] I.A. Katsoyiannis, A.I. Zouboulis, H. Althoff and H. Bartel, *Chemosphere.* 47, 325 (2002).

[55] I.A. Katsoyiannis and A.I. Zouboulis, *Water Res.* 38, 1922 (2004).

[56] I.A. Katsoyiannis and A.I. Zouboulis, *Water Res.* 38, 17 (2004).

[57] I.A. Katsoyiannis, A.I. Zouboulis and M. Jekel, *Ind. Eng. Chem. Res.* 43, 486 (2004).

[58] I.A. Katsoyiannis, A. Zikoudib and S.J. Hug, *Desalination.* 224, 330 (2008).

[59] N. Korte, *Environmental Geology and Water Science*. 18 (2), 137 (1991).

[60] L.C. Anderson and K.W. Bruland, *Environmental Science and Technology*. 25 (3), 420 (1991).

[61] M. Edwards, *J. AWWA*. 86, 64 (1994).

[62] H. Cheng, Y. Hu, J. Luo, B. Xu and J. Zhao, *J. Hazard. Mater*. 165, 13 (2009).

[63] H. Yang, D. Li, J. Zhang, R. Hao and B. Li, *J. Environ. Sci. Health A. Tox. Hazard. Subst. Environ. Engrg*. 39, 1447 (2004).

[64] E. Deschamps, V.S.T. Ciminelli, P.G. Weidler and A.Y. Ramos, *Clay Clay Miner*. 51, 17 (2003).

[65] E.O. Kartinen and C.J. Martin (Jr), *Desalination*. 103, 78 (1995).

[66] D.M. Sahabi, M. Takeda, I. Suzuki and J. Koizumi, *J. Hazard. Mater*. 168, 1310 (2009).

[67] M. J. Scott and J.J. Morgan, *Environ. Sci. Technol*. 29, 1898 (1995).

[68] E. Deschamps, V.S.T. Ciminelli, P.G. Weidler and A.Y. Ramos, *Clay Clay Miner*. 51, 197 (2003).

[69] Y. Khambhaty, K. Mody, S. Basha and B. Jha, *Chem. Eng. J*. 145, 489 (2009).

[70] I. Ruzic, *Marine Chem*. 53, 1 (1996).

[71] G. Sposito, *J.Soil Sci. Soc. Am*.44, 652 (1980).

[72] C.H. Giles, A.P. D'Silv and I.A. Easton, *J Colloid Interface Sci*. 47, 766 (1974).

[73] A.B. P´erez-Marın, V.M. Zapata, J.F. Ortuño, M. Aguilar, J. S´aez and M. Lor´ens, *J. Hazard. Mater*. 39, 122 (2007).

[74] W. Driehaus, R. Seith and M. Jekel, *Water Res*. 29, 297 (1995).

[75] B.A. Manning, S.E. Fendorf, B. Bostick and D.L. Suarez, *Environ. Sci. Technol*. 36, 976 (2002).

[76] S. Bajpai and M. Chaudhuri, *J. Environ. Eng*. 125, 782 (1999).

[77] D. Dong, Y.M. Nelson, L.W. Lion, M.L. Shuler and W.C. Ghiorse, *Water Res*. 34, 427 (2000).

[78] Y. Li, F. Yang, D. Dong, Y. Lu, and S. Guo, *Chemosphere*. 62, 1709 (2006).

[79] D.M. Sahabi, M. Takeda, I. Suzuki and J. Koizumi, *J. Environ. Eng*. 136, 496 (2009).

In: Arsenic
Editor: Andrea Masotti

ISBN: 978-1-62081-320-1
© 2013 Nova Science Publishers, Inc.

Chapter 13

ARSENIC ADSORPTION, REMOBILIZATION, AND REDOX TRANSFORMATION ON NANOCRYSTALLINE TITANIUM DIOXIDE IN THE PRESENCE OF SULFATE REDUCING BACTERIA

Ting Luo and Chuanyong Jing[*]

State Key Laboratory of Environmental Chemistry and Ecotoxicology,
Research Center for Eco-Environmental Sciences,
Chinese Academy of Sciences, Beijing, China

ABSTRACT

Arsenic (As) contamination poses severe health risks worldwide. Highly efficient removal of As is of great urgency and high priority. Recently, nanocrystalline titanium dioxide (TiO_2) has been successfully applied to remove As in groundwater and industrial wastewater. The molecular-level study of As removal using TiO_2 and the redox transformation and remobilization of adsorbed As in the presence of sulfate reducing bacteria (SRB) are crucial to understand the speciation and fate of As. The synchrotron-based techniques including EXAFS, NEXAFS, and STXM provide molecular-scale approaches to study the As biogeochemical processes at microbe–TiO_2 interfaces.

As contamination from copper smelting is one of the major anthropogenic sources. An average of 3890 ± 142 mg/L As(III) at pH 1.4 in copper smelting wastewater was reduced to 59 ± 79 μg/L with TiO_2. The As(III) adsorption on TiO_2 followed pseudo second-order rate kinetics. EXAFS results demonstrate that As(III) form bidentate binuclear surface complexes as evidenced by an average Ti-As(III) bond distance of 3.35 Å, and the dominant surface species was $(TiO)_2AsO$-. The pH edge and adsorption behavior of As on TiO_2 can be precisely described with the charge distribution multisite surface complexation model. The security of spent TiO_2 with elevated As(V) presents an obvious concern when the nanomaterial assemblage went through oxic to anoxic environment. Microbial activities are ubiquitous in the geochemical environment, and SRB play an essential role in redox transformation and remobilization of adsorbed As. The As(V) on spent TiO_2 was released to aqueous solution and subsequently reduced to As(III) in the biotic incubation with SRB. The sulfate reduction was also observed during the incubation. STXM combines the high spatial resolution (≤ 50 nm) image and the chemical sensitivity as characterized with NEXAFS. By imaging at different energies and measuring the variations in X-ray absorption, a certain amount of As(III) was observed on TiO$_2$ surface. Therefore, STXM illustrates that adsorbed As(V) could also be reduced to As(III) on TiO$_2$ surface in the presence of SRB. The insight of interactions at As-microbe-nanomaterial interfaces will shed further light on predicting As redox transformation and mobilization.

INTRODUCTION

Arsenic (As) as a toxic metalloid distributes in the earth's crust worldwide. As contamination in groundwater and industrial wastewater has attracted public concerns. Naturally-occurring groundwater As with elevated concentrations has been reported in the Bangladesh, India, USA, Argentina, China, Chile, Taiwan, Mexico, Canada, Vietnam and Japan [1]. The largest population at risk with known groundwater As contamination is reported in Bangladesh [2]. Moreover anthropogenic activities such as discharging industrial wastewater also lead to As contamination of air, water, and soil [3]. Prolonged As exposure may lead to chronic arsenicosis, cancers of the liver, bladder, and lung, and also causes cardiovascular disease and inhibits the mental development of children [4, 5].

To date, nano TiO$_2$, as a promising adsorbent, has been applied to remove As from groundwater and industrial wastewater [6-8]. The adsorption and desorption behaviors of As on TiO$_2$ surface will provide fundamental knowledge for engineering applications. In natural environment, the adsorbed As on nano TiO$_2$ will undergo the transition from oxidized arsenate [As(V)] form to reduced arsenite [As(III)] species. The redox potential and microorganisms play an important role in the process of As transformation and mobilization. The potential As release and accumulation in aqueous phase present a challenge for natural environment and human health.

In recent years, synchrotron based techniques including soft X-ray scanning transmission X-ray microscopy (STXM) have been developed for associated imaging and spectroscopic characterization of contaminant-microbe-metal oxides in the presence of water [9, 10]. STXM combines high spatial resolution (≤ 50 nm) with the chemical sensitivity of near-edge X-ray absorption fine structure (NEXAFS) spectroscopy. Therefore, it can characterize the elemental composition and distribution of minerals, and speciation of sorbed metals compared with conventional techniques [11-13]. This chapter investigates the As removal using nano TiO$_2$ and the redox transformation and mobility of adsorbed arsenic on nano TiO$_2$ with sulfate reducing bacteria, and offers new cognition for the application of arsenic removal technique.

RESULTS AND DISCUSSION

1. Mechanism of as Removal from Copper Smelting Wastewater Using Nano TiO$_2$

1.1. As(III) Removal from Copper Smelting Wastewater Using TiO$_2$

Arsenic from copper smelters has caused increasing public concern because of its high concentrations and apparent carcinogenicity [14, 15]. Conventional neutralization and precipitation are the most widely used technologies in these industries. However, due to the large volume and the instability of the treatment sludge, innovative technology for As-containing metallurgical industry wastewater is highly desired.

Figure 1 shows a novel As removal and recovery technique using TiO_2. The results of As removal from raw water obtained from a copper smelting company with three successive adsorption are shown in Figure 2. As(III) was the only As species in the wastewater, and the initial As(III) concentration at 3,310 mg/L was reduced to 27 μg/L after three consecutive adsorption reactions. The results reveal that TiO_2 is an effective adsorbent that could be used in the remediation of acidic metallurgical wastewater with As(III) concentrations at the gram per liter level.

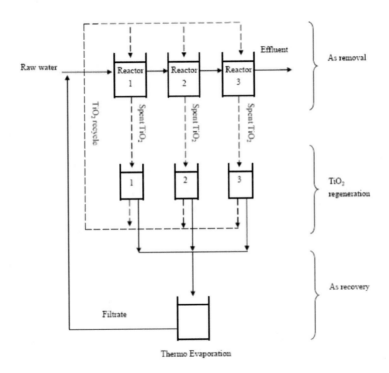

Figure 1. Schematic flowchart of As removal and recovery process from the copper smelter wastewater.

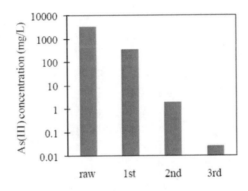

Figure 2. Concentrations of As(III) in raw water and in three consecutive adsorption reactions. Initial As concentration was 3,310 mg/L in raw water; TiO_2= 30 g/L in each adsorption; pH=7.

The XPS spectra of As 3d in spent TiO_2 samples are shown in Figure 3. The binding energy of the As 3d peak was at 44.6 eV, which is in agreement with previous reported values for As(III) at 44.6 ± 0.13 eV [16]. Conversely, no As(V) 3d peak at 45.6 ± 0.2 eV [17] was observed, indicating that As(III) is the only As species on spent TiO_2.

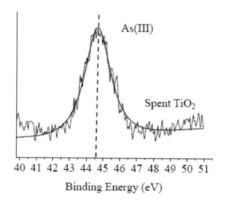

Figure 3. High-resolution XPS survey of As 3d for spent TiO_2. Vertical lines indicate As(III) binding energies at 44.6 ± 0.1 eV.

1.2. Adsorption Kinetics

The adsorption kinetics results in Figure 4 show that the adsorption rate of As(III) was rapid during the initial 2 min, and As(III) concentration kept relatively constant thereafter. A pseudo-second order kinetic model has been applied previously to describe As(III) adsorption on TiO_2 [18], and is expressed as $t/q_t = t/q_e + 1/k \cdot q_e2$, where q_t and q_e are the amounts (mg/g) of As(III) adsorbed at time t and at equilibrium, respectively, and k (g/mg·h) is the equilibrium rate constant of pseudo-second-order adsorption. The model parameters inserted in Figure 4 were obtained by linear regression of the integrated rate equation (R>0.999). The adsorption capacity q_e (104 mg/g) is an order of magnitude higher than that of a previous report (7.1 mg/g) in which a synthetic groundwater with 2 mg/L As(III) was adsorbed using 0.2 g/L TiO_2 [18]. The higher As adsorption capacity might due to the higher mass ratio of As:TiO_2 (1:10) in the present study than that in the previous report (1:100) [18]. The higher As load (1.7 mmol/g) than that in the previous report (0.13 mmol/g) could be able to occupy more surface sites on TiO_2 (6 mmol/g), which resulted in a higher adsorption capacity. This observation is in agreement with a previous report [19].

1.3. Surface Complexes of As(III) on TiO₂ and CD-MUSIC Model

Figure 5 shows the k^3-weighted As K-edge EXAFS spectra obtained for wastewater As(III) adsorption on TiO_2. The first shell in Figure 5B was the As(III)-O bond with a distance of 1.78 Å. The average coordination number (CN) of oxygen was calculated to be 2.9 for As(III). The second peak was attributed to Ti with interatomic As-Ti distances of 3.34 Å. Fitting the second peak was completed in both k-space and R-space using a single Ti shell, resulting in a CN of 1.6. The EXAFS results show that As(III) formed bidentate binuclear inner-sphere complexes on the surfaces of TiO_2. The As(III) surface configuration is in line with a previous EXAFS study of As(III) adsorption on TiO_2 with a bidentate binuclear configuration [18].

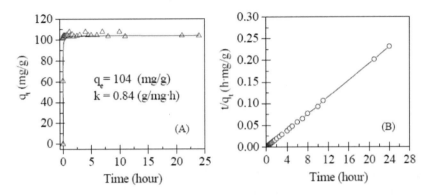

Figure 4. Adsorption kinetics of As(III) in raw water on TiO_2. Symbols are experimental data and the solid lines represent the pseudo-second-order kinetic model simulation (A). The linear regression (R>0.999) and the best-fit parameters is shown in (B). Initial As concentration was 3,310 mg/L in raw water; TiO_2=30 g/L; pH= 7.

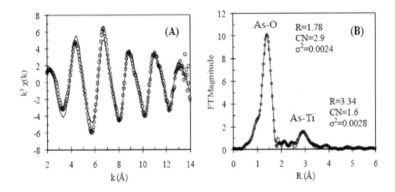

Figure 5. k^3-weighted observed (dotted line) and model calculated (solid line) As K-edge EXAFS spectra (A) and Fourier transform magnitude (B) resulting in a radial distance structure for As(III) adsorption on TiO_2. Initial As(III) concentration was 3,310 mg/L in raw water. TiO_2=30 g/L.

The results of the As(III) adsorption edge experiments, which determined the percentage of As(III) adsorbed as a function of equilibrium pH, and CD-MUSIC model calculations are shown in Figure 6. The experimental observations show that the As(III) adsorption increased from 78.8% to 98.2% when pH was increased from 3 to 9. Based on the constraint of As(III) bidentate binuclear surface configurations obtained from the EXAFS study of As(III) adsorption on TiO_2, the CD-MUSIC model described the As(III) adsorption behavior reasonably well. At high As(III) concentrations, bidentate surface configuration is preferable where an As atom shares two oxygens with two Ti atoms. This structure preference introduces less charge on the surface, and results in a more stable configuration than monodentate [20]. Additional batch adsorption experiments were performed by adding increasing amounts of TiO_2 into raw water at various pH values and the observed and CD-MUSIC results are shown in Figure 7. The experimental and model results suggest that a minimum dosage of 30 g/L TiO_2 was required to achieve more than 90% As(III) removal at pH 7.

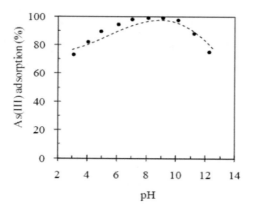

Figure 6. Experimental results (dots) and CD-MUSIC modeling (dashed line) of As(III) adsorption as a function of equilibrium pH. Initial As(III) concentration was 3,310 mg/L in raw water. TiO$_2$=30 g/L.

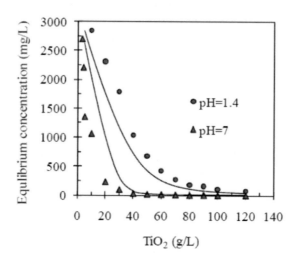

Figure 7. Adsorption of As(III) from raw water as a function of TiO$_2$ concentration at pH values of 1.4 and 7. Symbols are the experimental data and lines represent CD-MUSIC model results. Initial As concentration was 3,310 mg/L in raw water.

2. TiO$_2$ REGENERATION AND AS RECOVERY FROM NANO TiO$_2$

Spent TiO$_2$ can be reused in adsorption process after regeneration using NaOH extraction. We used 5 M NaOH to regenerate spent TiO$_2$ at a 100 mL NaOH to 30 g spent TiO$_2$ ratio and could extract more than 60% adsorbed As(III) on spent TiO$_2$ (data not shown). The incomplete extraction of As(III) may be attributed to the strong binding of As(III) on TiO$_2$ and possible formation of As(III) coprecipitates with coexisting cations at high pH. The regenerated TiO$_2$ was then reused to remediate a mixed solution of raw water and As recovery waste with an average As(III) concentration of 3890±142 mg/L. The As(III) desorption ratio in the 21 treatment cycles was in the range of 55-87% (Figure 8).

In these cycles the efficiency of regenerated TiO₂ did not decrease (Figure 9). The treatment process resulted in an average residual As concentration of 59 ± 79 μg/L, which is far less than the regulatory discharge limit of 500 μg/L [21]. The experimental and CD-MUSIC model results in Figure 9 suggest that the As(III) removal percentage was 81±6% in the first TiO₂ adsorption step, 98±2% in the second step, and 99±1% in the third step. The slightly reduced As(III) adsorption percentage in the first step may be attributed to the complicated composition of the raw water, with other species competing for the available surface sites. Besides the As(III) removal, heavy metals can be effectively removed in each cycle of the 21 treatment cycles, and the effluent Cu, Pb, and Cd concentrations were less than 0.02 mg/L (data not shown). The results suggest that TiO₂ can be used as an effective adsorbent for heavy metal removal from highly contaminated copper smelting wastewater.

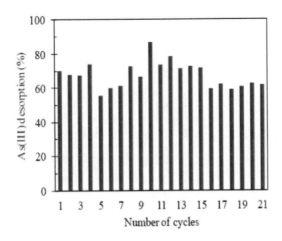

Figure 8. The desorption of As(III) from spent TiO₂ extracted with 5 M NaOH in 21 treatment cycles.

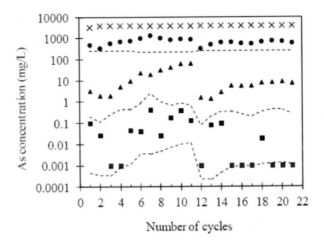

Figure 9. As(III) concentrations in raw water (✕), after the first adsorption (●), after the second adsorption (▲), and after the third adsorption (■) as a function of TiO₂ recycle times. Lines are results of CD-MUSIC simulations.

As(III) recovery was performed after regeneration of spent TiO_2. The average concentration of As(III) in the waste solution resulted from the TiO_2 regeneration process was 20,985 mg/L. As(III) was concentrated by heating at 70 °C and subsequently recovered by precipitation. The results demonstrate that As(III) was the primary As species, and little As(III) was oxidized to As(V) during the recovery process (Figure 10). After the solution cooled down to room temperature, 60~62% As(III) was recovered as precipitate based on the mass balance calculation.

The EDX analysis demonstrates that the predominant elements in the recovered solid were As, Na, and O (Figure 11A). The As 3d XPS spectrum shows a peak around 44.3 eV binding energy (Figure 11B), which is in good agreement with the peak of As 3d in $NaAsO_2$ at 44.2~44.3 [22, 23], suggesting the precipitation of sodium arsenite. The precipitate could be used as an intermediate for sodium arsenite chemical refinement. The residual liquid after recovery was recycled and mixed with raw water as influent in the TiO_2 adsorption process, and its contribution to the total As(III) concentration was less than 20%.

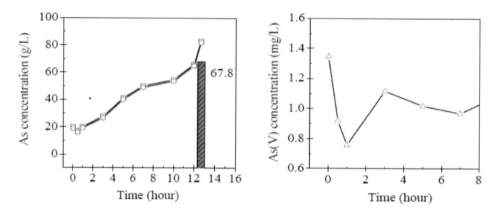

Figure 10. Concentrations of total As (□), As(III) (○), and As(V) (Δ) in the recovery process. The column shows the As(III) concentration in the residual liquid after recovery.

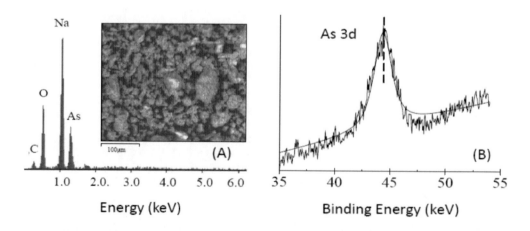

Figure 11. EDX spectrum with inserted SEM image (A) and XPS spectrum of As 3d (B) for recovered sodium arsenite solid sample.

3. ARSENATE REDUCTION AND MOBILITY ON NANO TiO$_2$ IN THE PRESENCE OF SULFATE REDUCING BACTERIA

After effectively adsorb As from water, the nano TiO$_2$ with elevated As content has a potential risk to release As and re-contaminate surface and groundwater. As cycling is controlled by microorganism, redox potential, pH, as well as adsorption, desorption, and coprecipitation on a variety of natural and engineered nano particles in natural and engineered environment [24]. SRB appears to be a dominant influence on determining whether As accumulates to hazardous levels in anaerobic conditions [25]. The sulfide produced during microbial sulfate reduction can form sulfide minerals, such as orpiment (As$_2$S$_3$) and realgar (AsS), which removes As from aquatic environment [26]. Therefore, the fate of As on the spent adsorbent in the natural environment, especially in the sulfate reducing condition, will be affected extensively by SRB. However, there is a paucity of data regarding biogeochemical processes leading to As release from the spent nano TiO$_2$ under reducing conditions, and the relevant mechanisms are not clear till now. To investigate the hypothesis that adsorbed As(V) on nano TiO$_2$ could be released and reduced to As(III) in the presence of SRB, we conducted anaerobic incubation experiments. The nano TiO$_2$ with pre-adsorbed As(V) was incubated with two SRB, *Desulfovibrio vulgaris* strain 7757 purchased from ATCC and *Desulfovibrio vulgaris* DP4 isolated from a soil sample collected at a naturally-occurring As contaminated site in China.

3.1. Sulfate Reduction and Sulfide Production

The sulfate concentrations decreased driven by SRB in aqueous solution over times, and sulfide accumulated gradually due to the reduction of sulfate (Figure 12). In the As load of 300 mg-As/g-TiO$_2$, sulfate contents decreased from about 160 mg/L to 10 mg/L with ATCC strain 7757 and 12 mg/L with strain DP4. In the As load of 5700 mg/g, sulfate content decreased from about 160 mg/L to 12 mg/L with strain 7757 compared with 11 mg/L with strain DP4. The sulfide was observed in aqueous solution, and levels increased to about 70 mg/L in these two As load systems with strain 7757 and strain DP4. Whereas sulfate contents in high As load (5700 mg/g) decreased more slowly than the low As load (300 mg/g). Kocar et al. [27] also reported that sulfate reduction occurred more slowly at high As concentrations (50% of the adsorption maximum) than the low As concentrations (10% of the adsorption maximum) with *Desulfovibrio vulgaris* (Hildenborough) in sulfate reducing environment.

3.2. Reduction and Desorption of Arsenate with Sulfate-Reducing Bacteria

The adsorbed As(V) on nano TiO$_2$ was reduced to As(III) with strain 7757 and strain DP4, while As(III) was not observed in abiotic controls (Figure 13A, B, C and D). The desorption of As(V) from nano TiO$_2$ was promoted by SRB relative to abiotic controls (Figure 13C and D). Stable dissolved As(V) concentrations in abiotic controls were around 8 mg/L in the As load of 300 mg/g compared with 16 mg/L in the As load of 5700 mg/g. The As(V) concentrations released to aqueous in biotic experiment were two times higher than that in abiotic controls. Compared with the As load 300 mg/g, the dissolved As(V) concentrations in As load 5700 mg/g were 27% higher (Figure 13 C and D).

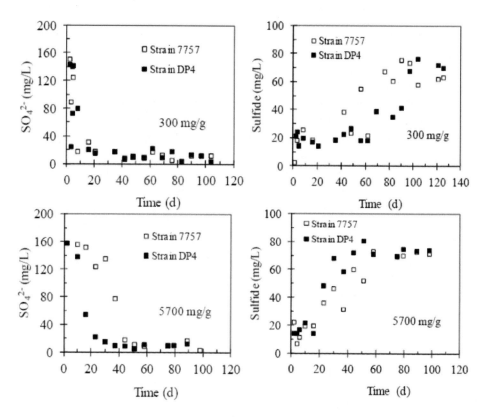

Figure 12. Dissolved sulfate and sulfide concentrations as a function of incubation time for low (300 mg/g) and high (5700 mg/g) As loads. Two sulfate reducing bacteria strains: *Desulfovibrio vulgaris* (strain 7757) obtained from ATCC, and *Desulfovibrio vulgaris* (strain DP4) isolated from an As contaminated soil.

The percentages of soluble As(III) and As(V) to total dissolved As are shown in Figure 13E and F. The change of percentage as a function of time apparently shows two different stages in terms of incubation time. In the initial stage (before 20 d), the decrease in As(V) coupled to the As(III) increase in low As load. The significant correlations between As(III) and As(V) concentrations were observed (R=0.87, p<0.05 for strain DP4, and R=0.75, p<0.05 for strain 7757) (Figure 14 A). In high As load, As(V) increased with the As(III) decrease in the initial stage. Statistically significant correlations between As(V) and As(III) were also detected (R= 0.92, p<0.01 for strain DP4, and R=0.86, p<0.05 for strain 7757) (Figure 14C). The increased As(V) percentage might be attributed to the faster rate of As(V) desorption than that of As(V) reduction. The results indicate that As(V) desorption before its reduction to As(III) in the aqueous phase was the dominant process, with a minor As(V) reduction on the solid phase in the initial stage. In the latter incubation stage, the reduction of As(V) to As(III) reached equilibrium because soluble As(V) and As(III) kept relative constant in low As load. Meanwhile no correlation (p>0.05) was detected between the As(III) and As(V) levels (Figure 14B). In high As load, the reduction of As(V) to As(III) continuously occurred as evidenced by the slight decrease of As(V) and increase of As(III). However, there was no statistically significant correlation (p>0.05) between As(V) and As(III) concentrations, indicating that concurrent redox reactions in the aqueous and solid phases regulated the As release in As-5700 in the latter stage (Figure 14D).

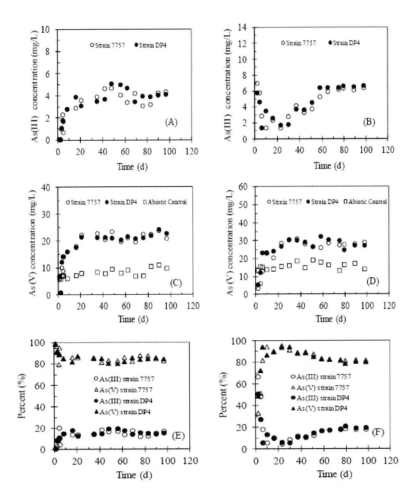

Figure 13. Dissolved concentrations of As(III) and As(V) during the incubation experiment. The As loads on TiO_2 were 300 mg/g (A, C and E), and 5700 mg/g (B, D and F). As(III) and As(V) as percentage of total dissolved As concentrations (E and F). Two sulfate reducing bacteria strains: *Desulfovibrio vulgaris* (strain 7757) obtained from ATCC, and *Desulfovibrio vulgaris* (strain DP4) isolated from an As contaminated soil.

The *Desulfovibrio vulgaris* can not directly reduce As(V) [28], whereas the biogenic sulfide can reduce As(V) to As(III) [29]. However, where the reduction of As(V) to As(III) occurs, in aqueous or solid phase, is a controversial issue [30, 31]. A previous study found that the As(V) adsorbed on iron oxyhydroxides was reduced to As(III) on solid surfaces [31]. To the contrary, Huang et al. found that the As(V) desorption from iron minerals occurred followed by reduction of dissolved As(V) with *Shewanella putrefaciens* strain CN-32 [30]. The author concluded that the reduction of adsorbed As(V) was insignificant or at least much slower than that of dissolved As(V). These studies concentrated on the As-Fe oxides-bacteria system. No investigation of As fate and transformation in As-Ti oxides-bacteria system has be reported.

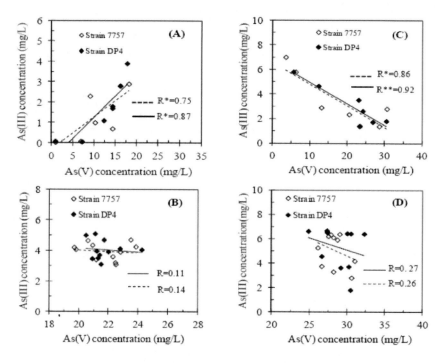

Figure 14. The correlation between dissolved As(III) and As(V) concentrations in low As load (300 mg/g) (A and B) and high As load (5700 mg/g) (C and D) in the presence of strain 7757 (◊) and DP4 (♦). The relationship between dissolved As(III) and As(V) concentrations in initial stage (A and C), and in latter stage (B and D). *: $p<0.05$; **: $p<0.01$. Solid (dashed) line represents regression for strain DP4 (strain 7757).

The As(III) concentrations were up to 7 mg/L in high As load compared with 0.06 mg/L in low As load after 2 d incubation (Figure 13A and B), because the initial levels of dissolved As(V) were higher in high As load than that in the low As load. The sharp drop of As(III) concentrations in high As load in the initial incubation stage (Figure 13B) might be attributed to the concurrent results of As transformation with biogenic sulfide production. Due to the presence of sulfide, the formation of As sulfide precipitates such as orpiment (As_2S_3) [32, 33] in the As-S aqueous solution is possible, which regulates the transport and fate of As under sulfate reducing conditions. At high ratio of As:S, As_2S_3 precipitates were observed in *Desulfotomaculum auripigmentum* cultures by Newman et al. [34]. They reported that the formation of As_2S_3 was occurred, and they further stated that As_2S_3 precipitation occurred only at high ratio of As(III):S^{2-} [34]. Previous research reported that iron oxides in aquifer sediments regulate As levels by formation of As-associated pyrite or arsenopyrite under reducing conditions in the presence of sulfide [35, 36]. However, limited iron source was available in our experiment. Thus, dissolved H_2S increased to high concentrations due to the lack of effective buffer by pyrite precipitation. With the further increase of H_2S concentrations, appreciable amount of As_2S_3 precipitates might become soluble because of the formation of arsenic-sulfide aqueous complexes [37].

Compared with the abiotic control, a substantial increase of dissolved As(V) was observed in the presence of SRB (Figure 13C and D). The increased desorption of As(V) might relate to bacteria-mineral surface interactions [30]. In agreement with our results, As(V) release from goethite was accelerated by *Shewanella putrefaciens* strain CN-32 [30]. The bacteria with highly ionizable functional groups such as phosphate and carboxylate can preferably attach to mineral surfaces to compete with As(V) for available surface sites. Moreover the terminal phosphate/phosphonate and phosphodiester groups on cell were involved in bacterial adhesion to metal-oxides through formation of innersphere P-O-Metal complexes [38].

3.2. STXM

Synchrotron-based STXM is one of the few methods capable of characterizing particles >30 nm in diameter in the presence of water [9, 10]. Soft X-ray spectromicroscopy has been used to explore the interfacial chemistry of microorganisms, trace and major element geochemistry [39]. Application of *in situ* X-ray spectroscopy methods will facilitate a significant advance in the current understanding of the bacterial-contaminant-metal oxides interactions at the molecular level. STXM can provide information on the impact of microbes on metal redox speciation using the $L_{2,3}$ edges of transition metals (e.g., Mn, Fe, Cr) or metalloids (e.g., As). Thus the high spatial resolution and spectra sensitivity are advantages when compared with conventional microscopic methods.

Currently, STXM beamlines at synchrotron radiation sources are available worldwide, including X1A1(250-500 eV) and X1A2 (250-1000 eV) at National Synchrotron Light Source (NSLS); BL 5.3.2 (250-700 eV) at Advanced Light Source (ALS); ID12 (250-600 eV) at Bessy II, Germany; ID 21 (2000-8000 eV) at European Synchrotron Radiation Facility (ESRF); 2-ID-B (1-4 keV) at Advance Photon Source (APS); 8A (250-1000 eV) at Pohang Accelerator Laboratory; and 08U (250-2000 eV) at Shanghai Synchrotron Radiation Facility (SSRF) [10].

In this chapter, we studied As speciation and distribution on TiO_2 in the initial incubation stage with STXM facility at SSRF. Based on the analysis of dissolved As speciation and concentration in aqueous solution, we deduced that the reduction of As(V) to As(III) induced by SRB occurred predominately in the aqueous phase. To exemplify the hypothesis that the adsorbed As(V) can also be reduced on the TiO_2 surface in the initial stage, STXM was employed in our study.

NEXAFS spectra of reference standards and STXM images are shown in Figure 15. The As L_3-edge NEXAFS spectra of $NaH_2AsO_4 \cdot 7H_2O$ and $NaAsO_2$ are depicted in Figure 15A. The As L_3-edge spectra of As(III) shows a single peak at 1339.4 eV, while an additional peak at 1342.2 eV can be observed for As(V). Based on the energy difference of As species in NEXAFS spectra, their spatial distribution in TiO_2-As-containing bacteria suspension samples could be evaluated by STXM image.

Detailed STXM images generated the spatial distribution of As(III) and As(V) in the TiO_2 suspension, by analyzing the variations in X-ray absorption below the As L_3-edge at 1335.2 eV, at the absorption peak for As(III) and As(V) (1339.4 eV), and at absorption peak only for As(V) (1342.2 eV).

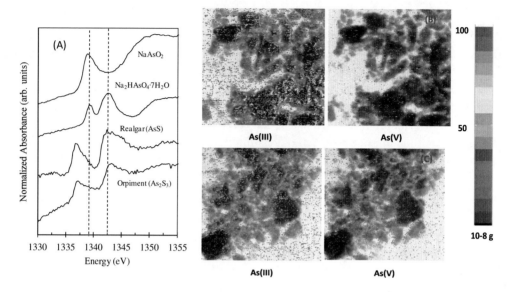

Figure 15. STXM analyses at As L_{III}-edge for As-TiO$_2$-SRB samples at 4 d incubation time. NEXAFS spectra of As(III) and As(V) reference at the As-L_{III} edge, and dashed lines indicate energy positions at around 1339.4 and 1342.2 eV (A). STXM image of low As load (300 mg/g) with strain DP4 (B), and high As load (5700 mg/g) with strain DP4 (C). The STXM images were taken at E =1339.4 eV for As(III) and As(V) and E=1342.2 eV for As(V) alone.

Images of TiO$_2$-As-containing bacteria suspension with two different As loads (300 mg/g with strain DP4, and 5700 mg/g with strain DP4) show that appreciable amount of As(III) were present on the TiO$_2$ surface as early as at 4 d incubation (Figure 15B and C). The results indicate that the reduction of As(V) to As(III) occurred also on the TiO$_2$ solid in the initial stage. It is an invalid assumption that the dissolved As(III) would readsorbed on TiO$_2$ because the As(III) mobility is greater than As(V) [40]. The negatively charged As(V) species can form bidentate binuclear complex on metal oxides surface, resulting in a higher affinity for metal oxides than the neutral As(III) species [41, 42].

There are extensive studies of As speciation with X-ray absorption spectroscopy techniques performed at the As K-edge of 11,867 eV [43-45]. However, investigation of interactions between As-mineral-microbes using the STXM technique at the As L$_3$-edge are rare [10, 46]. Benzerara et al. demonstrated the imaging of As speciation on Fe mineral in an acid mine drainage system with microbe using the STXM technology [46]. Here, this chapter shows that As L$_3$-edge provides clear information on As oxidation state in the samples of As containing TiO$_2$ with SRB. In such complex assemblages, simultaneous observation of As speciation and distribution on minerals accompanying microbes on the submicron scale may provide insightful knowledge about the As behavior on mineral interface.

3.3. As Solid Phase Transformations in the Reducing Condition

The As speciation on TiO$_2$ solid at the end of incubation experiment was determined by XANES analysis. XANES studies reveal that As(V) was the predominant species in TiO$_2$

solids at low As load (300 mg/g), and As(V), As(III), and orpiment were observed at high As load (5700 mg/g) (Figure 16).

The As XANES spectra were compared to those of reference arsenic oxide compounds in Figure 16. The XANES spectra of the low As load (300 mg/g) indicate that As(V) was the primarily As species in the solid, as evidenced by the same peak position as $Na_2AsO_4.7H_2O$ at 11874 eV, and no As(III) and orpiment peaks were observed. No significant orpiment formed in these samples at low As load. The XANES spectra of the high As load (5700 mg/g) show that orpiment mineral (As_2S_3) formed in the solids both with strain DP4 and strain 7757. By linear combination fitting of the XANES spectra, the solids contained 80% As(V), 5% As(III), and 15% orpiment for strain DP4, whereas 67% As(V), 4% As(III), and 29% orpiment for strain 7757.

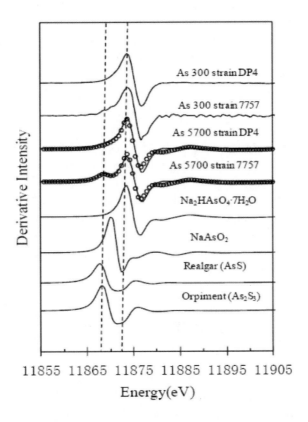

Figure 16. As k-edge first –derivative XANES spectra of experiment data (solid line) and model with linear combination fitting (dotted line). As 300: As load on TiO_2 was 300 mg/g; As 5700: As load on TiO_2 was 5700 mg/g.

CONCLUSION

A novel approach was investigated for the first time using TiO_2 for As adsorptive removal from wastewater and subsequent spent adsorbent regeneration and As recovery using NaOH. An average of 3890±142 mg/L As(III) at pH 1.4 in copper smelting wastewater was reduced to 59±79 μg/L with TiO_2. The As(III) adsorption on TiO_2 followed pseudo second-order rate kinetics. EXAFS results demonstrate that As(III) form bidentate binuclear surface complexes as evidenced by an average Ti-As(III) bond distance of 3.35 Å, and the dominant surface species was $(TiO)_2AsO^-$. The pH edge and adsorption behavior of As on TiO_2 can be precisely described with the charge distribution multisite surface complexation model. Approximately 60% As(III) in the waste solution after the TiO_2 regeneration process were recovered by thermo vaporization and subsequent precipitation of sodium arsenite, as suggested by the EDX and XPS analysis. Our innovative process could achieve adsorbent regeneration and arsenic recovery and has the advantage of producing "zero" amount of sludge. It resolves a difficult environmental problem about the secondary pollution from the huge volumes of sludge, which prevails in the smelting industry and causes a significant proportion of overall treatment costs.

In anaerobic incubation experiments, milligram levels of dissolved As(III) was observed in low As load (300 mg/g) and high As load (5700 mg/g) with *Desulfovibrio vulgaris* ATCC 7757 and *Desulfovibrio vulgaris* DP4. The process of As(V) release and reduction can be categorized into initial and latter stages which separated at 20 d. Adsorbed As(V) released from TiO_2 surface, and subsequently was reduced to As(III) in aqueous solution in initial stage. The reduction of As(V) reached equilibrium in low As load and had a minor effect on As cycling in high As load in the latter stage. STXM results reveal that As(V) reduction also occurred on the TiO_2 solid in the initial stage. The release of adsorbed As(V) from nano TiO_2 was promoted by SRB relative to abiotic controls. The dissolved As(V) concentrations in biotic experiment were two times higher than that in abiotic controls. For the solid samples at the end of incubation experiment, XANES spectra show that As(V) was the major component for low As load, and 15~29% orpiment was produced for high As load. The results provide fundamental knowledge about the As mobilization at the nano adsorbent and water interface and the security of adsorbed As on the nano material when the environment changed from oxic to anoxic. Further exploration on the interactions of As-nano material-bacteria at the molecular level should gain new insight into biogeochemical cycling of As.

REFERENCES

[1] Smedley, P. L.; Kinniburgh, D. G. (2002). A review of the source, behaviour and distribution of arsenic in natural waters. *Appl. Geochem.*, 17, 517-568.
[2] Fendorf, S.; Michael, H. A.; van Geen, A. (2010). Spatial and temporal variations of groundwater arsenic in south and southeast Asia. *Science*, 328, 1123-1127.
[3] Dudka, S.; Adriano, D. C. (1997). Environmental impacts of metal ore mining and processing: A review. *J. Environ. Qual.*, 26, 590-602.
[4] Chen, Y.; Ahsan, H. (2004) Cancer burden from arsenic in drinking water in Bangladesh. *American Journal of Public Health*, 94, 741-744.
[5] Wasserman, G. A.; Liu, X. H.; Parvez, F.; Ahsan, H.; Factor-Litvak, P.; van Geen, A.; Slavkovich, V.; Lolacono, N. J.; Cheng, Z. Q.; Hussain, L.; Momotaj, H.; Graziano, J. H. (2004). Water arsenic exposure and children's intellectual function in Araihazar, Bangladesh. *Environ. Health Perspect.*, 112, 1329-1333.

[6] Jegadeesan, G.; Al-Abed, S. R.; Sundaram, V.; Choi, H.; Scheckel, K. G.; Dionysiou, D. D. (2010). Arsenic sorption on TiO_2 nanoparticles: size and crystallinity effects. *Water Res.*, 44, 965-973.

[7] Luo, T.; Cui, J.; Hu, S.; Huang, Y.; Jing, C. (2010). Arsenic removal and recovery from copper smelting wastewater using TiO_2. *Environ. Sci. Technol.*, 44, 9094-9098.

[8] Bang, S.; Patel, M.; Lippincott, L.; Meng, X. G. (2005). Removal of arsenic from groundwater by granular titanium dioxide adsorbent. *Chemosphere*, 60, 389-397.

[9] Pecher, K.; McCubbery, D.; Kneedler, E.; Rothe, J.; Bargar, J.; Meigs, G.; Cox, L.; Nealson, K.; Tonner, B. (2003). Quantitative charge state analysis of manganese biominerals in aqueous suspension using Scanning Transmission X-ray Microscopy (STXM). *Geochim. Cosmochim. Acta*, 67, 1089-1098.

[10] Yoon, T. H.; Johnson, S. B.; Benzerara, K.; Doyle, C. S.; Tyliszczak, T.; Shuh, D. K.; Brown, G. E. (2004). In situ characterization of aluminum-containing mineral-microorganism aqueous suspensions using scanning transmission X-ray microscopy. *Langmuir*, 20, 10361-10366.

[11] Hunter, R. C.; Hitchcock, A. P.; Dynes, J. J.; Obst, M.; Beveridge, T. J. (2008). Mapping the speciation of iron in pseudomonas aeruginosa biofilms using Scanning Transmission X-ray Microscopy. *Environ. Sci. Technol.*, 42, 8766-8772.

[12] Dynes, J. J.; Tyliszczak, T.; Araki, T.; Lawrence, J. R.; Swerhone, G. D. W.; Leppard, G. G.; Hitchcock, A. P. (2006). Speciation and quantitative mapping of metal species in microbial biofilms using scanning transmission X-ray microscopy. *Environ. Sci. Technol.*, 40, 1556-1565.

[13] Obst, M.; Wang, J.; Hitchcock, A. P. (2009). Soft X-ray spectro-tomography study of cyanobacterial biomineral nucleation. *Geobiology*, 7, 577-591.

[14] Nordstrom, D. K. (2002). Worldwide occurrences of arsenic in ground water. *Science*, 296, 2143-2145.

[15] Lubin, J. H.; Moore, L. E.; Fraumeni, J. F. Jr.; Cantor, K. A. (2008). Respiratory cancer and inhaled inorganic arsenic in copper smelters workers: A linear relationship with cumulative exposure that increases with concentration. *Environ. Health Perspect.*, 116, 1661-1665.

[16] Lange, F.; Schmelz, H.; Knozinger, H. (1991). An X-ray photoelectron-spectroscopy study of oxides of arsenic supported on TiO_2. *J. Electron. Spectrosc. Relat. Phenom.* , 57, 307-315.

[17] Frau, F.; Addari, D.; Atzei, D.; Biddau, R.; Cidu, R.; Rossi, A. (2010). Influence of major anions on As(V) adsorption by synthetic 2-line ferrihydrite. Kinetic investigation and XPS study of the competitive effect of bicarbonate. *Water, Air, Soil Pollut.*, 205, 25-41.

[18] Pena, M. E.; Korfiatis, G. P.; Patel, M.; Lippincott, L.; Meng, X. G. (2005). Adsorption of As(V) and As(III) by nanocrystalline titanium dioxide. *Water Res.*, 39, 2327-2337.

[19] Liu, G.; Zhang, X.; Talley, J. W.; Neal, C. R.; Wang, H. (2008). Effect of NOM on arsenic adsorption by TiO_2 in simulated As(III)-contaminated raw waters. *Water Res.*, 42, 2309-2319.

[20] Hiemstra, T.; Van Riemsdijk, W. H. (1999). Surface structural ion adsorption modeling of competitive binding of oxyanions by metal (hydr)oxides. *J. Colloid Interface Sci.*, 210, 182-193.

[21] Integrated wastewater discharge standard, National Standard of the People's Republic of China. GB 8978-1996. Ministry of Environmental Protection of the People's Republic of China. 1996.

[22] Taylor, J. A. (1982). An XPS study of the oxidation of AlAs thin-films grown by MBE. *J. Vac. Sci. Technol.*, 20, 751-755.

[23] Wagner, C. D.; Riggs, W. M.; Davis, L. E.; Moulder, J. F.; Mulenberg, G. E. (1979). Handbook of X-ray photoelectron spectroscopy, Perkin-Elmer Corporation, Physical Electronics Division, Eden Prairie, Minn. 55344.

[24] Charlet, L.; Polya, D. A. (2006). Arsenic in shallow, reducing groundwaters in southern Asia: An environmental health disaster. *Elements*, 2, 91-96.

[25] Kirk, M. F.; Holm, T. R.; Park, J.; Jin, Q. S.; Sanford, R. A.; Fouke, B. W.; Bethke, C. M. (2004). Bacterial sulfate reduction limits natural arsenic contamination in groundwater. *Geology*, 32, 953-956.

[26] Rittle, K. A.; Drever, J. I.; Colberg, P. J. S. (1995). Precipitation of arsenic during bacterial sulfate reduction. *Geomicrobiol. J.*, 13, 1-11.

[27] Kocar, B. D.; Borch, T.; Fendorf, S. (2010). Arsenic repartitioning during biogenic sulfidization and transformation of ferrihydrite. *Geochim. Cosmochim. Acta*, 74, 980-994.

[28] Saalfield, S. L.; Bostick, B. C. (2009). Changes in iron, sulfur, and arsenic speciation associated with bacterial sulfate reduction in ferrihydrite-rich systems. *Environ. Sci. Technol.*, 43, 8787-8793.

[29] Rochette, E. A.; Bostick, B. C.; Li, G. C.; Fendorf, S. (2000). Kinetics of arsenate reduction by dissolved sulfide. *Environ. Sci. Technol.*, 34, 4714-4720.

[30] Huang, J. H.; Voegelin, A.; Pombo, S. A.; Lazzaro, A.; Zeyer, J.; Kretzschmar, R. (2011). Influence of arsenate adsorption to ferrihydrite, goethite, and boehmite on the kinetics of arsenate reduction by Shewanella putrefaciens strain CN-32. *Environ Sci Technol.*, 45, 7701-7709.

[31] Masscheleyn, P. H.; Delaune, R. D.; Patrick, W. H. (1991). Effect of redox potential and pH on arsenic speciation and solubility in a contaminated soil. *Environ Sci Technol.*, 25, 1414-1419.

[32] O'Day, P. A.; Vlassopoulos, D.; Root, R.; Rivera, N. (2004). The influence of sulfur and iron on dissolved arsenic concentrations in the shallow subsurface under changing redox conditions. *PNAS*, 101, 13703-13708.

[33] Moore, J. N.; Ficklin, W. H.; Johns, C. (1988). Partitioning of arsenic and metals in reducing sulfidic sediments. *Environ Sci Technol.*, 22, 432-437.

[34] Newman, D. K.; Beveridge, T. J.; Morel, F. M. M. (1997). Precipitation of arsenic trisulfide by Desulfotomaculum auripigmentum. *Appl. Environ. Microbiol.*, 63, 2022-2028.

[35] Savage, K. S.; Tingle, T. N.; O'Day, P. A.; Waychunas, G. A.; Bird, D. K. (2000). Arsenic speciation in pyrite and secondary weathering phases, Mother Lode Gold District, Tuolumne County, California. *Appl. Geochem.*, 15, 1219-1244.

[36] Kim, M. J.; Nriagu, J.; Haack, S. (2002). Arsenic species and chemistry in groundwater of southeast Michigan. *Environ. Pollut.*, 120, 379-390.

[37] Wilkin, R. T.; Wallschlager, D.; Ford, R. G. (2003). Speciation of arsenic in sulfidic waters. *Geochem. Trans.*, 4, 1-7.

[38] Parikh, S. J.; Chorover, J. (2006). ATR-FTIR spectroscopy reveals bond formation during bacterial adhesion to iron oxide. *Langmuir*, 22, 8492-8500.

[39] Tonner, B. P.; Droubay, T.; Denlinger, J.; Meyer-Ilse, W.; Warwick, T.; Rothe, J.; Kneedler, E.; Pecher, K.; Nealson, K.; Grundl, T. (1999). Soft X-ray spectroscopy and imaging of interfacial chemistry in environmental specimens. *Surf. Interface Anal.*, 27, 247-258.

[40] Pierce, M. L.; Moore, C. B. (1982). Adsorption of arsenite and arsenate on amorphous iron hydroxide. *Water Res.*, 16, 1247-1253.

[41] Manning, B. A.; Fendorf, S. E.; Goldberg, S. (1998). Surface structures and stability of arsenic(III) on goethite: spectroscopic evidence for inner-sphere complexes. *Environ Sci Technol.*, 32, 2383-2388.

[42] Sun, X. H.; Doner, H. E. (1996). An investigation of arsenate and arsenite bonding structures on goethite by FTIR. *Soil Sci.*, 161, 865-872.

[43] Pena, M.; Meng, X. G.; Korfiatis, G. P.; Jing, C. Y. (2006). Adsorption mechanism of arsenic on nanocrystalline titanium dioxide. *Environ Sci Technol.*, 40, 1257-1262.

[44] Onstott, T. C.; Chan, E.; Polizzotto, M. L.; Lanzon, J.; DeFlaun, M. F. (2011). Precipitation of arsenic under sulfate reducing conditions and subsequent leaching under aerobic conditions. *Appl. Geochem.*, 269-285.

[45] Johnston, S. G.; Keene, A. F.; Burton, E. D.; Bush, R. T.; Sullivan, L. A. (2011). Iron and arsenic cycling in intertidal surface sediments during wetland remediation. *Environ Sci Technol.*, 45, 2179-2185.

[46] Benzerara, K.; Morin, G.; Yoon, T. H.; Miot, J.; Tyliszczak, T.; Casiot, C.; Bruneel, O.; Farges, F.; Brown, G. E., Jr. (2008). Nanoscale study of As biomineralization in an acid mine drainage system. *Geochim. Cosmochim. Acta*, 72, 3949-3963.

Suggested review by
Dr. Xiaoguang Meng, Center for Environmental Systems,
Stevens Institute of Technology, Hoboken, NJ 07030, US.
Phone: 201 216 8014; Email: xiaoguang.meng@stevens.edu

In: Arsenic
Editor: Andrea Masotti

ISBN: 978-1-62081-320-1
© 2013 Nova Science Publishers, Inc.

Chapter 14

ARSENIC REMOVAL BY FUNCTIONAL POLYMERS COUPLED TO ULTRAFILTRATION MEMBRANES

Bernabé L. Rivas[*] and Julio Sánchez

Polymer Department, Faculty of Chemistry,
University of Concepción, Concepción, Chile

ABSTRACT

In northern Chile, arsenic is the most harmful environmental pollutant. The arsenic is of natural origin but is also released in the environment due to the exploitation of copper. The concentrations of arsenic in air, water, and soils in the region, exceed in some cases both the nationally and internationally established permissible limits. This may have consequent impacts on the health of the population exposed to arsenic.

Several methods are used to remove traces of arsenic from water including ion exchange resins, adsorption using especially modified chelating compounds, chemical precipitation, coagulation, and membrane processes. However, the redox- and pH-controlled diversity of arsenic species in water results in complex selectivity issues which have yet to be fully addressed.

The removal of As(III) is more difficult than As(V). As(III) in aqueous solution usually is not charged and is more difficult to remove from solution, and therefore is usually oxidized to As(V) prior to removal.

The present chapter shows that it is possible to remove arsenic from aqueous solutions as well as from natural water from northern Chile by an emerging technique called the liquid-phase polymer-based retention (LPR). The LPR technique using a ultrafiltration membrane enables the separation of arsenic ionic species, which are retained by the functional groups of water-soluble polymers. The toxic ions do not pass through the membrane and are separated from the aqueous solution.

This work shows also studies on nanocomposite electrode materials synthesized by incorporation of metal nanoparticles dispersed in conducting polymer matrix. That material presents strong electrocatalytic properties towards oxidation of As(III) to As(V). The nanocomposite films modified electrodes can be used for As(III) analysis as well as for exhaustive electrocatalytic oxidation of As(III) solutions. The use of water-soluble polymers acting both as supporting electrolyte and as extracting agent to remove As(V)

---

[*] E-mail: brivas@udec.cl.

species allowed to combine the electrocatalytic oxidation (EO) of As(III) with a LPR process to efficiently remove arsenic from water.

INTRODUCTION

The presence of arsenic at high concentration in aqueous environments is still a global environmental problem that needs to be solved. At the present, traces of arsenic in drinking water affect to people around the world, including countries like Bangladesh, India, China, Chile, Argentina, Mexico, Hungary, Taiwan, Vietnam, Japan, New Zealand, Germany, and United States among others [1].

The natural rocks weathering processes cause of high arsenic concentrations in the environment. Moreover, arsenic contamination of water and soil may result from human activities due to the disposal of industrial waste chemicals, the smelting of arsenic bearing minerals, the burning of fossil fuels, and the application of arsenic compounds in many products especially in the past few hundred years [2].

Arsenic poisoning episodes have been reported all over the world. Thousands and thousands of people are suffering from the toxic effects of arsenicals in many countries due to natural groundwater contamination as well as to industrial effluent and drainage problems [1].

The maximum permissible concentration in Europe and accepted by the World Health Organization (WHO) for arsenic in drinking water is 10 μg L^{-1} [3]. On the other hand, developing countries are struggling to find and implement systems to reach the standard of 50 μg L^{-1} in areas affected by the presence of arsenic [3, 4]. Background concentrations of arsenic in groundwater are less than 10 μg L^{-1} in most countries and sometimes substantially lower. However, values quoted in the literature show a very wide range from < 0.5 to 5000 μg L^{-1} [5].

ARSENIC IN CHILE

The water resources are limited in northern Chile. The most groundwater and surface waters are unsuitable for drinking water because they have high salinity and arsenic concentration.

The first health problems related to arsenic in Northern Chile were recognized in 1962 [5]. The symptoms included skin-pigmentation changes, keratosis, skin cancer, cardiovascular problems and respiratory disease [6, 7]. Since exposure to arsenic was mainly in the period since 1955 to 1970, this pointed to a long latency period of cancer mortality [5]. It was estimated that 7% of all deaths among 1989-1993 in northern Chile were due to past exposure to arsenic in drinking water at concentrations around of 500 μg L^{-1} [8]. Other studies related to Chilean case have shown that exposure to arsenic may give rise to several neurologic and reproductive diseases as well as lung cancer, liver cancer, bladder cancer and acute myocardial infarction, among others [9-15].

The concentrations of arsenic in air, water, and soils in the region, exceed in some cases the established permissible limits. The Chilean norm 409 for water allows a maximum concentration of 50 μg L^{-1} of arsenic in water until 2004. Currently, this rule has been modified and the goal is to achieve a concentration of 30 µg L^{-1} in 2010 and 10 µg L^{-1} in 2015 [16].

Elevated arsenic and boron concentrations as well as high salinity have been reported in surface waters and groundwaters from Region II of Northern Chile [17]. These high concentrations of arsenic are caused by local geochemistry, increased by evaporation and, in some cases, also by anthropogenic activities such as mining and mineral processing [7]. Additional arsenic exposure from smelting of copper ore has also been noted in northern Chile [5]. In this area, some plants based on the coagulation technology were implemented in 1970 in order to solve the arsenic contamination [18-20]. However, using this technology relatively large volumes of arsenic-containing sludges are formed, typically disposed off in landfills, and being a potential source of re-contamination [18].

REMOVAL OF ARSENIC

The complete arsenic removal from aqueous environments is not an easy task. Each aqueous system is different and requires a previous optimization for the removal of arsenic species.

In order to remove traces of arsenic from water various methods known like ion-exchange, adsorption (especially with reagents impregnated resins and metal-loaded chelating resins), chemical precipitation-coagulation, reverse osmosis, and complexation have been used [7, 18-23]. However, the complete extraction of arsenic from drinking water, wastewaters, and industrial effluents in order to reach acceptable levels still represents a real challenge. The redox and pH controlled diversity of arsenic species in water bring about complex selectivity issues which have not fully been addressed yet [24].

Arsenic is present in aqueous environments in different forms and mainly in the oxidation states -3, 0, 3, and 5. The most frequent arsenic species present in natural waters are the oxy-anions arsenate, As(V), and arsenite, As(III), which are related to arsenic acid (H_3AsO_4) and arsenous acid (H_3AsO_3), respectively [25-27]. However, the forms, concentrations, and relative proportions of As(V) and As(III) in water vary significantly depending on changes in input sources such as pH and oxidation potential [24]. At high redox potential, arsenic is stabilized as a series of pentavalent arsenic species: H_3AsO_4, $H_2AsO_4^-$, $HAsO_4^{2-}$, and AsO_4^{3-}; whereas under most reducing (acid and mildly alkaline) conditions and low redox potential, the trivalent arsenic species (H_3AsO_3, $H_2AsO_3^-$, $HAsO_3^{2-}$, and AsO_3^{3-}), become stable. Arsenate is less toxic than arsenite, but more abundant and more mobile in natural surface waters, whereas arsenite is found mostly in anaerobic environments such as groundwater [28].

The efficiency to remove arsenic varies according to many site-specific chemicals, geographic, and economic conditions [21]. The common technologies used for arsenic removal are efficient for arsenate, because it is present in the form of charged oxy-anions, while the arsenite is uncharged at pH below 9.2. This is the reason why many arsenic remediation methods use previously an oxidation step. However, oxidation without help of other physical or chemical processes does not remove arsenic from water [18].

The arsenic removal technologies include processes that can be used directly or in combination, such as oxidation, co-precipitation and adsorption, lime treatment, ion exchange resins and membrane technologies, among others [29, 30]. Usually, the removal of arsenic requires a previous step of oxidation in order to transform As(III) species to As(V) species. The oxidation process can be performed by direct aeration, but it is not efficient enough [31]. However, a number of compounds, including some gaseous chlorine, hypochlorite, ozone, permanganate, hydrogen peroxide, manganese oxides, and Fenton's reagent (H_2O_2/Fe^{2+}) can be employed to accelerate oxidation [32]. In all cases arsenic extraction efficiency strongly depends on the ability to convert As(III) species into the more easily extractable As(V) forms and these technologies should meet several basic technical criteria.

POLYMERIC MATERIALS FOR ARSENIC REMOVAL

Several technologies exist to remove arsenic using membrane or polymeric materials, including mainly reverse osmosis, nano, micro or ultrafiltration processes, ion exchange resins, fibers, and water-soluble polymers.

The membrane processes for remove arsenic that need low-pressure are microfiltration and ultrafiltration (UF). However, these technologies are not adequate because the arsenical species are very small and can pass through the membranes. Nanofiltration or reverse osmosis use high-pressure membranes [32]. Currently, available membranes are more expensive than other arsenic removal options but are more appropriate in municipal settings where very low arsenic levels are required.

The synthetic ionic exchange resins, generally polymers linked to charged functional groups, can be applied for arsenic removal. The quaternary amine functional groups are commonly used [21].

The ion-exchange fibers compared with traditional resins have many advantages such as higher specific surface area, smaller diameter and better elasticity. They may occur in the form of filaments and other textile articles and can be regenerated and reused [33].

Among the latest materials developed for the removal of arsenic are the water-soluble polymers, which combined with filtration membranes can remove arsenic species from aqueous solutions.

LIQUID PHASE POLYMER BASED
RETENTION (LPR) TECHNIQUE

The *liquid-phase polymer-based retention technique* (LPR) so called also as polymer assisted ultrafiltration (PAUF), polymer or polyelectrolyte enhanced ultrafiltration (PE-UF), is a hybrid method of membrane separation where the water-soluble polymer and arsenic ion are contacted on the feed-side in a filtration system.

The water-soluble polymer interacts with arsenic ions and these are bonded to the polymer resulting in polymer-arsenic ion macromolecules, which are then retained mainly by a size exclusion mechanism; whereas unbound arsenic species, with a diameter smaller than membrane cut-off diameter, pass through the membrane into the permeate stream (see Fig. 1) [34-36].

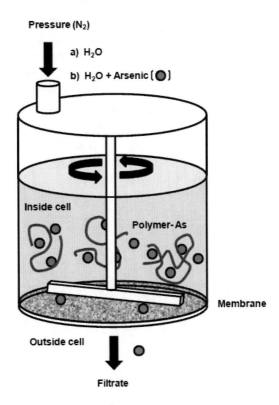

Figure 1. Procedure of arsenic removal using LPR technique. The different experiments: a) washing method, b) enrichment method.

In LPR technique, separation is a function of the interaction strength between target ions and the polymer functional groups. Thus, whereas in traditional UF, dissolved particles can be "retained" (larger than pore diameter) and "not retained" (smaller than pore diameter); in LPR, two kinds of target ions, both smaller than the membrane pore diameter, can be defined: "free target ions in solution" (not retained) and "bound to the polymer" (retained). In other words, in LPR, the retention of the species of interest depends on its interaction with the polymer and is independent of its size [36].

The great advantage of the LPR method is that it is performed in homogeneous media and largely avoids the phenomenon of mass transfer or diffusion that occurs in heterogeneous methods [35, 36].

In LPR, two kinds of experiments can be identified: a) *washing method*, which is an elution method based in the continuous diafiltration by addition of solvent at constant volume (see Figure 1a and b) *enrichment method*, which is a concentration method based in the continuous diafiltration by addition of solvent and target ions at constant volume (see Fig. 1b). This method is used to determine the maximum retention capacity of the water-soluble polymer.

The LPR experiments use fractions of polymer with high molecular weight (above to 100,000 Da) to interact with the ions and membranes, which have an exclusion limit of 10,000 Da. The most important physical properties of membranes are the interfacial properties, such as tension and interfacial adsorption. In this context, the van der Waals interactions, hydrogen bonds, electrostatic effects, charge transfer, and dipole moment play a critical role in membrane functioning. The membranes are usually made up of polycarbonate or cellulose esters, polyamides, and polysulfones, among others. The ultrafiltration system consists of a ultrafiltration cell, membrane, stirrer, flow switch, reservoir, and pressure source [36].

Previous studies report the ability of some polymers based ammonium salts to remove arsenate species from solution by the LPR technique [22, 24, 36-45]. These water-soluble polymers (see Fig. 2), that contain different counter-ions have been synthesized, characterized and studied, demonstrating different arsenic retention capability. Basically, the capacity of the polymer to interact and retain arsenate anions depends on the pH, counterion of quaternary ammonium group, ionic strength and the polymer concentration.

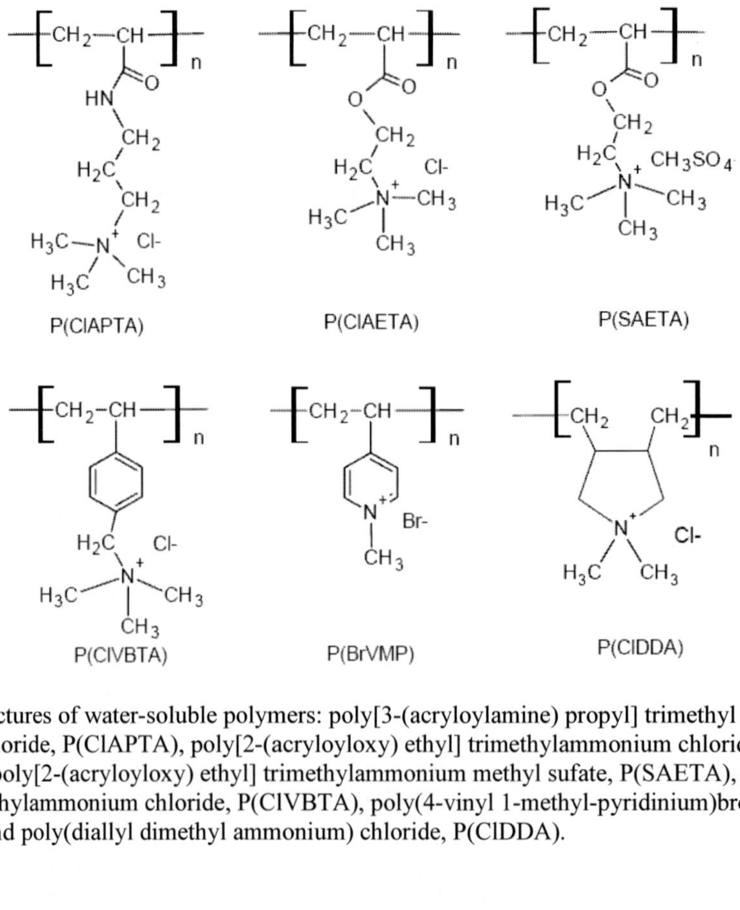

Figure 2. Structures of water-soluble polymers: poly[3-(acryloylamine) propyl] trimethyl ammoniumchloride, P(ClAPTA), poly[2-(acryloyloxy) ethyl] trimethylammonium chloride, P(ClAETA), poly[2-(acryloyloxy) ethyl] trimethylammonium methyl sufate, P(SAETA), poly(ar-vinyl benzyl) trimethylammonium chloride, P(ClVBTA), poly(4-vinyl 1-methyl-pyridinium)bromide, P(BrVMP), and poly(diallyl dimethyl ammonium) chloride, P(ClDDA).

OPTIMIZATION OF ARSENIC REMOVAL BY LPR, THE WASHING METHOD

In order to study the removal of arsenic ions from ionic solution using the LPR technique via washing method, two factors should be defined: 1) retention (R), which is the fraction of arsenic ions remaining in the cell and 2) filtering factor.

$$R = [As_{cell}] / [As_{init}] \tag{1}$$

where [As $_{cell}$] is the absolute amount of arsenic ions that are retained in the cell and [As $_{init}$] is the absolute amount of arsenic ions at the start of the experiment.

The filtration factor (Z) is the ratio between the total permeate volume (V_f) and the retentate volume (V_o):

$$Z = V_f / V_o \tag{2}$$

Depending on the experimental data, a graph (retention profile) in which R is represented as a function of Z, can be drawn.

A) The Effect of pH on Arsenic Removal

The first results showed the high affinity of the polymer to interact and remove As(V) species and the non-retention of As(III) species at pH between 3 to 9. This can be explained by the speciation of As(III) in aqueous media. Different As(III) species are present in solution according to the pH: $H_2AsO_3^-$, $HAsO_3^{2-}$ and AsO_3^{3-}, with $pK_{a1} = 9.2$, $pK_{a2} = 12.1$ and $pK_{a3} = 13.4$ respectively. Therefore, at pH 9 the As(III) species are in equilibrium between the non-dissociated salt and the mono arsenic oxy-anion. On the other hand, As(V) species coexist in an aqueous medium according to the pH: $H_2AsO_4^-$, $HAsO_4^{2-}$ and AsO_4^{3-}, $pK_{a1} = 2.2$, $pK_{a2} = 7.0$ and $pK_{a3} = 11.5$ respectively [39, 44].

The retention capacity of the water-soluble polymer is in direct relation with the anion exchange of the counterion corresponding to quaternary ammonium of polymer because in these systems the electrostatic interactions are predominant. According to the literature, it is suggested that the anionic exchange prefers more divalent ions that monovalent ions at the same conditions [38]. This can be confirmed by the high retention capacity of As(V) species and the non-retention of As(III) species using the water-soluble polymer at different pHs with the LPR technique.

The results about As(V) retention in function of pH are summarized in the figure 3. The arsenate retention (% R) was determined at filtration factor Z=10 using different cationic water-soluble polymers, at pH 3, 6, and 9. In these experimental conditions the total volume of permeate was 200 mL and the volume retentate in the cell was 20 mL. The polymer:As(V) mole ratio was 20:1 using 0.01 mmol absolute As(V) ion.

According to the results, As(V) is more easily retained at pHs between 6 and 9 in comparison with acid pHs. At pH 3, monovalent anionic species ($H_2AsO_4^-$) are in equilibrium with the coupled salt, and the polarity of the functional group is assumed to be a parameter to control the selectivity of ion exchange [36].

At pH 6, the monovalent ($H_2AsO_4^-$) and divalent ($HAsO_4^{2-}$) oxy-arsenic species exist in equilibrium. This is corroborated by the higher retention capacity of predominantly divalent species at pH 9 by the polymers. The polymer interaction capacity depends on the presence of a positively charged quaternary ammonium group because the interactions are produced by ion exchange between the chloride counter-ion of quaternary ammonium salt and the arsenate anions [38].

The polymers can remove between 55-100% of arsenate ions at pH 9, where arsenate is oxy-anionic divalent species. At pH 3, where the species of monovalent arsenate is in equilibrium with the coupled salt, clearance the removal is null.

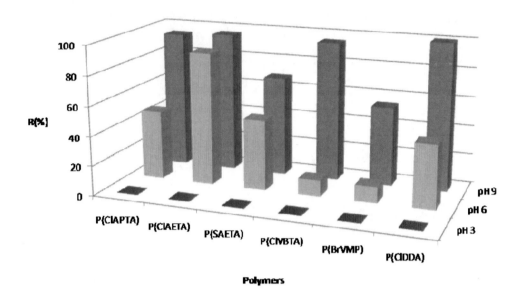

Figure 3. Retention profile of As(V) at Z=10 using water-soluble polymers, at different pH, with 0.2 mmol absolute polymer and 0.01 mmol absolute As(V) ion.

B) The Influence of Polymer-Counterion on Arsenate Removal

In order to study the influence of polymer anion exchanger, we compare the arsenic removal behavior of two polymers, poly[2-(acryloyloxy) ethyl]trimethylammonium chloride, P(ClAETA), and poly[2-(acryloyloxy) ethyl]trimethylammonium methylsulphate, P(SAETA) [36]. They have similar structures except the exchanger groups; chloride (Cl⁻) and methylsulphate ($CH_3OSO_3^-$) respectively.

All the removal experiments were performed by the washing method using a mole ratio 20:1 polymer: arsenate. In general, both polymers were capable to interact and remove arsenate species from aqueous solution at basic pH (see Fig. 4). This demonstrates that at wide range of pH, water-soluble polymers with chloride exchanger groups, such as P(ClAETA), showed a higher ability to remove arsenate than P(SAETA), which contains methylsulphate as anion exchanger group. These results can be attributed to the easier release of the chloride anion compared to the methylsulphate one, which is associated with the quaternary ammonium groups.

Monovalent ions, such as methylsulphate, are strongly retained by hydrophobic sites of quaternary ammonium groups due to the difference in size, solvation and polarity, as compare to chloride [36, 40]. It has been reported in the literature that larger and polarized ions produce a disruption in the local structure of water allowing an easy association with the quaternary ammonium group [40].

Specifically, monovalent ions can be polarized and retained better in comparison with chloride due to the high hydrophobicity of the anion exchanger site. The hydrophobic nature of the monovalent anion ($CH_3OSO_3^-$), which contains a methyl group, may explain this efficiency in the removal of arsenate due to the difficulty of exchanging this large group [36, 40].

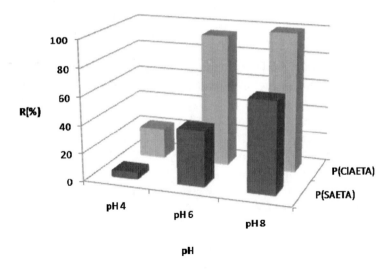

Figure 4. Retention profile of As(V) at Z=10 by P(ClAETA) and P(SAETA) at pH 4, 6 and 8, using 30 mg L^{-1} of As(V) and mole ratio of 20:1 polymer:As(V) (1.6×10^{-4} mol:8×10^{-6} mol).

C) Competitive Effect of Monovalent and Divalent Anions on Arsenate Retention

The water-soluble polymers with chloride counterion present the highest retention of arsenate species by the LPR technique when no other anions are present in the solution [45].

In order to determine the influence of other ions, different experiments in presence of divalent and monovalent anions, such as a sulfate and chloride, were performed using different concentrations of these salts at pH 8. In this study, we used the washing method at different ionic strength adding to both the reservoir and the ultrafiltration cell concentrations in the range of 1×10^{-3} M to 1×10^{-1} M NaCl and Na$_2$SO$_4$ in separate experiments with a P(ClAETA):As(V) mole ratio of 20:1 inside of ultrafiltration cell.

The arsenate retention is found to decrease with the increasing salt concentration and the increased charge of the added anion. The decrease in the retention was due to the presence of the added salts declining in the following order Na$_2$SO$_4$ > NaCl.

According to the literature [46], the order of interference in the arsenic retention is: trivalent ions > divalent ions > monovalent ions. The effect of added electrolytes on arsenic binding to the functional polymers can be understood as due to the competition between arsenate and other anions for binding sites on the polymer. The affinity of anions to bind onto the polymer is similar to the behavior observed in the ion-exchange resin containing ammonium groups when removing arsenic by ion exchange process [45, 46]. Another way to explain the effect is that the electrical double layer is compressed around the polymer as the ionic strength increases, thus reducing the electrical potential of the polymer. The divalent anions produce a higher reduction in arsenic retention than the monovalent anions because the divalent anions bind more strongly to the charged sites of the polymer and also compress the electrical double layer around the polymer more effectively than the monovalent anions [47].

It is reasonable that sulfate or chloride anions present different interference toward arsenate retention. The results prove the adsorption of the interfering ions at the same active sites on the polymer, especially in the case of sulfate, which like arsenate has a tetrahedral structure and divalent charge at basic pH. The results showed that arsenic retention decreased from 96% to 20% at Z = 10 when just 1×10^{-3} M of sodium sulfate was added. Moreover, arsenate retention dropped to zero when sulfate ion concentration increased to 5×10^{-3} M (see Figure 5). On the other hand, the competition between arsenate and monovalent chloride was lower than between sulfate and arsenate. In another separate experiment, when the minimum chloride concentration was added, corresponding to 1×10^{-3} M, the arsenate retention capacity of arsenate decreased from 96% to 55% at Z = 10 (see Fig. 5). This behavior shows that when the concentration of chloride was increased, it was blocking the polymer active sites and the retention of arsenate was decreasing gradually. These results proved that when the ionic strength increases the retention capacity of the polymer decreases due to the competition between ions in solution. This behavior depends directly of the type, charge, and concentration of interfering ion.

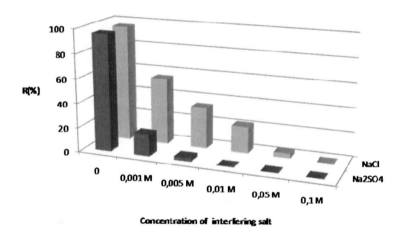

Figure 5. Retention profile of As(V) by P(ClAETA) in presence of different concentrations of NaCl and Na_2SO_4 in both the reservoir and ultrafiltration cell at pH 8, using molar ratio 20:1 polymer:As(V) (3.2×10^{-4} mol:1.6×10^{-5} mol).

D) Effect of Arsenate Concentration on Retention Capacity

The effect of arsenate concentration on arsenate removal in presence of NaCl was also studied. All the experiments were carried out at polymer: As(V), 20:1 molar ratio and pH 8. The arsenate concentration in the feed ranged from 2.46×10^{-6} M (5.5 mg L^{-1}) to 1.27×10^{-5} M (47.6 mg L^{-1}), and all were at constant ionic strength in presence of 1.54×10^{-3} M NaCl.

In comparison with P(SAETA), P(ClAETA) shows higher retention arsenate capacity in all the cases. At higher concentration (47.6 mg L^{-1}), arsenate retention by P(ClAETA) was 58% and this removal capacity increased gradually reaching 100% retention when the arsenate concentration in the cell was minimum (5.5 mg L^{-1}) (see Table 1).

The effect of the conformational changes of water-soluble polymer and the influence of ionic strength cannot be discarded. Indeed, it may be due to a conformational change on the polymer chains [48]. The filtration of arsenate ions and their subsequent release from the polymer induces an increase of the net charge on the polymer surface and then in an expansion of the chains in order to increase the total surface, minimizing the electrostatic repulsions at low arsenic concentration. Related with this, at high arsenic concentrations, the decrease on the surface charge density of the polymer induces a decrease in the strength of the interactions with the arsenic ions, and in consequence, their easier release into the solution from the polymer domain during filtration [45].

Table 1. Effect of As(V) concentration on the removal. Retention percentages of P(ClAETA) and P(SAETA)) polymer:As(V) ratio 20:1 at pH 8 and Z = 10 in presence of 1.54×10^{-3} M NaCl

Mol of polymer	Mol of As(V)	As(V) in feed (mg L^{-1})	R(%) of P(ClAETA)	R(%) of P(SAETA)
2.54×10^{-4}	1.27×10^{-5}	47.6	58	19
8.05×10^{-5}	4.02×10^{-6}	15.1	65	20
5.33×10^{-5}	2.66×10^{-6}	10.0	80	42
2.93×10^{-5}	2.46×10^{-6}	5.5	100	56

MAXIMUM ARSENATE RETENTION CAPACITY, THE ENRICHMENT METHOD

The maximum retention capacity (C) of arsenate by the water-soluble polymer was determined by the enrichment method. This method consists of adding the maximum concentration of the arsenate anion to the polymer solution so that the polymer can bind arsenate ions in order to reach the saturation. The maximum retention capacity is defined as:

$$C = (M\ V) / Pm \qquad (3)$$

where Pm is the amount of polymer (g), M is initial concentration of As(V) (mg L^{-1}), V is the volume of filtrate (volume set) containing As(V) (mL) that passes through the membrane. The maximum retention capacity (C) of arsenate was calculated for the total filtrate volume of 300 mL. Assuming a quantitative retention of As(V), the enrichment factor (E) is a measure of the binding capacity and it is determined as follows:

$$E = (P\ C) / M \qquad (4)$$

where P is the polymer concentration (g L^{-1}). Since the arsenate ion-polymer interactions are processes in equilibrium, a lower slope in the rate of increase of the arsenate concentration in the filtrate is normally observed [35]. The differences in the slopes can easily be used to calculate the amount of arsenate ions bound to the polymer and free in solution as well as the maximum retention capacity (see Fig. 6).

The maximum retention capacity (C) and enrichment factor (E) for some polymers are summarized in the Table 2. The highest retention capacity was found for polymers with counterion Cl$^-$, such as P(ClDDA) and P(ClAPTA), compared to P(SAETA) that contain CH$_3$OSO$_3^-$. The nature of counterion was a more important factor for the maximum retention of arsenate ions than the position of the quaternary nitrogen with respect to polymer chain, showing almost the same behavior for P(ClAPTA) and P(ClDDA) in all the experiments [24].

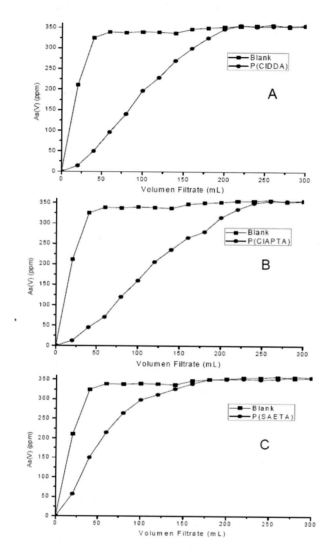

Figure 6. Maximum retention capacity of arsenate using (A) P(ClDDA), (B) P(ClAPTA), (C) P(SAETA) as a extracting polymer at pH 9. Mole ratio of 8×10^{-4} mol of polymer and 4×10^{-3} mol of As(V). The blank (■) is the experiment without polymer.

Table 2. Maximum retention capacity of arsenate and enrichment factor of water-soluble polymers

Polymer	Maximum retention capacity (C), mg As(V) / g polymer	Enrichment factor (E)
P(ClAPTA)	380	7.5
P(ClDDA)	369	9.4
P(SAETA)	79	2.5

DESORBING OF ARSENATE: THE CHARGE-DISCHARGE PROCESS

In order to study the maximum retention and elution capacity of arsenate, so called charge-discharge process, the enrichment and washing method were alternately used [45]. In these experiments P(ClAETA) and P(SAETA) were studied.

The first step of the experiment was the saturation of the polymers through the enrichment method. The enrichment method was performed at pH 8, using 8×10^{-4} mol of polymer into the ultrafiltration cell (20 mL) and adding a solution 4×10^{-3} M in As(V) from the reservoir. After reaching saturation, the polymer:As(V) solution was washed in the ultrafiltration cell with water buffered at pH 3, in a similar way to the washing method. It was assumed that the polymer activity can be recovered in the strongly acid conditions media and that this did not significantly affect the active sites of the polymer because acid pH was used in the radical polymerization. The same charge-discharge process was repeated twice for each polymer in order to determine the capacity of arsenate delivery and to regenerate the extracting ability of the water-soluble polymer [45].

Figure 7 shows the charge-discharge behavior for both polymers. Figure 7 (a) presents the enrichment process (charge) reaching the maximum retention capacity (C) for both polymers at pH 8. The values of C were 165 mg g^{-1} for P(ClAETA) and 79 mg g^{-1} for P(SAETA), and the total filtrate volume was 300 mL [45].

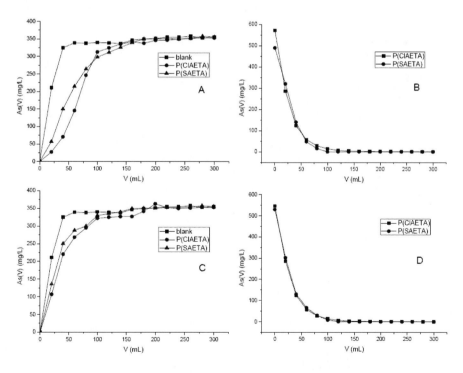

Figure 7. Charge–discharge process of arsenate ions using P(ClAETA) and P(SAETA). (a) first charge process of polymers through enrichment method at pH 8, (b) first discharge process of polymers using washing method at pH 3 with 1×10^{-1} M HCl. (c) recharge of polymers through enrichment method at pH 8, (d) second discharge process of polymers using washing method at pH 3 with 1×10^{-1} M HCl.

After the charge process, the discharge process was initiated changing the pH from basic to acid using buffered solution of 1×10^{-1} M HCl. Figure 7 (b) presents the discharge process of the arsenate ions from both polymers when the polymer-arsenate is in contact with acid solution (pH 3) from the reservoir. The first discharge of arsenate was effective and was carried out almost entirely in the first 100 mL of solution when a higher ion arsenate concentration is discharged from P(ClAETA) in comparison with P(SAETA) at the same volume. Both polymers discharge all the amount of arsenate at 300 mL of filtrate [45].

Figure 7 (c) shows that the second charge process did not improve the maximum retention capacity when compared with the first charge process. P(ClAETA) lost the capacity to remove arsenate, P(SAETA) was only slightly better at the same conditions. The values of C were 83 mg g^{-1} for P(ClAETA) and 47 mg g^{-1} for P(SAETA), and the total filtrate volume was 300 mL. This result is probably due to the presence of more species in the solution when the pH was adjusted from basic to acid in the discharge process and from acid to basic in the second charge process. Finally, the second discharge process (see Fig. 7 (d)) showed almost the same behavior in both polymers, releasing most of the arsenate ions into the filtrate in the first 100 mL in a similar manner [45].

ARSENIC REMOVAL FROM THE NORTHERN CHILE SAMPLES BY LPR

The town of Camarones is located in the Atacama Desert in the northern Chile. The water from the Camarones River is used principally for human consumption and agricultural activities in the area [49]. The Camarones River water presents natural arsenic contamination with total arsenic concentrations above 1000 μg L^{-1} that exists mainly in the form of As(V). The water samples were collected, characterized and reported in a study made by Cornejo et al. [45, 49].

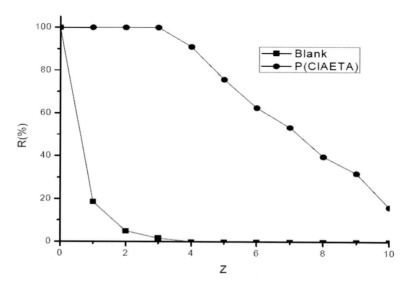

Figure 8. Retention profile of As(V) from Camarones river water by P(ClAETA) using mole ratio of 20: 1 polymer : As(V) and pH 9.

In the present study, we also include some preliminary results of the arsenic removal from the Camarones River water using LPR technique with P(ClAETA) in the already mentioned conditions: pH 9 and polymer: As(V) 20:1 mole ratio. The preliminary results of the As(V) removal from Camarones river water by P(ClAETA) are presented in figure 8. The water-soluble polymer showed a high performance (100%) for the first Z values and then decreased up to 16 % for Z = 10. This means that under these conditions the interaction between the polymer and arsenate is not strong enough, probably due to the presence of other ions and dissolved solids. The Camarones River water presents mainly pH 8.3, 154 mg L^{-1} of sulfate, 541 mg L^{-1} of chloride, 1650 mg L^{-1}of total dissolved solids, 15.68 mg L^{-1}of boron, among others [49].

In a future research we will try to optimize the conditions of the LPR technique in order to improve the arsenic retention from natural supplies.

ELECTROCHEMICAL OXIDATION COUPLED TO LPR TECHNIQUES TO REMOVE ARSENIC FROM AQUEOUS SOLUTIONS

The complete arsenic extraction efficiency, depends strongly on the ability to convert As(III) species into the more easily extractable As(V) forms as the first step. The main, unsolved obstacle to this conversion lies in high irreversibility of the electrochemical oxidation of As(III) into As(V) at bare electrodes. In particular, in aqueous media and for the entire pH range, the anodic oxidation of As(III) is hindered by the solvent oxidation, resulting in a very low As(V) yield. As(III) oxidation could be quantitatively applied to treat polluted water by using catalysts or electrode materials possessing appropriate catalytic properties [36]. Electrochemical oxidation of arsenite combined with adsorption [31] or membrane ultrafiltration procedures [24, 50-52] has also emerged as a remediation technology, to remove arsenic from contaminated aqueous solutions. We have demonstrated that treatment by the LPR technique of aqueous arsenic solutions previously submitted to an electrocatalytic oxidation at bulk platinum [50] or onto iridium oxide film modified carbon electrodes [51] to convert As(III) to As(V) species quantitatively, allowed removing hazardous arsenic. Arsenic removal efficiency was directly linked to the oxidation of As(III) derivatives into more efficiently extractable As(V) species.

ELECTROCATALYTIC OXIDATION (EO) OF AS(III) TO AS(V) BY POLYMER-METAL NANOCOMPOSITE

Lately, we have reported that the electrocatalytic oxidation of arsenic (III) to arsenic (V) can be efficiently performed at platinum metal-polymer nanocomposites film modified electrodes [52].

The As(III) to As(V) conversion is likely electrocatalyzed by the formation of platinum hydroxide on the electrode surface [24, 52]:

$$Pt(OH) + As(OH)_3 \leftrightarrows Pt(OH)As(OH)_3 + H^+ + e^- \tag{5}$$
$$Pt(OH)As(OH)_3 \rightarrow Pt + OAs(OH)_3 \tag{6}$$

Noble metals and metal oxides can be used for this purpose especially in the form of particles of nanometer size, particularly as electrode materials formed from the dispersion of nanoparticles in functionalized polymer matrices [52].

The efficiency of electro-oxidation of As(III) to As(V) was evaluated using the carbon/polymer/Pt nanocomposites-modified electrodes previously synthesized and reported [52]. The functionalized polypyrrole films were grown by oxidative polymerization on carbon felt (RVC 2000, 65 mg cm^{-3}, from Le Carbone Lorraine) electrodes (20×20×4 mm) using a polymerization charge of 5 Coulomb. This led to the deposition onto the carbon felt of a polymeric material containing about 8-9 μmol of ammonium groups (polymerization yield 35-40%). The precipitation of platinum metal in the polymer was performed by electroreduction at - 0.2 V vs. Ag|AgCl in 10^{-4} M K$_2$PtCl$_4$; a charge of 5 Coulomb was passed, which corresponds to deposition of 5 mg of platinum. Glassy carbon electrodes (3 mm diameter) were also previously modified in order to obtain the sensor nanocomposite material for As(III) monitoring by anodic oxidation [24, 52].

The possibility of using water-soluble polymers as supporting electrolytes in the As(III) oxidation process was also evaluated. At first, conductivity measurements were performed in aqueous solutions containing increasing concentrations of different water-soluble polymers used in this study. As expected, both the concentration and molecular weight of polymers significantly influence solution conductivity. Increasing polymer size and concentration simultaneously diminished the conductivity values but all the investigated polymers can overall be regarded as efficient electrolytes whose conductivity values fall in the range of commonly used aqueous electrolytes [51]. The electrocatalytic oxidation of As(III) to As(V) at carbon/polymer/Pt nanocomposites-modified macroelectrodes was tested directly in solutions containing a water-soluble polymers, P(ClAPTA), P(SAETA), and P(ClDDA), as supporting electrolyte, which are able to complex As(V) oxy-anionic species [24].

In a typical experiment, a solution of arsenite and cationic soluble polymer at a 20:1 polymer:As(III) molar ratio (As(III) 7.5 × 10^{-4} M, polymer 15 × 10^{-3} M) was submitted to electrocatalytic oxidation. The solution of As(III) was electrolyzed at + 0.7 V vs. Ag/AgCl until the complete conversion of As(III) to As(V) was achieved. The electro-oxidation was conducted in a one-compartment cell, without separator between the working modified electrode and the auxiliary electrode (a platinum basket). During the process, the solution was stirred at 1000 rpm. The theoretical charge for the complete oxidation was 4.3 Coulombs, calculated according to Coulomb law (Q = number electrons × 96500 × As(III) concentration). The pH of solution was above 9 before the As(III) oxidation using P(ClDDA) and P(ClAPTA) and around 3 for P(SAETA). At the end of As(III) oxidation, the pH decreased to acidic for all the cases due the oxidation processes correspond to acidic catalysis. However, when P(ClDDA) was used, the pH did not decrease below 5 in the course of the electrolysis, probably because P(ClDDA) gives buffer properties towards As(III) oxidation in comparison with the other polymers. Moreover, the protons generated were subsequently reduced to hydrogen at the Pt-basket counter electrode. As expected, the use of a Pt counter electrode (*i.e.*, a Pt-basket set around the working electrode) improved the system due to the good catalytic activity of platinum for proton reduction [52]. The progress of the electrolysis was monitored *in-situ* using an analytical modified electrode (glassy carbon/polymer/Pt) [24] as an amperometric sensor to determine the remaining arsenite (see Figure 9 a, b, c). Compared to our previous work [50] performed in bulk platinum electrodes, the use of a

(Carbon/Polymer/Pt) modified electrode allows catalytic oxidation of As(III) at a lower potential.

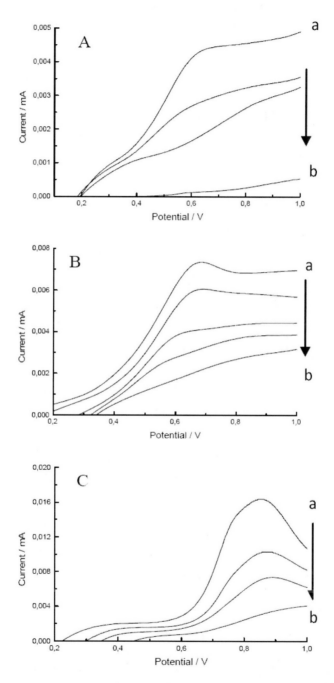

Figure 9. Exhaustive oxidation experiments of 7.5×10^{-4} M of As(III) at a carbon/polymer/Pt carbon felt modified macroelectrode, using (A) P(ClDDA), (B) P(ClAPTA) and (C) P(SAETA) as supporting polyelectrolyte. The anodic voltammetry shows the advancement of electrolyses (carried out at E = + 0.7 V vs. Ag / AgCl), monitored *in-situ* by recording at a carbon/polymer/Pt analytical modified electrode. The decrease of the anodic CV peak corresponding to the remnant As(III) species.

The water-soluble polymers show excellent behavior in both, for exhaustive oxidation and for analytical measurements. Using (A) P(ClDDA), the (B) P(ClAPTA) containing Cl⁻ counter-ion shows the oxidation potential between + 0.6 V to + 0.7 V (see Figure 9 a, b). On the other hand, the use of P(SAETA) as a support in the electrochemical process increases the current response of arsenite measurements at + 0.8 V (see Fig. 9 c). The counterion Cl⁻ on the water-soluble polymer was important factor in order to obtain a lower oxidation potential during the electrolysis, it was due to better charge transport by this class of support water-soluble polymers. The results showed a complete conversion of As(III) to As(V), reaching the theoretical charge calculated for the process under these conditions. The exhaustive oxidation of the As(III) solution was completed in less than 2 hours (see Table 3).

Table 3. Charge applied, peak current of As(III) remnant, pH and time measured during the advancement of the electrolysis of 7.5 × 10⁻⁴ M of As(III) to As(V) at the preparative scale using C|poly1-Pt⁰ carbon felt modified working macroelectrode (E = + 0.7 V vs. Ag/AgCl). Retention percentage (R%) of As(V) previously electro-oxidized

Water-soluble polymer	Total charge applied / C	Peak current / μA	pH during electrolysis	Total time of electrolysis /min
P(ClDDA)				
	0	4.2	9.8	0
	1.6	2.68	7	20
	2.6	1.65	5.8	40
	4.3	0.12	5.6	60
P(ClAPTA)				
	0	7.04	9.2	0
	1	5.66	6.1	20
	2	4.02	4.9	40
	3	2.62	4.2	60
	4.3	1.37	3.9	120
P(SAETA)				
	0	16.2	3.3	0
	2.2	10.1	3.2	20
	3.2	7.12	3.1	42
	4.3	3.25	3	52

LPR TO REMOVE ELECTRO-OXIDIZED ARSENIC SPECIES

After consumption of the charge required for the exhaustive oxidation of As(III) to As(V), the pH of the electrolyzed solution was adjusted to 3, 6, and 9 with HNO₃ or NaOH. Then, the resulting solution (polymer:As(V) of 20:1 mole ratio) was assayed with the LPR-technique by the washing method and the arsenic concentration in the filtrate was determined by atomic absorption spectrometry. In these experimental conditions, at basic pH, the arsenic retention achieved was between 70% to 100%, and was maximum for water-soluble polymers with chloride as a counter ion: 100% for P(ClDDA) (see Figure 10 a), 94% for P(ClAPTA) (see Figure 10 b), and 70% for P(SAETA) (see Fig. 10 c) at Z = 10. However, at pH 3, the species of monovalent arsenate prevail and the clearance is bellow 5%. This behavior is similar to the previous results obtained in the study of the influence of the counterion in the arsenate retention. Finally, we checked that arsenic recovery was 0% for a solution containing water-soluble polymer and As(III) in a 20:1 mol ratio that has not been submitted to electrolysis (see Figure 10 curve (■)) [24].

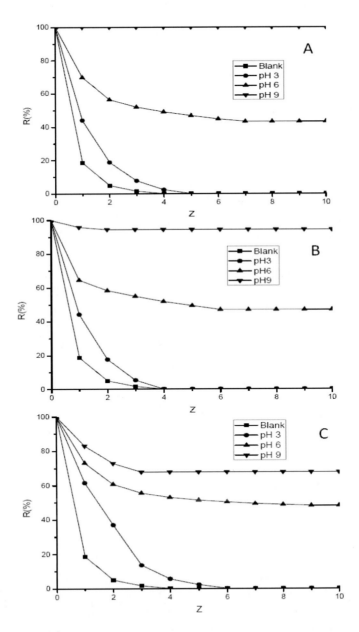

Figure 10. Retention profile of As(III) after exhaustive oxidation of the solution and without oxidation (blank; ■) using (A) P(ClDDA), (B) P(ClAPTA), (C) P(SAETA) as a extracting agent, at different pH, with 2×10^{-4} mol absolute polymer and 1×10^{-5} mol absolute As(V) ion.

The permeate flux was different for each experiment at 3.5 bar of pressure and pH 9 showing almost the same permeate flux for polymers with chloride counterion such as P(ClDDA) and P(ClAPTA) (0.5 mL min^{-1}) and lower for P(SAETA) (0.4 mL min^{-1}). Moreover, the permeate flux also depends on the pH, showing higher permeate flux for experiments at acid pH and lower at basic pH. For example, P(ClDDA) presented permeate flux of 0.8, 0.6 and 0.5 mL min^{-1} for pH 3, 6 and 9, respectively [24].

The results confirm that the combination of the LPR with the electrocatalytic oxidation of arsenic at a polymer-platinum nanocomposite film modified electrode might be a useful technique for arsenic removal from contaminated solutions.

CONCLUSION

The removal capacity of As(V) by water-soluble cationic polymers, that have quaternary ammonium salt groups and different ion exchangers was demonstrated. These polymer materials were synthesized and characterized in our laboratory, showing an acceptable yield of synthesis, water solubility, high thermal resistance, electrical conductivity and molecular weights suitable to use as an extracting agent of As(V) through the LPR technique.

The study of As(V) removal was optimized by analyzing variables such as pH, arsenic concentration, influence of the polymer counterion and presence of interfering ions in solution. Besides, the maximum removal capacity and recovery of the polymer through charge-discharge cycles were also determined.

On the other hand, conducting functional polymers have shown benefits as to their metallic counterparts in bulk as the polymers present electrocatalytic activity towards the oxidation of arsenite oxy-anions. Not only has it been possible to use these materials for analytical purposes but also as a convenient tool for the complete removal of arsenic species from water. This is because it is possible to use this material in the quantitative conversion of As(III) species to As(V) in aqueous solution.

After proving the possibility to use these water-soluble polymers as the supporting electrolyte in the electrolysis and as the extracting agent in ultrafiltration, the next challenge was to couple off line electrocatalytic oxidation with ultrafiltration. The results showed an almost complete removal of species of As(III), which were previously electro-oxidized on the electrode modified with polymer films containing metal nanoparticles. The dual use of water-soluble polyelectrolyte was the key to couple both processes because it avoids the use of external agents as support or oxidants. It was also possible to follow the oxidation process *in situ* by monitoring the residual concentration of As(III) using the modified electrodes in analytical scale.

Finally, it is necessary to set new challenges and improve the use of these electrocatalytic processes and ultrafiltration together. It would be ideal to perform the electrolysis directly in the ultrafiltration cell, or to explore the possibility of realizing the two processes circuit connected. Another important point is to be able to conduct monitoring of As(III) as well as the total removal of arsenic species in real water samples, either in naturally contaminated water or industrial waste water, to reach permissible levels. Each aqueous system is different and requires a previous optimization in order to use adequate technologies to remove arsenic species. For this, much remains to be done in this area.

ACKNOWLEDGMENTS

The authors are grateful for grants from FONDECYT (Grant No 1110079), PIA (Grant Anillo ACT 130), and CIPA. The authors also thank to the Department of Molecular Chemistry, University Joseph Fourier, Grenoble 1, France.

REFERENCES

[1]   B. K. Mandal and K. T. Suzuki, *Talanta* 58, 201 (2002).
[2]   M. Bissen and F. H. Frimmel, *Acta Hydroch. Hydrob.* 31, 9 (2003).
[3]   WHO. Arsenic Compounds, Environmental Health Criteria 224, 2nd ed., Geneva: World Health Organization, (2001).
[4]   U. S. Environmental Protection Agency. Interim Primary Drinking Water Standards. Fed. Reg. 40, (1975).
[5]   P. L. Smedley and D. G. Kinniburgh, *Appl. Geochem.* 17, 517 (2002).
[6]   R. Zaldivar. *Beitr. Zur Pathol.* 151, 384 (1974).
[7]   S. Karcher, L. Cáceres, M. Jekel, R. Contreras, *J. Chart. Inst. Water Environ. Manag.* 13, 164 (1999).
[8]   A. H. Smith, M. Goycolea, R. Haque, M. L. Biggs, *Am. J. Epidemiol.* 147, 660 (1998).
[9]   Y. Yuan, G. Marshall, C. Ferreccio, C. Steinmaus, S. Selvin, J. Liaw, M. N. Bates, A. H. Smith, *Am. J. Epidemiol.* 166,1381 (2007).
[10]   C. Ferreccio and A. M. Sancha, *J. Health Popul. Nutr.*, 24,164 (2006).
[11]   J. M. Borgoño, P. Vicent, H. Venturino, A. Infante, *Environ. Health Persp*, 19, 103 (1977).
[12]   A. H. Smith, A. P. Arroyo, D. N. Guha Mazumder, M. J. Kosnett, A. L. Hernandez, M. Beeris, M. M. Smith, L. E. Moore, *Environ. Health Persp*, 108, 617 (2000)
[13]   C. Hopenhayn-Rich, S. R. Browning, I. Hertz-Picciotto, C. Ferreccio, C. Peralta, H. Gibb, *Environ. Health Persp*, 108, 667 (2000).
[14]   G. Marshall, C. Ferreccio, Y. Yuan, M. N. Bates, C. Steinmaus, S. Selvin, J. Liaw, A. H. Smith, *J. Natl. Cancer Inst.*, 99, 920 (2007).
[15]   J. Liaw, G. Marshall, Y. Yuan, C. Ferreccio, C. Steinmaus, A. H. Smith, *Cancer Epidemiol Biomarkers Prev.,* 17, 1982 (2008).
[16]   NCh 409/1.Of 2005 drinking water. Pt.1.Requirements. Santiago: Instituto Nacional de Normalización Chile, (2005).
[17]   L. Cáceres, E. Gruttner, R. Contreras, *Ambio*, 21, 138 (1992).
[18]   M. I. Litter, M. E. Morgada, J. Bundschuh, *Environ. Pollut.*, 158, 1105 (2010).
[19]   A. M. Sancha, *International Water Association, Water Supply*, 18, 621 (2000).
[20]   A. M. Sancha, *Journal of Health, Population and Nutrition*, 24, 267 (2006).
[21]   R. Johnston and H. Heijnen, *Safe water technology for arsenic removal*, (2001) in: M. Feroze Ahmed, M. Ashraf Ali, A. Zafar (Eds.), *Technologies for Arsenic Removal from Drinking Water,* 2001 (Bangladesh), pp. 1-22.
[22]   N. Kabay, J. Bundschuh, B. Hendry, M. Bryjak, K. Yoshizuka, P. Bhattacharya, S. Anac, *The Global Arsenic Problem: Challenges for Safe Water Production*, CRC Press, Taylor and Francis Group (2010).

[23]   J. Bundschuh, M. Litter, V. S. T. Ciminelli, M. E. Morgada, L. Cornejo, S. Garrido Hoyos, J. Hoinkis, M. T. Alarcon-Herrera, M. A. Armienta, P. Bhattacharya, *Water Res.* 44, 5828 (2010).

[24]   J. Sánchez and B. L. Rivas, *Chem. Eng. J.* 165, 625 (2010).

[25]   J. F. Ferguson and J. Gavis, *Water Res.* 6, 1259 (1972).

[26]   R. W. Cullen and K. Reimer, *Chem Rev.* 89, 713 (1989).

[27]   A. H. Welch, D. B. Westjohn, D. R. Helsel, R. B. Wanty, Arsenic in ground water of the United States: occurrence and geochemistry. *Ground Water*, 38, 589 (2000).

[28]   D. Melamed, *Monitoring arsenic in the environment: a review of science and technologies for field measurements and sensors.* EPA 542/R-04/002, (2004).

[29]   P. Ravenscroft, H. Brammer, K. Richards, *Arsenic pollution: a global synthesis.* Oxford, U. K., Wiley-Blackwell, (2009).

[30]   V. K. Sharma and M. Sohn, *Environ Int.* 35, 743 (2009).

[31]   M. Bissen and F. H. Frimmel, *Acta Hydroch. Hydrob.* 31, 97 (2003).

[32]   M. Pirnie, *Technologies and costs for removal of arsenic from drinking water.* USEPA Report 815-R-00-028, (2000).

[33]   X. Zhang, K. Jiang, Z. Tian, W. Huang, L. Zhao, *J. Appl. Polym. Sci.* 110, 3934 (2008).

[34]   B. Ya. Spivakov, K. Geckeler, E. Bayer, *Nature*, 315, 313 (1985).

[35]   B. L. Rivas, E. D. Pereira, I. Moreno-Villoslada, *Prog. Polym. Sci.* 28, 173 (2003).

[36]   B. L. Rivas, E. D. Pereira, M. Palencia, J. Sánchez, *Prog. Polym. Sci.* 36, 294 (2011).

[37]   K. E. Geckeler and K. Volchek, *Environ. Sci. Technol.* 30, 725 (1996).

[38]   B. L. Rivas, M. C. Aguirre, E. Pereira, *J. Appl. Polym. Sci.* 102, 2677 (2006).

[39]   B. L. Rivas, M. C. Aguirre, E. Pereira, *J. Appl. Polym. Sci.* 106, 89 (2007).

[40]   B. L. Rivas, M. C. Aguirre, E. Pereira, J.-C. Moutet, E. Saint-Aman, *Poly. Eng. Sci.* 47, 1256 (2007).

[41]   B. L. Rivas and M. C. Aguirre, *J. Appl. Polym. Sci.* 112, 2327 (2009).

[42]   B. L. Rivas and M. C. Aguirre, *Water Res.* 44, 5730 (2010).

[43]   B. L. Rivas and M. C. Aguirre, *Polym. Bull.* 67, 441 (2011).

[44]   B. L. Rivas, J. Sánchez, S. A. Pooley, L. Basaez, E. Pereira, C. Bucher, G. Royal, E. Saint Aman, J.-C. Moutet, *Macromol. Symp.* 296, 416 (2010).

[45]   J. Sánchez and B. L. Rivas, *Desalination*, 270, 57 (2011).

[46]   A. Berdal, D. Verrie, E. Zaganiaris, *Removal of arsenic from potable water by ion exchange resins*, (2000) in: J.A. Greig, (Ed), Ion Exchange at the Millennium, Proceedings of IEX. Imperial College Press, 2000, London, pp.101-108.

[47]   R. A. William, *Colloid and Surface Engineering: Applications in the Process Industries*, Butterworth-Heinemann, Oxford, (1992).

[48]   B. L. Rivas, I. Moreno-Villoslada, *J. Phys. Chem. B*, 106, 9708 (2002).

[49]   L. Cornejo, H. Lienqueo, M. Arenas, J. Acarpi, D. Contreras, J. Yáñez, H. D. Mansilla, *Environ. Pollut.* 156, 827 (2008).

[50]   B. L. Rivas, M. C. Aguirre, E. Pereira, C. Bucher, G. Royal, D. Limosin, E. Saint-Aman, J.-C. Moutet, *Water Res.* 43, 515 (2009).

[51]   B. L. Rivas, M. C. Aguirre, E. Pereira, C. Bucher, J.-C. Moutet, G. Royal, E. Saint-Aman, *Polym. Adv. Technol*, 22, 414 (2011).

[52]   J. Sánchez, B. L. Rivas, A. Pooley, L. Basaez, E. Pereira, I. Pignot-Paintrand, C. Bucher, G. Royal, E. Saint-Aman, J.-C. Moutet, *Electrochim. Acta,* 55, 4876 (2010).

In: Arsenic
Editor: Andrea Masotti

ISBN: 978-1-62081-320-1
© 2013 Nova Science Publishers, Inc.

Chapter 15

ARSENIC: ENVIRONMENTAL IMPACT REDUCTION USING NATURAL AND MODIFIED ADSORBENTS

Dimitris Papoulis[,1], Dionisios Panagiotaras[2] and Georgios Panagopoulos[3]*

[1]Department of Geology, University of Patras, Patras, Greece
[2]Department of Mechanical Engineering, Laboratory of Chemistry,
Technological Educational Institute (T.E.I.) of Patras, Patras, Greece
[3]Department of Mechanical and Water Resources Engineering,
Technological-Educational Institute (T.E.I.) of Messolonghi,
Nea Ktiria, Messolonghi, Greece

ABSTRACT

Arsenic is a ubiquitous inorganic element that can enter the groundwater as a result of geological processes such as rock-weathering and volcanic activity, and anthropogenic processes such as mining activities, leaching of wastes and wastewaters and burning of coals in power facilities. Arsenic pollution has become one of the most serious environmental problems worldwide today.

Recent epidemiological studies refer high toxicity and fatal effects of As on human health and for this reason WHO proposed a reduction of maximum allowable arsenic concentration in drinking water from 50 μg/L to 10 μg/L. The removal of arsenic from drinking water is a recurring challenge, especially in developing countries. Cost considerations can make it expedient to use local materials, from soils or produced in agricultural or industrial operations, as adsorbents for arsenic. These materials may not always be optimal, but their availability at various locations all over the world and as a consequence their low cost often makes them attractive choices. This review presents a compilation of the reported up to date adsorption techniques and their evaluation based on the availability and cost of the materials that their capacity for the removal of arsenic chemical species from waters/wastewaters seems to be affective or at least promising.

[*] E-mail address: papoulis@upatras.gr.

INTRODUCTION

Arsenic (As) is a common inorganic element found widely in the environment. The presence of arsenic in the environment can pose a risk to human health [1, 2, 3]. Historical and current industrial use of arsenic has resulted in soil and groundwater contamination that may require remediation [4, 5]. Some industrial wastes and wastewaters currently being produced require treatment to remove or immobilize arsenic [6]. In addition, arsenic must be removed from some sources of drinking water before they can be used. Arsenic is widely known for its adverse effects on human health, affecting millions of people around the world [7, 3]. There is a growing need for cost-effective arsenic treatment. Because arsenic readily changes valence states and reacts to form species with varying toxicity and mobility, effective, long-term treatment of arsenic can be difficult. In some disposal environments arsenic has leached from arsenic-bearing wastes at high concentrations [8, 9, 10].

This review presents a compilation of the reported up to date Arsenic adsorption techniques from waters/wastewaters as well as their evaluation. It should be noted that in the literature data for too many arsenic adsorbents are published. In this review we present selected adsorbents for every category of adsorbents as we classify them (based on their origin).

ARSENIC GEOCHEMISTRY

Arsenic occurs naturally in rocks, soil, water, air, plants, and animals. Natural activities such as volcanic activity, erosion of rocks, and forest fires, can release arsenic into the environment.

The man-made sources of arsenic in the environment include mining and smelting operations; agricultural activities; burning of fossil fuels and wastes; pulp and paper production; cement manufacturing; and former agricultural uses of arsenic [11].

Arsenic is widely distributed in more than 320 minerals [12], the most common ones being arsenopyrite (FeAsS) [13, 14, 15], orpiment (As_2S_3), realgar (As_2S_2), while it also participates as solid solution in pyrite (FeS_2) [16, 17]. Arsenic is also found in sedimentary environments adsorbed by Fe(III) and Mn(IV) after weathering of sulfide minerals [18,19, 20, 15].

Arsenic bearing minerals, together with a once widespread use of arsenic in pigments, insecticides and herbicides, represent the major sources of arsenic in natural waters. About 70% of all arsenic used is in pesticides, which principally are the following [21].

- Monosodium methane arsenate (MSMA) – $HAsO_3CH_3Na$;
- Disodium methane arsenate (DSMA) – $Na_2AsO_3CH_3$;
- Dimethylarsinic acid (cacodylic acid) – $(CH_3)_2 AsO_2H$;
- Arsenic acid – H_3AsO_4.

The other uses of arsenic and its compounds are in wood preservatives, glass manufacture, alloys, electronics, catalysts, feed additives and veterinary chemicals [8, 10].

Many studies document the adverse health effects in humans exposed to inorganic arsenic compounds. A discussion of those effects is available in the following documents:

- National Primary Drinking Water Regulations; Arsenic and Clarifications to Compliance and New Source Contaminants Monitoring (66 FR 6976 / January 22, 2001) [18].
- The Agency for Toxic Substances and Disease Registry (ATSDR) ToxFAQsTM for Arsenic [22].

Arsenic is a metalloid or inorganic semiconductor that can form inorganic and organic compounds. It occurs with valence states of -3, 0, +3 (arsenite), and +5 (arsenate). However, the valence states of -3 and 0 occur only rarely in nature. This discussion of arsenic chemistry focuses on inorganic species of As(III) and As(V). Inorganic compounds of arsenic include hydrides (e.g., arsine), halides, oxides, acids, and sulfides [23].

ARSENIC TOXICITY

The World Health Organization (WHO) has recognized arsenic as the cause of the largest mass poisoning of population in history [24].

A large number of diseases and cancers related to the chronic arsenic toxicity (arsenicosis) can be found in the literature [25-27]. Milton [28], Mazumder and Dasgupta [29] and Chakraborti [30] give reviews about the health effects of arsenic toxicity in humans. Skin diseases attributed to arsenicosis are pigmentation and keratosis [31-32]. Prolonged exposure to arsenic can cause chronic lung and liver diseases as chronic bronchitis, bronchiectasis and non cirrhotic portal fibrosis [33-36]. Cardiovascular diseases such as hypertension, nonpitting edema of feet/hands and anemia [37-39]; diabetes [40-43]; adverse pregnancy outcomes [44-47] and decrease in children's intellectual function [48-51] are well documented as a result of arsenicosis. Cancers of skin, lung, and urinary bladder are the most important cancers associated with this toxicity [52-58].

Fresh water contains As in concentrations ranging from 0.5 ppb to more than 5000 ppb, but typical values are less than 10 ppb and frequently less than 1 ppb [59]. Taking into account the aforementioned epidemiological studies, many authorities have lowered the maximum allowed value of As in the drinking water. WHO reduced the guideline value for As from 50 ppb [60] to 10 ppb [61].

US. EPA, E.C. and Japan have adopted the limit of 10 ppb while the Canadian interim maximum limit is 25 ppb. On the other hand, many developing countries still keep the maximum limit of 50 ppb, mainly because they cannot afford facilities to measure extremely low concentrations of arsenic [59].

West Bengal (India) and Bangladesh constitute hot spots in terms of population drinking water highly contaminated in arsenic [29, 59]. Other well known high-As groundwater areas have been reported in all over the world, i.e. Latin America (Argentina, Chile, Mexico) [62-64]; SW USA [65]; Asia (China, Vietnam, Taiwan, Nepal, Myanmar, Cambodia) [66-71] and Europe (Hungary, Romania, Italy, Greece) [72-74].

Sources of Arsenic in Groundwater

Arsenic is a ubiquitous element that can enter the environment as a result of both geological and anthropogenic processes. The geological origin of As is mainly associated with the presence of volcanic rocks, especially volcanic glass and ashes [64, 73, 75-82]. High concentrations of As have been found in sedimentary alluvial aquifers as well [67, 83-88], because arsenic is present in the crystal structure of many rock-forming minerals [59]. Furthermore, high values of this metal are often observed in areas with geothermal activity due to the reducing action of acid gases of the deep geothermal fluids [73, 89-90]. On the other hand, ground and surface waters enriched in As have often man-made origin which is associated with mining and smelting operations; agricultural applications; burning of fossil fuels and wastes; pulp and paper production; cement manufacturing; and former agricultural uses of arsenic [59, 91-96].

Natural and Modified Arsenic Adsorbents

Arsenic can be found in many industrial products, wastes, and wastewaters, and is a contaminant of concern at many remediation sites. Arsenic contaminated soil, waste, and water must be treated either by removing the arsenic or immobilizing it. The removal of arsenic from drinking water is a recurring challenge all over the world and especially in developing countries [97, 98].

Cost considerations can make it expedient to use local materials, from soils or produced in agricultural or industrial operations, as adsorbents for arsenic. These materials may not always be optimal, but their availability at various locations all over the world and as a consequence their low cost often makes them attractive choices. This review presents a compilation of the reported up to date adsorption techniques as well as a general evaluation.

The Arsenic adsorbents are too many (Tables 1-7), for that reason it is useful to categorize them and present data for selected adsorbents for each category. We present eleven categories even though in literature one can find different classifications as well as lower number of categories [e.g. 97]. These adsorbents can be categorized into the following categories:

1. commercial activated carbons
2. synthetic activated carbons
3. agricultural product and by-products
4. industrial by-products/wastes
5. the clay minerals (e.g. smectite minerals like montmorillonite), including soils
6. other minerals (e.g. zeolites) and synthetic zeolites.
7. oxides (e.g. iron oxides),
8. hydrotalcites
9. phosphates
10. metal-based methods
11. biosorbents

It should be noted that categories 5, 6 and 7 could be unified in one category because oxides and other minerals can be found in soils. Despite that we did recognise three different categories because oxides and other minerals are usually found in relatively low amounts in soils. For example feldspars and goethite can be found in many soils in relatively low amounts and in some cases in relatively high amounts. In reverse gibbsite is a soil constituent but also an oxide so it could be categorized in both categories. The presence of many different arsenic adsorbents (in varying amounts) from different categories in soils shows that the arsenic adsorption capacity for each soil would depend from its mineralogical composition and it would vary for each soil.

Selected adsorbents from every category used, up to date (as tracked from references), for removal of Arsenic (III) and Arsenic (V) from waters/wastewaters are presented in Tables 1, 2, 3, 4, 5, 6 and 7. From Tables 1-7 it is evident that many adsorbents have been tested and most of them with good results over the last decades.

EVALUATION OF ARSENIC ADSORBENTS

The efficiency of the different adsorbents is not easily comparable because many tests (e.g. contamination concentration, particle diameter) were conducted under different conditions (some of them presented in Tables 1-7). Despite this inconsistency it is possible to find criteria in order to evaluate the different adsorbents. The evaluation of the adsorption techniques can be based generally on the availability and as a consequence the cost of the materials that their capacity for the removal of arsenic chemical species seems to be effective or at least promising. An important factor is Arsenic contamination concentration in water and as a result how effective adsorbents do we need. For example if the Arsenic amount in water is only 10-20% above limits we could use a less affective adsorbent than in case that Arsenic amount is more than 100% above limits. Another important factor is pH values of the water. It is evident (Tables 1-7) that different adsorbents have different optimum pH values.

In order to evaluate the adsorbents we have to present two different cases. The first case is for relatively high Arsenic contamination concentration (about 10-20% above limits) and the second is for very-extremely high Arsenic contamination concentration.

In the case of relatively high Arsenic contamination concentration the most attractive materials are the natural materials due to its low cost and its availability. From the above categories the most abundant materials are by far the clay minerals. Clay minerals are more abundant in the earth's crust (in amount) as well as the most widespread (common minerals in most of the soils all over the world). Taking into consideration that clay minerals are common, widespread in surface occurrences and deposits and therefore cheap as well as efficient arsenic adsorbents for wide pH range values we would evaluate these materials as the first choice in most of the cases. Generally our second choice would be zeolites and third choice the oxides. Exceptionally, zeolites or oxides would be our first choice in areas with naturally occurring zeolite deposits or laterites (less common than clays) respectively.

Other adsorbents in terms of availability and cost can not be evaluated as first choices at least in most of the cases. Some of them could be our first choice in cases of contaminated water under extreme conditions like very low or very high pH values. In cases of low or very low pH values our first choice would be iron compounds (e.g. iron oxides).

Table 1. Selected Commercial and synthetic activated carbons

Adsorbent	Method /type of water	Optimum pH	Contamination concentration (mg/L)	Capacity (mg/g) As (III)	As (V)	References
Commercial activated carbons	Drinking water	7	5-200	0.21-1.4	3-31	[99]
Commercial activated carbons	Aqueous solution	6.4 - 7.5	157-992	29.9	30.48	[99]
Commercial activated carbons	Waste water		300		25-2860	[100]
Activated synthetic carbons	Column	4-9	1.33	40.5	27	[101]

Table 2. Selected Agricultural and industrial products and by-products used as Arsenic adsorbents

Adsorbent	Method /type of water	Optimum pH	Contamination concentration (mg/L)	Capacity (mg/g) As (III)	As (V)	References
Rice husk	Column	6.5-8.0	50-500	20	7	[102]
Chars	Drinking water	3.5	10-100	0.0012-12		[103]
Chars	Aqueous solution	2-3	157-992	89	35	[104]
Red mud	Aqueous solution /Batch	7.25/3.5	33-37-400.4	0.884	0.941	[105]
Bauxsol, activated	water	4.5	2.04-156.7 mol/L	0.541	7.642	[106, 107]

Table 3. Selected Soils and soil constituents used as Arsenic adsorbents

Adsorbent	Method /type of water	Optimum pH	Contamination concentration (mg/L)	Capacity (mg/g) As (III)	As (V)	References
Bentonite (modified)	Batch	6.0/9.0	0.2-1	0.82	1.48	[108]
montmorillonite	Batch	5-6	20μM	~0.2 mmol/g	~0.4 mmol/g	[109]
Illite	Batch	9	20μM	~0.2mmol/g	~0.5 mmol/g	[109]
Kaolinite	Batch	3-8	20μM	~0.3mmol/g	~0.5 mmol/g	[109]
Kaolinite, surfactant modified	Batch/column	5.0-6.5	0.2-14	4.3 mmol/kg	9.0 mmol/kg	[110]
Gibbsite	Wastewater	5.5	10-1000	3.30	4.60	[111]
Soil, Sharkey	Soil	5-6	5-100		0.74	[112]

Table 4. Other minerals (e.g. zeolites) and synthetic zeolites used as Arsenic adsorbents (selected)

Adsorbent	Method /type of water	Optimum pH	Contamination concentration (mg/L)	Capacity (mg/g) As (III)	As (V)	References
Zeolite, surfactant modified	Batch/column	7.2-7.5	0.2-14	1.6 mmol/kg	7.2 mmol/kg	[113] and [114]
Zeolites	Batch	4.0	0.1-4.0	0.017	0.1	[115]
Malachite	Batch	4.0	5.000		57.1 mg/g	[116]
Feldspar	Water/wastewater	4.2	133.49 μmol/L		0.18	[117]
Siderite	Batch and column	7	250-2000	1040 μg/g	516 μg/g	[118] and [119]
Fe-Mn mineral material	Batch/column	3/3,5.5	0.47 mmol/L	14.7	6.7	[120]

Table 5. Selected Oxides used as Arsenic adsorbents

Adsorbent	Method /type of water	Optimum pH	Contamination concentration (mg/L)	Capacity (mg/g) As (III)	As (V)	References
$Al_2O_3/Fe(OH)_3$	Batch	6.1 ± 0.3/8.0 ± 0.3	0.1-0.4	0.12 mmol/g	36.7	[121]
Fe-Mn binary oxide	Batch	4.8	0.20 mmol/L	1.77 mmol/g	0.93 mmol/g	[122] and [123]
Goethite	Batch wastewater	5.5	10-1000	7.50	12.5	[122] and [124]
Ferrihydrite	Batch/natural		325 μg/L		0.25	[125]
Ferric hydroxide, granular	Column drinking water	8-9	5-100	2.3		[126]
TiO_2	Batch	8.5/7.3	0.4-80	32.4	41.4	[127] and [128]

Table 6. Selected Hydrotalcites, phosphates and metal-based methods used as Arsenic adsorbents

Adsorbent	Method /type of water	Optimum pH	Contamination concentration (mg/L)	Capacity (mg/g) As (III)	As (V)	References
Synthetic hydrotalcite	Ground water	7.0	400		105	[129]
Layered double hydroxides, calcined	Wastewater	4.2-5.4	20-200		5.61	[130]
$FePO_4$ (amorphous)	Drinking water	7-9/6-6.7	0.5-100	21	10	[131], [132]
$FePO_4$ (cryst.)	Drinking water	7-9/6-6.7	0.5-100	16	9	[131], [132]
Fe/NN-MCM-41	Drinking water	6.0	~0-1500	119.8		[133]
Cu/NN-MCM-48	Drinking water	7.0	~0-1500		37.46	[133]

Table 7. Selected Biosorbents used as Arsenic adsorbents

Adsorbent	Method /type of water	Optimum pH	Contamination concentration (mg/L)	Capacity (mg/g) As (III)	As (V)	References
Chitosan	Batch/column	9.0	1000 mg/L	2.0		[134]
Chitosan	Wastewater	4.0	400		58	[135]
Cellulose (cotton)	Batch/column	7.1	1 mg/L		35.0	[136]
Cellulose (bead) with iron oxyhydroxide	Ground water	7.0	1–100 mmol/L	33.2	33.2	[137]
Biomass, yeast, methylated	Surface and ground water	6.5	0.5–2.5 mM		3.75	[138]
Biomass, immobilized	Ground water	6.0	50–2500	704.1		[139]

In the case of very-extremely high Arsenic contamination concentration and with no significant problems in increasing the cost to reasonable levels the most attractive materials are the most widely used, the iron compounds, usually iron oxides. In order to deal with water characterized by extremely high Arsenic contamination concentration we would have to use the most effective adsorbents. In this case we would have to choose biosorbents like biomass or some other relatively low cost adsorbent from agricultural wastes or industrial wastes. If the cost for this treatment would be higher than can be afford then we would have to choose other affective but cheaper adsorbents, like oxides.

The above evaluation of the Arsenic adsorbents is general. In order to decide which adsorbent to use one has to consider many parameters as presented by Mohan and Pittman, 2007 [97]. These parameters are:

1) The range of initial arsenic concentrations
2) Other elements and their concentration in water
3) Optimization of adsorbent dose
4) Filtration of treated water
5) Adjustment of pH in water
6) Post treatment difficulties
7) Handling of waste and
8) Proper operation and maintenance [97].

The above parameters show that in fact every case is different, and that is how it should be considered. The uniqueness of every case is evident considering for example the parameter 2 (other elements and their concentration in water). Clay minerals adsorb many inorganic as well as organic pollutants. In some cases, that would be the optimum but the ability of clay minerals to adsorb cationic, anionic and neutral metal species and organic molecules lowering their sorption efficiency to arsenic. For example the adsorbance of cadmium (if present), an inorganic pollutant, would be desirable procedure while the adsorbance of calcium would just lowering the sorption efficiency of the adsorbent to arsenic.

REFERENCES

[1]   W. R. Cullen and K. J. Reimer, *Chem. Rev.*, 89, 713 (1989).
[2]   D. Dermatas, D. H. Moon, N. Menounou, X. Meng and R. Hires, *J. Hazard. Mater.* 116, 25 (2004).
[3]   K. A. Hudson-Edwards, S. L. Houghton and A. Osborn, *Trend. Anal. Chem.* 23, 745 (2004).
[4]   J. Matschullat, *Sci. Total Environ.* 249, 297 (2000).
[5]   E. Miteva, D. Hristova, V. Nenova and S. Manava, *Scien. Horticult.* 105, 343 (2005).
[6]   A. Kabata-Pendias and D. C. Adriano, Trace metals. *In: Soil amendments and environmental quality.* Rechcigl, (1995), J. E. (Ed.). CRC press. Boca Raton. USA. (1995) pp. 139-167.

[7]   V. Matera and I. LeHecho, I. 2001. Arsenic behavior in contaminated soils: mobility
       and speciation. *In: Heavy metals release in soils*. H. M. Selim and D. L. Sparks (Eds).
       CRC Press. Boca Raton, FL. (2001) pp. 207-235.
[8]   U.S. EPA. Land Disposal Restrictions: Federal Register, Volume 65, Number 118, June
       19, 2000 at http://www.epa.gov/osw/hazard/tsd /ldr/lrrp-fr.pdf.
[9]   U.S. EPA. National Primary Drinking Water Regulations: Federal Register, Volume 66,
       Number 14, January 22, 2001 at http://www.epa.gov/sbrefa/documents/pnl14f.pdf.
[10]   U.S. EPA. Arsenic Treatment Technology for Soil, Waste, and Water (542-R-02-004),
       September 2002 at www.epa.gov/tioclu-in.org/arsenic.
[11]   U.S. EPA. National Primary Drinking Water Regulations: Federal Register, Vol 65,
       Number 121, June 22, 2000 at http://www.epa.gov /safewater/ars/arsenic.pdf.
[12]   M. Fleischer, *Glossary of Mineral Species*, The mineral record Inc, Tucson, Arizona
       (1983).
[13]   D. R. Boyle, R. J. W. Turner and G. E. M. Hall, *Environ. Geochem. Health* 20, 199
       (1998).
[14]   P. L. Smedley, W. M. Edmunds and K. B. Pelig-Ba, Mobility of arsenic in groundwater
       in the Obuasi gold-mining area Ghana: some implications for human health. In:
       Appleton, J.D., Fuge, R., McCall, G.J.H. (Eds.), *Environ. Geochem. Health*, 113
       (1996), Geological Society Special Publication, London, pp. 163–181.
[15]   J.K. Myoung, J. Nriagu and S. Haack, *Environ. Poll.* 120, 379 (2002).
[16]   T. R. Chowdhury, G. K. Basu, B. K. Mandal, B. K. Biswas, G. Samanta, U. K.
       Chowdhury, C. R. Chanda, D. Lodh, S. L. Roy, K. C. Saha, S. Roy, S. Kabir, Q.
       Quamruzzaman and D. Chakraborti, *Nature* 401, 545 (1999).
[17]   B. K. Mandal, T. R. Chowdhury, G. Samanta, D. P. Mukherjee, C. R. Chanda, K. C.
       Saha and D. Chakraborti, *Sci. Total Environ.* 218, 185 (1998).
[18]   S. K. Acharyya, P. Chakraborty, S. Lahiri, B. C. Raymahashay, S. Guha, and A.
       Bhowmik, *Nature* 401, 545 (1999).
[19]   R. Nickson, J. McArthur, W. Burgess, K. M. Ahmed, P. Ravenscroft, M. Rahman,
       Nature 395, 338 (1998).
[20]   R. T. Nickson, J. M. McArthur, P. Ravenscroft, W. G. Burgess and K. M. Ahmed, *Appl.
       Geochem.* 15, 403 (2000).
[21]   M. Kumaresan and P. Riyazuddin, *Current Science*, 80, 837 (2001).
[22]   The Agency for Toxic Substances and Disease Registry (ATSDR): ToxFAQs [TM] for
       Arsenic, 12, July, 2001 at http://www.atsdr.cdc. gov/tfacts2.html.
[23]   K. Othmer. "Arsenic and Arsenic Alloys." *The Kirk-Othemer Encyclopedia of
       Chemical Technology*, 3, John Wiley and Sons, New York. (1992).
[24]   A.H. Smith, E.O. Lingas and M. Rahman, Bull. WHO 78, 1093 (2000).
[25]   M.P. Waalkes, J. Liu and B.A. Diwan, *Tox. Appl. Pharmac.* 222, 271 (2007).
[26]   L. Li, E.C. Ekström, W. Goessler, B. Lönnerdal, B. Nermall and M. Yunus, *Environ.
       Health Perspect.* 116, 315 (2008).
[27]   E.J. Tokar, W. Qu and M.P. Waalkesm, *Tox. Sci.* 120, S192 (2011).
[28]   A.H. Milton, "Health effects of arsenic: toxicity, clinical manifestation and health
       management", Arsenic contamination: Bangladesh Perspective, 2003 (Dhaka,
       Bangladesh) pp281–295).
[29]   D.N.G. Mazumder and U.B. Dasgupta, *J. Med. Sci.* 27, 360 (2011).
[30]   D. Chakraborti, Encycl. *Environ. Health*, 165 (2011).

[31] D.N.G. Mazumder, R. Haque, N. Ghosh, B.K. De, A. Santra and D. Chakraborty, *Int. J. Epidemiol.* 27, 871 (1998).

[32] M. Tondel, M. Rahman, A. Magnuson, O.A. Chowdhurt, M.H. Faruquee and S.A. Ahmad, *Environ. Health Perspect.* 107, 727 (1999).

[33] D.N.G. Mazumder, J.D. Gupta, A. Santra, A. Pal, A. Ghose and S. Sarkar, *J. Indian Med. Assoc.* 96, 4 (1998).

[34] D.N.G. Mazumder, C. Steinmaus, P. Bhattacharya, O.S. von Ehrenstein, N. Ghos and M. Gotway, *Epidemiol.* 16, 760 (2005).

[35] S.A.Ahmad, M. Sayed, S.A. Hadi, M.H. Faruquee, M.H. Khan and M.A. Jalil, Intern. J. *Environ. Health Res.* 9, 187 (1999).

[36] A.H. Smith, G. Marshall, Y. Yuan, C. ferreccio, J. Liaw and O. Ehrenstein, *Environ. Health Perspect.* 114, 1293 (2006).

[37] C.J. Chen, Y.M. Hsueh, M.S. Lai, M.P. Shyu, S.Y. Chen, M.M. Wu, T.L. Kuo, and T.Y. Tai, *Hypertension* 25, 53 (1995).

[38] H.Y. Chiou, W.I. Huang, C.L. Su, S.F. Chang, Y.H. Hsu and C.J. Chen, *Stroke* 28, 1717 (1997).

[39] S.L. Wang, J.M. Chiou, C.J. Chen, C.H. Tseng, W.L. Chou, C.C. Wang, T.N. Wu and L.W. Chang, Environ. Health Perspect. 111, 155 (2003).

[40] W.P. Tseng, *Angiology* 40, 547 (1989).

[41] M.S. Lai, Y.M. Hsueh, C.J. Chen, M.P. Shyu, S.Y. Chen and T.L. Kuo, *Am. J. Epidemiol.* 139, 484 (1994).

[42] C.H. Tseng, C.P. Tseng, H.Y. Chiou, Y.M. Hsueh, C.K. Chong and C.J. Chen, *Toxicology Letters* 133, 69 (2002).

[43] L.M. Del Razo, G.G. García-Vargas, O.L. Valenzuela, E. Castellanos, L.C. Sánchez-Peña, J.M. Currier, Z. Drobná, M. Stýblo, Environmental Health: A Global Access Science Source 10, art. No. 73 (2011).

[44] S.A. Ahmad M.H. Sayed, S. Barua, M.H. Khan, M.H. Faruquee, A. Jalil, S.A. Hadi, and H.K. Talukder, *Environm. Health Perspect.* 109, 629 (2001).

[45] C.H. Wang, J.S. Jeng, P.K. Yip, C.L. Chen, L.I. Hsu, Y.M. Hsueh, H.Y. Chiou, M.M. Wu and C.J. Chen, *Circulation* 105, 1804 (2002).

[46] C.H. Tseng, C.K. Chong, C.P. Tseng, Y.M. Hsueh, H.Y. Chiou, C.C. Tseng and C.J. Chen, *Toxicology Letters* 137, 15 (2003).

[47] A.H. Milton, W. Smith, B. Rahman, Z. Hasan, U. Kulsum and K. Dear, *Epidemiol.* 16, 82 (2005).

[48] G.A. Wasserman, X. Liu, F. Parvez, H. Ahsan, P. Factor-Litvak, A. Geen, V. Slavkovich, *Environ. Health Perspect.* 112, 1329 (2004).

[49] O.S. von Ehrenstein, D.N.G. Mazumder, M. Hira-Smith, N. Ghosh, Y. Yuan, G. Windham, A. Ghosh, *Am. J. Epidemiol.* 163, 662 (2006).

[50] O.S. von Ehrenstein, S. Poddar, Y. Yuan, D.G. Mazumder, B. Eskenazi, A. Basu, M. Hira-Smith, *Epidemiol.* 18, 44 (2007).

[51] M.N. Asadullah and N. Chaudhury, *Econ. Education Rev.* 30, 873 (2011).

[52] IARC, "Arsenic and arsenic compounds (Group 1)", IARC monographs on the evaluation of carcinogenic risk to humans (supplement 7), 1987, (Lyon, France) p100.

[53] IARC, "Some drinking-water disinfectants and contaminants including arsenic, Monographs on the evaluation of carcinogenic risks to humans, WHO 2004 (Lyon, France) p84.

[54] C.J. Chen, T.L. Kuo and M.M. Wu, Lancet 1, 414 (1988).

[55] H.R. Cluadia, M.L. Biggs and A.H. Smith, Intern. *J. Epidemiol.* 27, 561 (1998).

[56] K.H. Morales, L. Ryan, T.L. Kuo, M.M. Wu and C.J. Chen, *Environ. Health Perspect.* 108, 655 (2000).

[57] C.L. Chen, L.I. Hsu, H.Y. Chiou, Y.M. Hsueh, S.Y. Chen, M.M. Wu and C.J. *Chen, J. Am. Med. Assoc.* 292, 2984 (2004).

[58] ATSDR, Toxicological Profile for Arsenic, U.S. Department of Helalth and Human Services, Atlanta.

[59] P.L. Smedley and D.G. Kinniburgh, *Appl. Geochem.* 17, 517 (2002).

[60] WHO, "Guidelines for drinking-water quality. Volume 1: Recommendations, 2nd ed." World Health Organization, Geneva (1993).

[61] WHO, "Environmental Health Criteria 224: Arsenic compounds 2nd edition", World Health Organisation, Geneva (2001).

[62] H.B. Nicolli, J.W. García, C.M. Falcón, P.L. Smedley, P.L. Environ. Geochem. Heanlth in press (2011).

[63] A.H. Smith, M. Goycolea, R. Haque and M.L. Biggs, Am. J. Epidemiol. 147, 660 (1998).

[64] L.M. Del Razo, M.A. Arellano and M.E. Cebrián, *Environ. Pollut.* 64, 143 (1990).

[65] F.N. Robertson, Environ. *Geochem. Health* 11, 171 (1989).

[66] M. Currell, I. Cartwright, M. Raveggi and D. Han, *Appl. Geochem.* 26, 540 (2011).

[67] N.M. Phuong, Y. Kang, K. Sakurai, M. Sugihara, C.N. Kien, N.D. Bang, H.M. Ngoc, *Environ. Monit. Assess.* in press (2011).

[68] L. Winkel, M. Berg, M. Amini, S.J. Hug, C. Annette Johnson, *Nature Geosci.* 1, 536 (2008).

[69] I.C. Yadav, U.P. Dhuldhaj, D. Mohan and S. Singh, *Environ. Rev.* 19, 55 (2011).

[70] S.W. Wang, Y.M. Kuo, Y.H. Kao, C.S. Jang, S.K. Maji, F.J. Chang and C.W. Liu, *J. Hydrol.* 408, 286 (2011).

[71] K.H. Cho, S. Sthiannopkao, Y.A. Pachepsky, K.W. Kim, J.H. Kim, Water Res. 45, 5535 (2011).

[72] H.A.L. Rowland, E.O. Omoregie, R. Millot, C. Jimenez, J. Mertens, C. Baciu, S.J. Hug and M. Berg, *Appl. Geochem.* 26, 1 (2011).

[73] A. Aiuppa, W. D'Alessandro, C. Federico, B. Palumbo and M. Valenza, *Appl. Geochem.* 18, 1283 (2003).

[74] M. Aloupi, M.O. Angelidis, A.M. Gavriil, M. Koulousaris, S.P. Varnavas, *Environ. Monit. Assess.* 151, 383 (2009).

[75] H.B. Nicolli, J.M. Suriano, M.A.G. Peral, L.H. Ferpozzi and O.A. Baleani, *Environ. Geol. Water Sci.* 14, 3 (1989).

[76] H.B. Nicolli, A. Tineo, J.W. Garcia, C.M. Falcon and M.H. Merino, "Trace-element quality problems in groundwater from Tucuman, Argentina", Proceedings of the 10th international symposium on water-rock interaction WRI-10, Jul. 2001 (Villasimius, Italy) pp993–996.

[77] L. Cáceres, E. Gruttner and R. Contreras, *Ambio* 21, 138 (1992).

[78] S. Karcher, L. Cáceres, M. Jekel and R. Contreras, J. Chart. Inst. *Water Environ. Manag.* 13, 164 (1999).

[79] P.L. Smedley, H.B. Nicolli, D.M.J. Macdonald, A.J. Barros and J.O. Tullio, *Appl. Geochem.* 17, 259 (2002).

[80] M. Claesson and J. Fagerberg, "Arsenic in groundwater of Santiago del Estero. Argentina. Sources, mobilization, controls and remediation with natural materials", Minor Field Studies Scholarship Programme MFS, Royal Institute of Technology (KTH), Stockholm (2003).

[81] R. Vivona, E. Preziosi, B. Madé and G. Giuliano, Hydrogeol. J. 15, 1183 (2007).

[82] L. Achene, E. Ferretti, L. Lucentini, P. Pettine, E. Veschetti and M. Ottaviani, Toxicol. *Environ. Chem.* 92, 509 (2010).

[83] L. Wang and J. Huang, "Chronic arsenism from drinking water in some areas of Xinjiang, China", Arsenic in the Environment, Part II: Human Health and Ecosystem Effects, John Wiley, New York, pp.159–172 (1994).

[84] C. Zhai, G. Dai, Z. Zhang, H. Gao, and G. Li, "An environmental epidemiological study of endemic arsenic poisoning in Inner Mongolia" Abstr. 3rd Internat. Conf. Arsenic Exposure and Health Effects, San Diego, 1998, p.17.

[85] M. Berg, H.C. Tran, T.C. Nguyen, H.V. Pham, R. Schertenleib and W. Giger, *Environ. Sci. Technol.* 35, 2621 (2001).

[86] BGS, DPHE, "Arsenic contamination of groundwater in Bangladesh", British Geological Survey (Technical Report, WC/00/19. 4 Volumes), British Geological Survey, Keyworth (2001).

[87] E.S. Gurzau and A.E. Gurzau, "Arsenic in drinking water from groundwater in Transylvania, Romania" Arsenic Exposure and Health Effects IV, Elsevier, Amsterdam, pp.181–184 (2001).

[88] D.S. Vinson, J.C. McIntosh, G.S. Dwyer and A. Vengosh, *Appl. Geochem.* 26, 1364 (2011).

[89] R. Cidu and S. Bahaj, Geothermics 29, 407 (2000).

[90] T.W. Horton, J.A. Becker, D. Craw, P.O. Koons and C.P. Chamberlain, *Chem. Geol.* 177, 323 (2001).

[91] A. Ure and M. Berrow, Environmental Chemistry, H.J.M. Bowen, Royal Society of Chemistry, London (1982).

[92] W.H. Ficklin and E. Callender, "Arsenic geochemistry of rapidly accumulating sediments, Lake Oahe, South Dakota", US Geol. Surv. Water Resour. Investig. Rep. 88–4420, U.S. Geol. Surv. Toxic Substances Hydrology Program—Proc. Tech. Meeting, Sept. 1988 (Phoenix, Arizona) pp.217–222.

[93] J.M. Azcue and J.O. Nriagu, J. Geochem. Explor. 52, 81 (1995).

[94] M. Williams, F. Fordyce, A. Paijitprapapon and P. Charoenchaisri, *Environ. Geol.* 27, 16 (1996).

[95] M. Williams, "Mining-related Arsenic Hazards: Thailand Case-study", Summary Report. Brit. Geol. Surv. Tech. Rep., WC/97/49 (1997).

[96] L. Altas, M. Isik and M. Kavurmaci, J. Environ. Manag. 92, 2182 (2011).

[97] D. Mohan and C.U. Pittman, *J. Hazard. Mater.* 142, 1 (2007)

[98] S.K.R.Yadanaparthi, D. Graybill and R.von Wandruszka, *J. Hazard. Mater.* 171, 1 (2009)

[99] T. Budinova, N. Petrov, M. Razvigorova, J. Parra, P. Galiatsatou, *Ind. Eng. Chem. Res.* 45, 1896 (2006)

[100] P. Navarro and F.J. Alguacil, *Hydrometallurgy* 66, 101 (2002).

[101] L.V. Rajakovic, *Sep. Sci. Technol.* 27, 1423 (1992).

[102] M.N. Amin, T. Kaneco, T. Kitagawa, A. Begum, H. Katsumata, T. Suzuki, K. Ohta, *Ind. Eng. Chem. Res.* 45, 8105 (2006).

[103] D. Mohan, C.U. Pittman, M. Bricka, F. Smith, B. Yancey, J. Mohammad, P.H. Steele, M.F. Alexandre-Franco, V.G. Serrano, H. Gong, J. Colloid Interf. Sci. 310, 57 (2007).

[104] J. Pattanayak, K. Mondai, S. Mathew, S.B. Lalvani, *Carbon* 38, 589 (2000).

[105] H.S. Altundogan, S. Altundogan, F. Tumen and M. Bildik, *Waste Manag.* 20, 761–767 (2000).

[106] H. Genc-Fuhrman, J.C. Tjell and D. McConchie, *J. Colloid Interface Sci.*, 271, 313–320 (2004).

[107] H. Genc-Fuhrman, J.C. Tjell and D. McConchie, *Environ. Sci. Technol.*, 38, 2428–2434 (2004).

[108] J. Su, H-G. Huang, X-Y. Jin, X-Q. Lu and Z-L. Chen, *J. Hazard. Mater.* 185, 63 (2011).

[109] S. Goldberg, *Soil Sci. Soc. Am. J.* 66, 413 (2002).

[110] H. Matsunaga, T. Yokoyama, R.J. Eldridge and B.A. Bolto, *React. Funct. Polym.*, 29, 167 (1996).

[111] A.C.Q. Ladeira and V.S.T. Ciminelli, *Water Res.* 38, 2087 (2004).

[112] H. Zhang and H.M. Selim, *Environ. Sci. Technol.*, 39, 6101 16 (2005).

[113] Z. Li, R. Beachner, Z. McManama and H. Hanlie, *Micropor. Mesopor. Mater.* 105, 291 (2007).

[114] E.J. Sullivan, R.S. Bowman and I.A. Legiec, *J. Environ. Qual.* 32, 2387 (2003).

[115] M.P. Elizalde-Gonzalez, J. Mattusch, W.-D. Einicke and R. Wennrich, *Chem. Eng. J. (Lausanne)*, 81, 187 (2001).

[116] J. Saikia, B. Saha and G. Das, *J. Hazard. Mater.* 186, 575 (2011).

[117] D.B. Singh, G. Prasad and D.C. Rupainwar, *Colloid Surf. A*, 111, 49–56 (1996).

[118] S.M.I. Sajidu, I. Persson, W.R.L. Masamba, E.M.T. Henry and D. Kayambazinthu, *Water SA*, 32, 519 (2006).

[119] S.K. Maji, A. Pal and T. Pal, *J. Environ. Sci. Health, A*, 42, 453 (2007).

[120] E. Deschamps, V.S.T. Ciminelli and W.H. Höll, *Water Res.*, 39, 5212 (2005).

[121] J. Hlavay and K. Polyak, *J. Colloid Interface Sci.*, 284, 71 (2005).

[122] G. Zhang, J. Qu, H. Liu, R. Liu and R. Wu, *Water Res.*, 41, 1921 (2007).

[123] S.-Z. Gao, H.-Q. Jiu, J.L. HuI, P.L. Rui and T.L. Guo, *Environ. Sci. Technol.* 41, 4613 (2007).

[124] A.C.Q. Ladeira and V.S.T. Ciminelli, *Water Res.*, 38, 2087–2094 (2004).

[125] R. Tobias, C. Jordi, A. Carlos, J.-L. Cortina and J. De Pablo, *Environ. Sci. Technol.*, 40, 6438 (2006).

[126] B. Daus, R. Wennrich and H. Weiss, *Water Res.*, 38, 2948–2954 (2004).

[127] F.-S. Zhang and H. Itoh, *Chemosphere*, 65, 125 (2006).

[128] S. Bang, M. Patel, L. Lippicott and X.G. Meng, *Chemosphere*, 60, 389 (2005).

[129] Y. Kiso, Y.J. Jung, T. Yamada, M. Nagai and K.S. Min, *Water Sci. Technol. Water Supply*, 5, 75 (2005).

[130] L. Yang, Z. Shahrivari, P.K.T. Liu, M. Sahimi and T.T. Tsotsis, *Ind. Eng. Chem. Res.*, 44, 6804 (2005).

[131] S.H. Abdel-Halim, A.M.A. Shehata and M.F. EI-Shahat, *Water Res.* 37, 1678 (2003).

[132] V. Lenoble, C. Laclautre, V. Deluchat, B. Serpaud and J.-C. Bollinger, *J. Hazad. Mater.* 123, 262 (2005).

[133] H. Yoshitake, T. Yokoi and T. Tatsumi, *Chem. Mater.* 15, 1713 (2003).

[134] D.D. Gang, B. Deng and L. Lin, *J. Hazad. Mater.* 182, 156 (2010).

[135] B.J. Mcafee, W.D. Gould, J.C. Nedeau and A.C.A. da Costa, *Sep. Sci. Technol.* 36, 3207 (2001).

[136] Y. Zhao, M. Huang, W. Wu and W. Jin, *Desalination* 249, 1006 (2009).

[137] X. Guo and F. Chen, *Environ. Sci. Technol.* 39, 6808 (2005).

[138] H. Seki, A. Suzuki and H. Maruyama, *J. Colloid Interface Sci.* 281, 261 (2005).

[139] V.K. Gupta, C.K. Jain, I. Ali, M. Sharma and V.K. Saini, *Water Res*, 37, 4038 (2003).

Reviewed by

Basilios Tsikouras, Geology Department,
University of Patras, Greece, 26504.

In: Arsenic
Editor: Andrea Masotti

ISBN: 978-1-62081-320-1
© 2013 Nova Science Publishers, Inc.

Chapter 16

ARSENIC REMOVAL IN CONTINUOUS SYSTEMS USING ZERO-VALENT IRON FIXED BEDS

Fernando S. García Einschlag[1] *and Juan M. Triszcz[2]*

[1]Instituto de Investigaciones Fisicoquímicas Teóricas y Aplicadas (INIFTA), CCT-La Plata-CONICET, Departamento de Química, Facultad de Ciencias Exactas, Universidad Nacional de La Plata. CP, Argentina
[2]Laboratorio de Ingeniería Sanitaria (LIS), Departamento de Hidráulica, Facultad de Ingeniería, Universidad Nacional de La Plata, CP, Argentina

ABSTRACT

The aim of this chapter is to describe fundamental concepts associated with the removal of soluble arsenic species using continuous systems based on the zero-valent iron technology. The chapter is divided into four sections. The introduction describes the basic chemical reactions involved in the formation of iron corrosion products that are closely related to arsenic removal.

In addition, a brief description of the procedures commonly used to characterize arsenic uptake in batch systems is given. The second section deals with the main ideas associated with the removal of contaminants by packed-bed sorption systems since several concepts can usually be applied to understand the behavior of continuous systems that use fixed beds containing iron as reactive material. The third section focuses on ZVI systems designed to continuously remove arsenic from drinking water. After a brief comparison between reactive ZVI-based beds and sorptive beds, a survey of the data reported on arsenic removal by means of ZVI-based fixed-bed columns is given. The analysis includes column configurations, mechanisms and kinetics of arsenic uptake as well as bed clogging effects and ZVI reactivity losses during long-term tests. In addition, some results obtained by using two continuous ZVI systems operated in our laboratory at very different dissolved oxygen concentrations are discussed. Finally, the fourth section presents the concluding remarks of this chapter.

INTRODUCTION

Groundwater contamination by arsenic has become an issue of worldwide concern since arsenic concentrations in natural waters above its maximum contaminant level (i.e., 0.01 ppm) have been reported in many regions of the world. Long-term exposure to high levels of arsenic in drinking water may cause skin alteration, damage to major body organs and many types of cancer. Most areas exposed to arsenic-contaminated groundwater are situated in developing countries.

Several techniques have been proposed for arsenic removal from water including precipitation, coagulation and filtration, reverse osmosis, electrodialysis, ion-exchange and adsorption. Arsenic removal by adsorption is gaining importance due to its technical simplicity and easy applicability in rural areas.

The predominant oxidation states of inorganic arsenic in water are As(III) (arsenite) and As(V) (arsenate). The pH of most natural waters lies between 5 and 9, the important As species being $H_2AsO_4^-$, $HAsO_4^{-2}$ and H_3AsO_3 [1, 2]. It has been reported that soluble arsenic species are efficiently removed by iron corrosion products owing to the formation of strong surface complexes, the uptake being virtually irreversible at typical pH values of drinking water. The use of zero-valent iron (ZVI) technology for arsenic removal has recently gained attention due to its applicability under different conditions, operational simplicity and low-cost maintenance.

It is worth mentioning that both As(III) and As(V) can efficiently be removed from aqueous solution in ZVI-based plants. Most iron-based treatment methods are more effective in removing arsenic in its pentavalent state rather than in its trivalent state, therefore As(III) species usually require oxidation as a pretreatment step. In comparison with other methods, ZVI can simultaneously remove As(V) and As(III) without previous oxidative treatment and does not require the use of additional chemicals, since metallic iron is used for the sustained production of colloidal hydrous ferric oxides (HFO) [3, 4].

Small plants based on the ZVI technique are advantageous since low-cost and easy to obtain materials, such as iron wool or iron fillings, can be used as the source of fresh and effective sorbent phases for inorganic As(III) and As(V) removal. In addition, they are usually built with standard plastic materials (PVC and polypropylene pipes and fittings used in plumbing), do not use electric power, may be operated by people without technical knowledge, and the residues do not cause environmental problems.

Iron Corrosion and Mechanisms for Arsenic Removal

The mechanism for arsenic removal by ZVI is rather complex since different processes are involved. It is generally accepted that As uptake from aqueous systems at near neutral pH is based on adsorption and co-precipitation phenomena coupled with the continuous generation of iron oxyhydroxides [5, 7]. However, depending on the operating conditions different processes, such as reactions associated with iron corrosion, arsenic speciation, co-precipitation and adsorption, may have an important role in controlling the overall effectiveness of arsenic removal.

The spontaneous chemical oxidation of ZVI in the presence of dissolved oxygen (i.e., corrosion) is a complex process involving a variety of metastable ferrous–ferric intermediate species, which are ultimately transformed into different stable iron oxides. It is known that Fe(II), which is the primary product of ZVI oxidation, may stay at the interface, where it may form ferrous precipitates (i.e., carbonate, hydroxide, etc.) or oxidize to form Fe(III) species. In addition, Fe(II) may be transported away from the surface, where it is subject to homogeneous oxidation and precipitation of Fe(III) [3, 8, 9]. During the oxidation of Fe(II) in aerobic environments, the transformation of arsenite to arsenate can be promoted by iron species [10]. The latter process is followed by the uptake of As(V) in the presence of the forming hydrous ferric oxides (HFO).

Previous studies have shown that adsorption, surface precipitation and co-precipitation appear to be the predominant mechanisms for arsenic removal by ZVI. The relative importance of each process is strongly dependent on operating conditions such as dissolved oxygen concentration, pH and aqueous matrix composition. The stratified nature of the corrosion coating formed on the metallic iron surface involves different stable and metastable phases of iron oxides and green rusts [5, 11, 12]. The formation of corrosion products on the surface of ZVI results in the creation of sites for both As(III) and As(V) uptake. The suggested mechanism for arsenic removal comprises the formation of strong inner-sphere bidentate As(III) and As(V) complexes with iron corrosion products ([13] and references therein).

In spite of the physicochemical complexity associated with the removal of arsenic by ZVI systems, reactions R1–R6 [2-6, 14-18] roughly represent the most relevant processes that may be involved in the formation of iron species associated with arsenic uptake at near neutral pH values.

$$Fe(0) + 2\,H_2O \rightarrow Fe(II) + H_2 + 2\,HO^- \qquad \textbf{R1}$$

$$Fe(0) + H_2O + 1/2\ O_2 \rightarrow Fe(II) + 2\,HO^- \qquad \textbf{R2}$$

$$Fe(0) + 2\,Fe(III) \rightarrow 3\,Fe(II) \qquad \textbf{R3}$$

$$Fe(II) + 1/2\,H_2O + 1/4\ O_2 \rightarrow Fe(III) + HO^- \qquad \textbf{R4}$$

$$Fe(III) + 3\,H_2O \rightarrow Fe(OH)_3 + 3\,H^+ \qquad \textbf{R5}$$

$$Fe(II) + CO_3^{-2} / 2\,HO^- \rightarrow \text{Ferrous precipitates} \qquad \textbf{R6}$$

As stated above, arsenic removal may involve parallel processes [17, 18] such as adsorption onto iron corrosion products, co-precipitation and sequestration in the matrix of iron corrosion products. It should be pointed out that the analysis of the uptake mechanism is rather complex since, given the dynamic nature of the Fe(0)/H$_2$O interface, these removal processes take place in a changing environment.

Arsenic Uptake Studies in Batch Systems

The knowledge of the factors that govern both kinetic and equilibrium aspects of arsenic uptake is critical for the design of efficient treatment plants. In order to evaluate the performance of a given material for arsenic uptake, studies are usually carried out in batch systems. Since the removal of soluble arsenic species by ZVI systems is due to their interaction with solid surfaces, quantitative adsorption models are usually applied to describe arsenic uptake. The amount of arsenic removed per unit mass of a given solid (q) and the adsorption efficiency (%R), determined as the removal percentage relative to the initial concentration in batch systems, are given by Eq. 1 and Eq. 2, respectively

$$q = \frac{(C_0 - C)}{w_{Solid}} \times V \qquad (1)$$

$$\%R = \frac{(C_0 - C)}{C_0} \times 100 \qquad (2)$$

where C_0 is the initial concentration, V is the reactor volume and w_{Solid} is the mass of solid material used for arsenic removal. The values of q and %R values can be used in both equilibrium studies (for the evaluation of q_e and $\%R_e$ by using $C = C_e$) and kinetic studies (for the evaluation of q_t and $\%R_t$ by using $C = C_t$).

Kinetic Analysis

The study of removal kinetics derives significance as it ultimately controls the residence time of solute uptake at the solid–solution interface. Several researchers have analyzed the rates of arsenic removal in the presence of different sorbent materials. In general, batch experiments are conducted to determine the reaction time required to reach sorption equilibrium and the experimental data are usually evaluated using different simple kinetic models. Table 1 summarizes simple models commonly applied to analyze uptake kinetics ([19, 20] and references therein). However, it should be emphasized that, owing to surface heterogeneities and the influence of transport phenomena, simple kinetic models may not be able to accurately describe sorption processes that involve porous materials. For example, in batch experiments it is rather frequent to observe an initial fast phase followed by a much slower one. The adsorption of a solute by a porous sorbent involves three consecutive steps: (i) transport of the adsorbate from the solution bulk to the external surface of the sorbent (film transport), (ii) transport of the sorbate within the pores of the adsorbent (intra-particle diffusion), and (iii) adsorption on the sorbent surface.

Table 1. Simple equations for describing adsorption kinetics

Model	Kinetic equation	Parameters
Pseudo-first order	$q_t = q_e[1 - \exp(-k_1.t)]$	q_e: sorption capacity at equilibrium. q_t: sorption capacity used at time t. k_1: pseudo-first-order rate constant.
Pseudo-second order	$q_t = \dfrac{k_2.q_e^2.t}{1 + k_2.q_e.t}$	q_e: sorption capacity at equilibrium. q_t: sorption capacity used at time t. k_2: pseudo-second-order rate constant.
Pore Diffusion	$q_t = k_{id}.t^{0.5}$	q_t sorption capacity used at time t. k_i: pore diffusion rate constant.

Since the relative importance of each step may depend both on operating conditions (such as mixing speed, temperature and sorbate concentration) and sorbent characteristics (such as particle size, sorbent-sorbate affinity and porosity of the particles), simple models only give a rough description of uptake rates.

Adsorption Isotherms

At a constant temperature, the equilibrium amount of solute adsorbed by a certain sorbent depends on the solute concentration, the resulting function being the adsorption isotherm. Among the models most frequently used to describe the adsorption equilibrium are the Langmuir, Freundlich and Dubinin-Radushkevich adsorption isotherms shown in Table 2 ([19-21] and references therein)

Table 2. Simple adsorption isotherms

Model	Isotherm	Parameters
Langmuir	$q_e = \dfrac{b\, q_m\, C_e}{1 + b\, C_e}$	qm: saturation sorption capacity. b: adsorption equilibrium constants. C_e: equilibrium solution concentration. q_e: amount of solute adsorbed at equilibrium.
Freundlich	$q_e = k\, C_e^{1/n}$	C_e: equilibrium solution concentration. q_e: amount of solute adsorbed at equilibrium. k: adsorption capacity. n: adsorption intensity.
Dubinin-Radushkevich	$q_e = q_m \times \exp\!\left(-\beta.\varepsilon^2\right)$	qm: theoretical saturation sorption capacity. ε is the Polanyi potential $\varepsilon = RT.ln[1+1/Ce]$ and β is related to the mean free energy of sorption (i.e., $E = [2\beta]^{-0.5}$).

The Langmuir isotherm assumes monolayer chemisorption. The adsorption occurs uniformly on the active sites of the adsorbent, and once an adsorbate occupies a site, no further adsorption can take place at this site. The most useful aspect of the Langmuir isotherm is for obtaining the sorbent saturation capacity. The Freundlich isotherm is an empirical model that is widely used to describe adsorption on surfaces having heterogeneous energy distribution, although there is no saturation and the equation does not reduce to Henry's law at low concentrations. The Dubinin-Radushkevich is a semiempirical equation used to describe adsorption on microporous materials. It is more general than the Langmuir isotherm, because it does not assume a homogeneous surface. Based on the Polanyi potential theory, the Dubinin-Radushkevich equation can be used to estimate the mean free energy of sorption [22, 23].

FIXED-BED SORPTION SYSTEMS

Contaminant Uptake in Packed Column Beds

Many of the reported studies for arsenic removal by sorption processes have been conducted in batch operation mode in order to evaluate sorbent strength, sorption capacities and sorption kinetics [20, 24]. Despite the fact that results in batch conditions may be used for a preliminary characterization of sorption systems, equilibrium and kinetic parameters obtained in batch conditions do not necessarily give accurate scale-up data [25]. Most separation and purification processes that employ the sorption technology use continuous-flow columns [26]. The fixed-bed operation constitutes the appropriate configuration mode for large-scale applications such as water and wastewater treatment. This operating mode ensures the highest possible concentration difference driving force. Starting at the inlet, the zone of saturated sorbent gradually extends throughout the column, with the sorbate eventually breaking through the column [27]. The process is performed until contaminant concentration in the effluent exceeds the allowable level of stream pollution. This moment is defined as a breakthrough point of the bed since it loses its protective capabilities. Given that the main aim in designing sorption columns is to predict how much effluent the bed can treat or how long the bed will last, it is necessary to carry out fixed-bed continuous flow tests both at laboratory and pilot scales [25, 28].

Sorptive Column Processes

In packed-bed sorptive systems, the flow of contaminated water creates a wave front, known as the mass transfer zone (MTZ), which moves from the inlet to the outlet as time increases. Behind the MTZ the sorbent is completely exhausted, within the MTZ the degree of sorbent saturation varies from 100 percent to effectively zero, and beyond the leading edge of the MTZ lays unused sorbent. The MTZ travels through the column until it reaches the outlet end, after which the bed reaches its breakthrough point (Fig. 1a). During the displacement of the MTZ across the bed, the ratio C/C_0 is a function of the relative distance from the inlet to the outlet (Figure 1b).

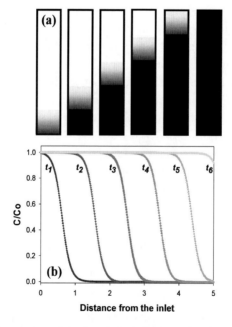

Figure 1. (a) Schematic representation of the fraction of used sorbent bed at different service times. (b) Profiles of contaminant concentration inside the fixed bed at different service times.

The dynamic behavior of fixed-bed continuous-flow systems is described in terms of breakthrough curves [26, 27], obtained by plotting the values of C/C_0 at the outlet against the treatment time (or the volume passed through the column). Breakthrough curves usually display a typical S-shaped profile. For a given contaminant, both the time for the breakpoint appearance and the slope of the breakthrough curve depend on the properties of the fixed bed as well as on the operating conditions. Figure 2 depicts a typical effluent concentration profile and also indicates both the breakthrough and exhaustion points.

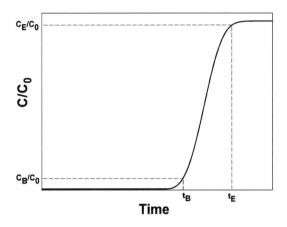

Figure 2. Typical shape of a breakthrough curve indicating the breakthrough and exhaustion points.

Simple Mathematical Models for Describing Breakthrough Curves

The nature of breakthrough curves depends not only on the type of adsorption isotherm governing the static equilibrium, but also on the mass transfer to and throughout the sorbent in the column [29]. Since sorption from water on a fixed bed is a process of unsteady-state mass transfer between the liquid and solid phases, process kinetics may be controlled by several independent processes such as film diffusion, intra-particle diffusion and chemisorption [30]. The quantitative description of the mass transfer zone (MTZ) problem involves sets of partial differential equations and usually requires complex numerical methods to solve. Therefore, various simple mathematical models have been developed to predict the dynamic behavior of fixed-bed columns. The main assumptions and the equations describing the breakthrough profiles for frequently used simplified models are listed in Table 3 ([24, 27, 31] and references therein).

With the aim of illustrating several issues concerning breakthrough curves, a short explanation of the Bohart-Adams approach will be given in the following paragraphs [32, 33]. In order to describe how the MTZ progresses through a single fixed bed of adsorbent, the method assumes that the adsorption rate is proportional to both the residual adsorbent capacity (a) and the remaining adsorbate concentration (C_b). The mass transfer rates for the fixed adsorbing material and liquid mobile phase obey the following equations, respectively:

$$\frac{\partial a}{\partial t} = -k_{AB}\, a\, C_b \tag{3}$$

$$\frac{\partial C_b}{\partial z} = -\frac{k_{AB}}{v_{Filt}}\, a\, C_b \tag{4}$$

where k_{AB} ($[mg/L]^{-1}.min^{-1}$) is the kinetic constant of the bed and v_{Filt} (cm/min) is the filtration rate. Equation 3 describes the decrease in the unused sorption capacity with increasing service time. Given that "a" and "C_b" are equal to zero behind and beyond the MTZ, respectively, Eq. 4 shows that the slope of the liquid phase concentration gradient is proportional to k_{AB}/v_{Filt} within the MTZ and zero elsewhere. Taking into account that the mass transfer zone is defined as the region where the ratio C_b/C_0 varies from 1 to effectively 0, the length of the MTZ decreases with an increasing adsorption rate and increases with an increasing influent flow rate.

Table 3. Simplified breakthrough equations[††]

Model	Breakthrough curve	Main assumptions
Bohart-Adams	$$\frac{C}{C_0} = \frac{1}{1+\exp\left(\frac{k_{AB}.N_o.Z}{v_{Filt}} - k_{AB}.C_0.t\right)}$$	No axial dispersion. Sorption rate controlled by surface reaction between sorbate and residual sorbent capacity.
Thomas	$$\frac{C}{C_0} = \frac{1}{1+\exp\left(k_T\,(q_o m_c - C_0 V_{eff})/Q\right)}$$	Plug flow conditions. Langmuir isotherm and second-order kinetics. Film mass transfer resistance and intra-particle diffusion absent.
Yoon-Nelson	$$\frac{C}{C_0} = \frac{\exp[k_{YN}\times(t-\tau)]}{1+\exp[k_{YN}\times(t-\tau)]}$$	The rate of decrease in the probability of sorption is proportional to the probability of sorbate sorption and sorbate breakthrough.
Wolborska	$$\frac{C}{C_0} = \exp\left(\frac{\beta_a\,C_0}{N_o}\,t - \frac{\beta_a\,Z}{v_{Filt}}\right)$$	General mass transfer equation. Kinetic coefficient dependent on flow rate.
Clark	$$\frac{C}{C_0} = \left(\frac{1}{1+A\,e^{-rt}}\right)^{1/n-1}$$	Mass transfer coefficient in combination with Freundlich isotherm.

[††] The parameters k_{AB}, k_T, k_{YN}, β_a and r are related to the kinetics of the uptake process, whereas N_0, $q_0.m$, τ, N_0 and A are related to the total uptake capacity of each simplified model. In addition, Z, v_{Filt} and Q represent the bed height, the filtration velocity and the volumetric flow rate, respectively.

After integration of Eqs. 3 and 4, the following expression for describing breakthrough curves is obtained:

$$\frac{C}{C_0} = \frac{1}{1 + \exp\left(\dfrac{k_{AB}.N_o.Z}{v_{Filt}} - k_{AB}.C_0.t\right)} \tag{5}$$

where C is the concentration in the column effluent, N_0 is the maximum sorption capacity (mg/L) and Z is the bed depth (cm). Equation 5 may be rearranged to yield

$$\frac{C}{C_0} = \frac{1}{1 + \exp(EBCT.k_{AB} \times [N_o - C_0.N_{BV}])} \tag{6}$$

where EBCT (= V_{Column}/Q) is the empty bed contact time and N_{BV} (= V/V_{Column}) is the number of bed volumes passed through the column. Figure 2 shows the effects of the operating parameters (i.e., EBCT and C_0) and the fixed-bed properties (i.e., N_0 and k_{AB}) on the breakthrough curves calculated by using Eq. 6.

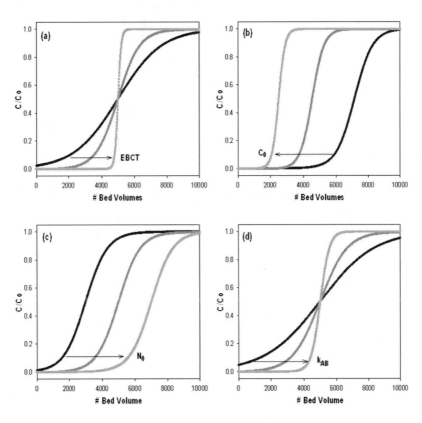

Figure 3. Influence of operating conditions and sorbent properties on the shape of breakthrough curves. (a) Effect of EBCT. (b) Effect of initial influent concentration. (c) Effect of sorbent removal capacity. (d) Effect of kinetic constant.

According to Figure 3a, the slope of the breakthrough curve decreases with decreasing EBCT. In addition, although the number of bed volumes needed to reach 50% of the inlet concentration is independent of the EBCT (i.e., $N_{BV}^{50\%} = N_0/C_0$), the number of bed volumes treated before the breakpoint diminishes with decreasing EBCT. Since the rate of migration of a given species involves advection and diffusion [28], the length of the MTZ is a function of the influent flow rate and the rate of adsorption [34]. Inspection of Eq. 4 suggests that as v_{Filt} increases, the length of the MTZ also increases. For any predetermined treatment objective, minimizing the length of the MTZ maximizes sorption capacity to be utilized.

The increase in inlet concentration increases the slope of the breakthrough curve and reduces the number of bed volumes treated before the breakpoint (Figure 3b). These trends can be explained taking into account that as the inlet concentration increases, the length of the MTZ is reduced and the rate of sorbent usage increases due to the higher contaminant loading. Figure 3c shows that, as expected, the number of treatable bed volumes increases with increasing column capacity (N_0). Finally, the trend observed in Fig. 3d is very similar to the one discussed for Fig 3a since the length of the MTZ is proportional to the ratio k_{AB}/v_{Filt}.

Experimental Trends in Breakthrough Curves

Several authors have observed the general trends predicted by the Bohart-Adams model: t_B and V_B were found to increase with increasing Z [25, 35, 36] and to decrease with increasing C_0 [29, 35, 37] or with increasing flow rate [38, 39]. In addition, the steepness of the time-based breakthrough curves increased with C_0.

Theoretically, the values of k_{AB} and N_0 should be independent of the operating conditions [40]. However, it should be taken into account that mass transport in fixed beds may be quite complex since uptake rates may be limited by film mass transfer resistance, intra-particle diffusion or adsorption.

The adsorption step is typically fast, whereas the film mass transfer and intra-particle steps are dependent on the flow rate and the particle size, respectively. Hence, the slope of the breakthrough curves may depend not only on C_0 values, but also on the flow rate, column length and bed properties.

For example, although the steepness of the time-based breakthrough curves (Eq. 5) should be independent of either the value of Z or the flow rate, it has been reported that the slope decreases with increasing Z [29], while an increase in the flow rate may either result in a decrease [29, 35] or an increase [40] of the slope. In addition, even though the total sorption capacity should be a constant independent of the operating conditions, it has been reported that N_0 may decrease [25, 36, 41], remain constant [29] or increase [37] with an increasing flow rate. The latter results show that simplified models may be used for describing general trends but may fail to give accurate descriptions of breakthrough curves in a wide range of operating conditions.

Rapid Small-Scale Column Tests (RSSCTs)

Rapid small-scale column tests (RSSCTs) may be used to simulate the performance of pilot-scale treatment systems at the bench scale [42, 43]. In the RSSCT approach, developed by Crittenden et al. [44, 45], the dynamic behavior of the full size or pilot column is simulated by using a smaller column packed with adsorbent particles of smaller size. Due to the similarity of mass transfer processes and hydrodynamic characteristics, the breakthrough curves of the small column and pilot-scale systems are correlated [46]. Hence, bench-scale systems could be quickly and inexpensively operated in order to reduce time and costs associated with pilot studies [42]. In particular, the use of RSSCTs may be very helpful for assessing the optimal operating conditions required for different water qualities.

RSSCTs, based upon dimensionless scaling of hydraulic conditions and mass transport processes, give the relationship between empty bed contact time (EBCT) and particle radius (R) at both scales:

$$\frac{EBCT_{SC}}{EBCT_{Pilot}} = \left[\frac{R_{SC}}{R_{Pilot}}\right]^{2-x} \tag{7}$$

where x, the diffusivity factor, is given by

$$\frac{D_{s,SC}}{D_{s,Pilot}} = \left[\frac{R_{SC}}{R_{Pilot}}\right]^{x} \quad \Rightarrow \quad x = \ln\left(\frac{D_{s,SC}}{D_{s,Pilot}}\right) \Big/ \ln\left(\frac{R_{SC}}{R_{Pilot}}\right) \tag{8}$$

where Ds are the surface diffusion coefficients. Scaling equations have been obtained assuming both constant diffusivity (CD, x=0) and proportional diffusivity (PD, x=1):

$$\textbf{CD:} \quad \frac{EBCT_{SC}}{EBCT_{Pilot}} = \left[\frac{R_{SC}}{R_{Pilot}}\right]^{2} \tag{9a}$$

$$\textbf{PD:} \quad \frac{EBCT_{SC}}{EBCT_{Pilot}} = \left[\frac{R_{SC}}{R_{Pilot}}\right] \tag{9b}$$

Recently reported results [47] suggest that the PD model is more appropriate than the CD model for scaling arsenic uptake and simulating arsenic breakthrough curves by porous metal oxide adsorptive media.

In addition to the scaling relationship for EBCTs, the filtration rates have to fulfill the following condition:

$$\frac{v_{Filt,SC}}{v_{Filt,Pilot}} = \frac{R_{Pilot}}{R_{SC}} \times \frac{\text{Re}_{SC,Min}}{\text{Re}_{Pilot}} \tag{10}$$

where $\text{Re}_{SC,Min}$ is the minimum value of the Reynolds number required to establish, in the small column, the minimum velocity that will not over overstress the effects of dispersion and external mass transfer.

ZVI-BASED CONTINUOUS SYSTEMS

Contaminant Removal by ZVI

In recent years, filtration experiments in continuous operation mode have been conducted by several investigators to evaluate the effectiveness of Fe(0) for arsenic removal. Fixed-bed sorption filters filled with an iron-based sorbent have been used to remove chlorinated organic compounds, heavy metals and inorganic anions such as chromate, perchlorate, nitrate and soluble arsenic species [48, 49]. The low costs and operational simplicity of ZVI filters make them suitable for point of use, individual wellhead, or other small-scale treatment systems [50]. Metallic iron is widely available and contains three times more iron per weight than most common iron salts [30]. In order to prevent bed clogging, Fe(0) is usually mixed with sand or other inert materials. ZVI/sand filters have been tested in the past two decades mainly as permeable reactive barriers for underground water treatment or as household filters for drinking water in rural areas. Contaminant removal by iron filters is based on iron corrosion (reactive filtration), whereas sand removes suspended particles that may be present in the influent water or may be formed as a result of iron corrosion (media filtration) [51].

Reactive Column Processes

As previously described by Noubactep [51], a conventional filter based on iron oxides for arsenic removal contains an inert bed with fixed sorption capacity. Hence, breakthrough is observed when the MTZ reaches the end of the column and the bed capacity is exhausted. In contrast, ZVI-based filters continuously produce iron corrosion products capable of removing soluble arsenic species. Metallic iron may be oxidized by H_2O (or by H^+ at pH values below 4) or by dissolved oxygen, the rate of the latter process being much higher in aerobic environments at near neutral pH. It has been reported that water-mediated oxidation occurs uniformly in the whole bed, whereas corrosion induced by dissolved oxygen may be located at the entrance or distributed along the bed depending on parameters such as flow rate, inlet oxygen concentration, water composition and column age [52]. Therefore, although a reaction front may exist, filtration in ZVI-based beds is a deep-bed filtration process since contaminants are not only removed at a reaction front but in the whole bed [51, 53, 54]. Figure 4a schematically shows the behavior of a ZVI-based bed at different service times.

It should be noted that, although there is a moving front of ZVI reactivity that progresses through the column, there is a slow and progressive reactivity loss across the entire column [54-56]. In addition, the zone behind the moving front is not completely inactive and therefore the pollutant concentration profiles shown in Figure 4b display a slow decrease until the moving front is reached.

Despite the above-discussed spatiotemporal differences in the activity profiles between purely sorptive beds and ZVI-based reactive beds, the breakthrough plots of ZVI systems designed to remove different pollutants display the typical S-shaped profile [57, 58]. Therefore, several simplified mathematical models developed to describe sorptive systems have been routinely used for describing the dynamic performance of ZVI-based reactive beds.

ZVI Column Configuration

The removal of different pollutants by ZVI-based beds in continuous systems has been reported by several authors [3, 5, 6, 59]. Studies include laboratory and field tests under different operating conditions. Laboratory experiments are usually performed using columns in an upflow configuration (i.e., against gravity) in order to minimize channeling and gas entrapment. Typical values of column height and diameter range from 15 to 50 cm and from 1.5 to 5 cm, respectively. Glass, stainless steel, Plexiglas, acrylic or plastic are frequently used as building materials. Glass wool, gravel and/or sand are placed at both ends of the columns to ensure optimum flow distribution across the cross-sectional area and prevent transfer of fine particles. Iron fillings are mainly used as active material, although iron chips and steel wool have also been used. Silica sand is frequently mixed with ZVI to prevent clogging and channeling due to the formation of ZVI corrosion products. In some cases, side sampling ports are used to monitor the spatial distribution of the fixed-bed activity. In addition, initial porosity (n) and initial hydraulic conductivity (K) are determined to characterize the columns [7, 42, 60-64].

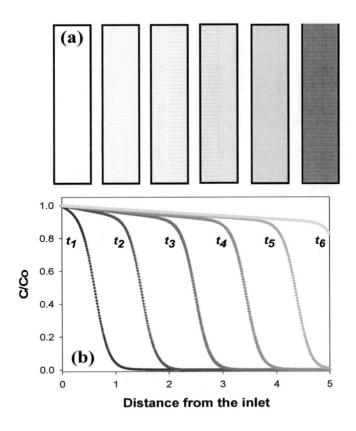

Figure 4. (a) Schematic representation of the accumulation of corrosion products in a ZVI-based bed at different service times. (b) Profiles of contaminant concentration inside the fixed bed at different service times.

Mechanisms for Arsenic Uptake

ZVI-based fixed beds have successfully removed soluble arsenic species in field and laboratory scale studies [10, 18, 42, 60, 65, 66]. Arsenic may be removed by ZVI/air systems without using flocculants or any other reagents, since the continuous release of Fe(II) in a column of corroding iron triggers various paths for arsenic uptake and also provides conditions for the oxidation of As(III) [10]. Arsenic species may be removed by adsorption, precipitation or co-precipitation processes that are linked to the in situ formation of corrosion products. The column performance may be characterized by measurements of arsenic concentration, pH, Eh, Fe(II) and Fe(III). Figure 5 shows a schematic representation of a multiparametric sensor developed in our laboratory [67].

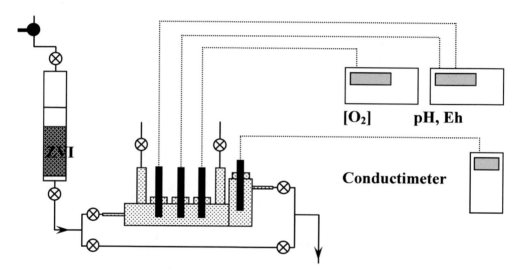

Figure 5. Schematic representation of the multiparametric sensor designed for on-line monitoring of the ZVI-fixed bed behavior.

In order to analyze the effect of dissolved oxygen concentration on the performance of ZVI-based columns designed for arsenic removal we have tested the corrosion rates as well as the efficiency of arsenic uptake at two different regimes [52, 67]. Prototype A was operated at high flow rates and under continuous air supply at the inlet end of the column, whereas Prototype B was operated at low flow rates and without added oxygen. The results showed that the effluent composition of column A was mainly characterized by high concentrations of Fe(III) species, high Eh values and negligible pH changes [52]. On the other hand, the effluent of column B was characterized by a relatively high Fe(II)/Fe(III) ratio, negligible dissolved oxygen concentration, Eh values close to the standard reduction potential of the couple Fe(II)/Fe(0), and a significant increase in pH [67]. In addition, it was observed that an oxygenation step between the outlet end of the column and the filtration stage was required/ for achieving complete arsenic removal in Prototype B.

The above-discussed results suggest that at relatively low dissolved oxygen concentrations, the uptake mechanism mainly involves co-precipitation with ferric species formed during the oxygenation stage applied to the column effluent. On the other hand, at high dissolved oxygen concentrations, arsenic uptake seems to be controlled by sorption onto ferric species formed inside the column bed. The comparison of the efficacies of both prototypes shows that, for high flow rates and high dissolved oxygen concentrations, a higher amount of ZVI per volume of treated water is required. Hence, the necessary iron to arsenic ratio is highly variable since bed capacities, iron corrosion rates and arsenic removal rates are strongly dependent on operating conditions.

Kinetics of Arsenic Uptake

Arsenate removal kinetics by corroding iron in batch tests has been studied by several authors [50, 68]. Farrell et al. found that at relatively high arsenic concentrations zeroth-order removal was observed, whereas at relatively low arsenic concentrations first-order removal was obtained.

Zeroth order is expected when the uptake is limited by the rate of formation of corrosion products capable of removing arsenic species. On the other hand, first order is observed when the uptake process is limited by the rate of mass transport of arsenic species to corrosion products. Column experiments showed a switch between pseudo-first- and pseudo-zeroth-order kinetics with increasing influent arsenic concentration [50]. The latter behavior was attributed to a shift from an excess to a shortage of adsorption sites as arsenic concentration increased. The analysis of arsenic uptake rates is rather complex since differences in corrosion rates and the nature of the corrosion products along the length of the column may give rise to variations in the apparent rate constants of both zeroth and first order [69].

Decrease of ZVI Reactivity

In drinking water treatment, a hydraulic retention times of less than 10 min is are usually required due to the high flow rates and space limitations in the treatment facilities [70]. For high water flow velocities, an elevated dissolved oxygen (DO) content in influent water may be necessary for an efficient arsenic removal by Fe(0) filters. High DO concentrations are likely to induce changes in ZVI reactivity over the entire flowthrough thickness [54]. In addition, ZVI corrosion in the presence of DO results in the formation of solid phases including $Fe(OH)_2$, $Fe(OH)_3$, $FeOOH$, Fe_3O_4, Fe_2O_3 and green rust [68, 70]. As discussed before, arsenic species may be trapped by adsorption, precipitation or co-precipitation in changing interfaces subject to aging processes, making the uptake process virtually irreversible [42].

It is obvious that, as service time increases, ZVI within the column is not only consumed but also progressively covered by minerals, resulting in a substantial decrease in the overall corrosion rate [71].

In addition, the resultant thicker oxide layers may lead to greater mass transfer resistance for arsenic removal [50]. However, precipitate formation appears to be a self-regulating process, with the precipitates causing reduced ZVI reactivity, which in turn reduces the rate of precipitate formation [54, 72]. Hence, the decrease of ZVI activity in fixed beds depends on several factors and further research in this field is certainly needed for the proper design and optimization of long-term ZVI-based water treatment systems.

Porosity Loss and Reactivity Decrease

Besides the decrease of corrosion activity due to the exhaustion of Fe(0) material, a major drawback for ZVI beds is the formation of corrosion products, which leads to a decrease in the permeability of Fe(0)/sand filters as the operation time increases [73]. Clogging and the development of preferential flow channels may arise due to the volumetric expansive nature of iron corrosion [73-75]. The progressive occupation of void spaces within the fixed bed reduces the hydraulic conductivity of the column filter [73]. Furthermore, amorphous and gelatinous ferric oxyhydroxides can bridge or cement iron particles together, effectively blocking a significant fraction of the pore spaces [76]. Finally, it should be taken into account that, besides the contribution of solid phases, hydrogen produced by anaerobic corrosion may also lead to porosity losses.

Strategies for Bed Porosity Sustainability

One of the major problems with ZVI media for arsenic removal from waters with elevated DO content is the high rate of formation of corrosion products. High corrosion rates may rapidly reduce the porosity and the hydraulic conductivity of the fixed beds resulting in inoperable systems even for beds with large amounts of residual iron. Columns packed with a high proportion of active (ZVI) media may undergo porosity losses and flow channeling due to cementation of neighboring granules by corrosion products [75]. Hence, mixing Fe(0) with inert additives such as sand is a prerequisite for column sustainability [74]. In typical Fe(0)/sand filters, the fraction of ZVI used usually represents 30–60 vol.% of the mixture, although beds containing more than 50% of ZVI are not advisable since they are likely to get clogged before the exhaustion of the active material ([74] and references therein). In addition, it has been recently shown that partial replacement of quartz sand by other porous materials may delay filter clogging due to the storage of corrosion products in the internal volume of porous materials [73].

Finally, it is recommended that inert material particles with a size similar to that of iron particles be used, otherwise the smaller particles are likely to fill the pore spaces between the larger particles. Moreover, large particles can afford larger pore spaces and less particle-to-particle contact and therefore are harder to bridge together [76].

CONCLUSION

Treatment plants based on the zero-valent iron technique are advantageous since low-cost and easy to obtain iron-based materials can be used as the source of fresh and effective sorbent phases for inorganic As(III) and As(V) removal. ZVI-based filters continuously generate iron corrosion products capable of removing soluble arsenic species. Adsorption, surface precipitation and co-precipitation appear to be the predominant removal mechanisms, although the contribution of each process may be difficult to assess given the dynamic nature of the Fe(0)/H$_2$O interface.

The dynamic behavior of fixed-bed continuous-flow systems may be described in terms of breakthrough curves, which usually display a typical S-shaped profile. Active filtration using ZVI-based beds is a deep filtration process since soluble arsenic species can be removed not only at a reaction front but in the whole bed. Hence, although simple models can be used to predict general trends, a rigorous analysis of arsenic uptake rates may be rather complex due to the spatial and temporal variations associated with both the corrosion rates and the nature of the corrosion products.

The two major drawbacks of ZVI-based filtration systems are the passivation of the reactive media and the decrease of hydraulic conductivity. As service time increases, the progressive accumulation of corrosion products onto the Fe(0) surface and inside the void space of the fixed bed results in substantial decreases in the overall corrosion rate and column porosity, respectively. In order to overcome these problems, mixing Fe(0) particles with inert material particles of similar size is recommended. In addition, since the efficiency of As removal by ZVI-based systems depends on the interplay of several factors, it is necessary to perform small-scale column tests in order to select the optimal operating conditions for site-specific water qualities.

REFERENCES

[1]   J.F. Ferguson, J. Garvis, *Water Res.* 6, 1259-1274 (1972).
[2]   S. Bang, M.D. Johnson, G.P. Korfiatisa, X. Meng, *Water Res.* 39, 763-770 (2005).
[3]   O.X. Leupin, S.J. Hug, *Water Res.* 39, 1729-1740 (2005).
[4]   R.B. Johnston, P.C. Singer, *Chemosphere* 69, 517-525 (2007).
[5]   N. Melitas, J.P. Wang, M. Conklin, P. O'Day, J. Farrell, *Environ. Sci. Technol.* 36, 2074-2081 (2002).
[6]   J.A. Lackovic, N.P. Nikolaidis, G.M. Dobbs, *Environ. Eng. Sci.* 17, 29-39 (2000).
[7]   C. Su and R.W. Puls, *Environ. Sci. Technol.* 37, 2582-2587 (2003).
[8]   A.N. Pham, A.L. Rose, A.J. Fetiz, T.D. Waite, *Geochim. Cosmochim. Acta* 70, 640-650 (2006).
[9]   N. El Azher, B. Gourich, C. Vial, M. Belhaj Soulami, M. Ziyad, *Chem. Eng. Process.* 47, 1877–1886 (2008).
[10]   O.X. Leupin, S.J. Hug, *Water Res.* 39, 1729–1740 (2005).
[11]   T.C. Zhang, Y.H. Huang, *Water. Res.* 40, 2311-2320 (2006).
[12]   K. Tyravola, N.P. Nikolaidis, N. Veranis, N. Kallithrakas-Kontos, P.E. Koulouridakis, *Water. Res.* 40, 2375-2386 (2006).
[13]   J. Farrell, J. Wang, P. O´Day, M. Conklin, *Environ. Sci. Technol.* 35, 2026-2032 (2001).
[14]   W. Stumm and F. Lee, *Ind. Eng. Chem.* 53, 143 (1961).
[15]   S. Bang, G.P. Korfiatis, X. Meng, *J. Hazard. Mater.* 121, 61-67 (2005).
[16]   H.L. Lien, R.T.Wilkin, *Chemosphere* 59, 377–386 (2005).
[17]   C. Noubactep, *J. Hazard. Mater.* 168, 1626–1631 (2009).
[18]   J.M. Triszcz, A. Porta, F.S. García Einschlag, *Chem Eng. J.* 150, 431-439 (2009).
[19]   S. Ayoob, A.K. Gupta, P.B. Bhakat, *Colloid. Surf. A: Physicochem. Eng. Aspects* 293, 247–254 (2007).

[20] N. Jain, H.C. Joshi, S. C. Dutta, S. Kumar, H. Pathak, Journal of Scientific and Industrial Research, 67, 154-160 (2008).
[21] T. Basu, K. Gupta, U.C. Ghosh, *J. Chem. Eng.* 55, 2039–2047 (2010).
[22] W.J.Weber Jr., Physico-chemical Process for Water Quality Control,Wiley Interscience Publication, New York (1972).
[23] O. Ceyhan, D. Baybas, *Turk. J. Chem.* 25, 193-200 (2001).
[24] S. Kundu, A.K. Gupta, *Chem. Eng. J.* 129, 123–131 (2007).
[25] A. Maiti, S. DasGupta, J.K. Basu, S. De, *Ind. Eng. Chem. Res.* 47, 1620-1629 (2008).
[26] B. Volesky, *Hydrometallurgy* 71, 179–190 (2003).
[27] A. Wolborska, *Chem. Eng. J.* 73, 85-92 (1999).
[28] Chenxi Li, Batch and bench scale fixed bed columna evaluations of heavy metal removals from aquous Solutions and synthetic landfill leachate using low cost natural adsorbnets. Master of Science Thesis. Queen's University. Kingston, Ontario, Canada. (January, 2008).
[29] M. Sarkar, A. Banerjee, P.P. Pramanick, A.R. Sarkar, *Chem. Eng. J.* 131, 329–335 (2007).
[30] E.A. Deliyanni, E.N. Peleka, K.A. Matis, *J. Hazard. Mater.* 172, 550–558 (2009).
[31] M. Trgo, N.V.Medvidovié, J. Perié, *Indian J. of Chem Toxicol.* 18, 123-131 (2011).
[32] A.C. Texier,Y. Andres, C. Faur-Brasquet, P. LeCloirec, *Process Biochem* 47, 333–342 (2002).
[33] G. Bohart G, E.Q. Adams, *J. Am. Chem. Soc.* 42, 523–544 (1920).
[34] M. Siegel, A. Aragon, H. Zhao, S. Deng, M. Nocon, M. Aragon. Prediction of Arsenic Removal by Adsorptive Media: Comparison of Field and Laboratory Studies. http://www.sandia.gov/water/docs/SAND2007-1923P-Siegel-ACS.pdf
[35] V.C. Srivastava, B. Prasad, I.M. Mishra, I.D. Mall, M.M. Swamy, *Ind. Eng. Chem. Res.* 47, 1603-1613 (2008).
[36] D. Ranjan, M. Talat, S.H. Hasan, *Ind. Eng. Chem. Res.* 48, 10180–10185 (2009).
[37] K. Komnitsas, G. Bartzas, K. Fytas, I. Paspaliaris, *Minerals Eng.* 20, 1200–1209 (2007).
[38] V.C. Srivastava, B. Prasad, I.M. Mishra, I.D. Mall, M.M. Swamy, *Ind. Eng. Chem. Res.* 47, 1603-1613 (2008).
[39] R.A. Hutchins, *Chem. Eng.* 80, 133-138 (1973).
[40] D.C.K. Ko, J.F. Porter, G. McKay, *Ind. Eng. Chem. Res.* 38, 4868-4877 (1999).
[41] Z. Aksu, Ferda Gönen, *Process Biochem.* 39, 599–613 (2004).
[42] N.P. Nikolaidis, G.M. Dobbs, J.A. Lackovic, *Water Res.* 37, 1417–1425 (2003).
[43] V.L. Nguyen, W-H Chen, T. Young, J. Darby, *Water Res.* 45, 4069-4080 (2011).
[44] J. C. Crittenden, J.K. Berrigan, D.W Hand, *Journal Water Pollution Control Federation* 58 312-319 (1986).
[45] J. C. Crittenden, J.K. Berrigan, D.W Hand, B. Lyskin. Journal of *Environmental Engineering* 113, 243-259. (1987).
[46] J. C. Crittenden, P.S. Reddy, H. Arora, J. Trynoski, D.W. Hand, D.L. Perran, S. Summers, *Journal of American Water Works Association* 83, 77-87 (1991).
[47] K. Hristovski ,A. Baumgardner, P. Westerhoff, *J. Hazard. Mater.* 147, 265–274 (2007).
[48] Y.H. Huang, T.C. Zhang, Water Res. 40, 3075–3082 (2006).
[49] H. Sun, L. Wang, R. Zhang, J. Sui, G. Xu, *J. Hazard. Mater.* 129, 297–303 (2006).

[50] N. Melitas, J. Wang, M. Conklin, P. O ' Day, J. Farrell, *Environ. Sci. Technol.* 36, 2074-2081 (2002).

[51] C. Noubactep, *Chem. Eng. J.* 165, 740–749 (2010).

[52] J.M. Triszcz, N. Luchessi, B. Sosa, E.M. Rosales, F.S. García Einschlag, *Ingeniería Sanitaria y Ambiental (AIDIS Argentina)* 93, 64-69 (2008).

[53] C. Noubactep, *Chem. Eng. J.* 171, 393– 399 (2011).

[54] I. Kouznetsova, P. Bayer, M. Ebert, M. Finkel, *J. Cont. Hydrol.* 90, 58–80 (2007).

[55] G.J. Okwi, N.R. Thomson, R.W Gillham, *Ground Water Monitoring and Remediation*, 35, 123-128 (2005).

[56] K.H. Chu, M.A. Hashim, *J. of Environ. Sci. China*, 19, 928-932 (2007).

[57] U. Kumar, M. Bandyopadhyay, *J. of Hazard. Mater.* 129, 253-259 (2006).

[58] M. Calero de Hoces, G. Blázquez García, A. Ronda Gálvez, M.A. Martín-Lara, *Ind. Eng. Chem. Res.* 49, 12587–12595 (2010).

[59] M. Biterna, L. Antonoglou, E. Lazou, D. Voutsa, *Chemosphere* 78, 7–12 (2010)

[60] K. Tyrovola, E. Peroulaki, N.P. Nikolaidis, *Euro. J. Soil Biology* 43, 356-367 (2007).

[61] J. Farrell, M. Kason, N. Melitas, T. Li, *Environ. Sci. Technol.* 34, 514-521 (2000).

[62] C. Noubactep, S. Caré, *J. Hazard. Mater.* 189, 809–813 (2011).

[63] A. Maiti, J.K. Basu, S. De, *Ind. Eng. Chem. Res.* 49, 4873–4886 (2010).

[64] M. Badruzzamana, P. Westerhoff, D.R.U. Knappe, *Water Res.* 38, 4002–4012 (2004).

[65] M. Biterna, A. Arditsoglou, E. Tsikouras, D. Voutsa, *J. Hazard. Mater.* 149, 548–552 (2007).

[66] K. Karschunke, V.L. Cáceres, M. Jekel, 26TH WEDC Conference, Water, Sanitation and Hygiene: Challenges of the millennium, Dhaka Bangladesh (2000).

[67] J.M. Triszcz, N. Luchessi, F.S. García Einschlag, *Ingeniería Sanitaria y Ambiental (AIDIS Argentina)* 99, 52-57 (2008).

[68] H-L Lien, R.T. Wilkin, *Chemosphere* 59, 377–386 (2005).

[69] N. Melitas, O. Chuffe-Moscoso, J. Farrel, *Environ. Sci. Technol.* 35, 3948-3953 (2001).

[70] S. Bang, G.P. Korfiatis, X. Meng, *J. of Hazard. Mater.* 121, 61–67 (2005).

[71] J.S. O, J. S-W. Jeen, R.W. Gillham, L. Gui, *J. Contam. Hydrol.* 103, 145–156 (2009).

[72] Y. Zhang, R.W. Gillham, *Ground Water*, Vol. 43, Iss.1, 113-121 (2005).

[73] C. Noubactep, S. Caré, *Chem. Eng. J.* 162, 635–642 (2010).

[74] C. Noubactep, S. Caré, *Chem. Eng. J.* 163, 454–460 (2010).

[75] D. Mishra, J. Farrell, *Environ. Sci. Technol.* 39, 9689-9694 (2005).

[76] P.D. Mackenzie, D.P. Horney, T.M. Sivavec, *J. Hazard. Mater.* 68, 1–17 (1999).

In: Arsenic   ISBN: 978-1-62081-320-1

Editor: Andrea Masotti   © 2013 Nova Science Publishers, Inc.

Chapter 17

CLOSED CYCLE PROCESS INVESTIGATIONS FOR ARSENIC REMOVAL FROM WATERS USING ADSORPTION ON IRON-CONTAINING MATERIALS FOLLOWED BY WASTE IMMOBILIZATION

L. Lupa, M. Ciopec, A. Negrea and R. Lazău*

University "Politehnica" Timisoara, Faculty of Industrial Chemistry
and Environmental Engineering, Timisoara, Romania

ABSTRACT

The chapter deals with adsorption process evaluation for removing arsenic from water using different iron containing materials, including an industrial waste. Adsorption is one of the most widely used techniques for arsenic removal from water and iron-based adsorbent materials proved to be very efficient. Therefore, kinetic, equilibrium, thermodynamic and column studies were performed for Fe_2O_3, $Fe_2O_3:SiO_2$ mixtures and an industrial sludge mostly containing Fe_2O_3 from hot dip galvanizing industry. Competing anions effect upon adsorption performance was also followed. All the studied materials showed good adsorption performance in both synthetic solutions and natural waters. The most suitable amongst the tested materials for arsenic adsorption remains the iron containing waste sludge (IS), due to economic reasons and high adsorption capacity. Thus, it has also been explored the possibility of immobilization of the exhausted adsorbent (IS) in vitreous matrices. This has been successfully done in frits/ ceramic decorative glazes. Considering the chosen solutions, the results are in full agreement with the sustainable development principles and comply with the closed-cycle technologies.

INTRODUCTION

The presence of dissolved arsenic in groundwater has created significant concern on a global scale, because of its toxic and carcinogenic effects on human beings [1, 2].

---

* E-mail address: lavinia.lupa@chim.upt.ro (L. Lupa).

Unfortunately, there is no known cure for As poisoning and therefore providing As free drinking water is the only way to eliminate its adverse health effects [3]. The World Health Organization (WHO) recommends a maximum admissible concentration of arsenic of 10 μg/L in drinking water [4]. It is well known from the chemistry of arsenic that As(III) is more toxic and mobile than As(V) [5]. The removal of As(III) from aqueous solution is usually more difficult compared to that of As(V), by almost all of the methods developed. It is because the predominant As(III) species are of neutral charge, whilst the As(V) species are negatively charged in the pH range of 4-10 [2, 5]. So sometimes pre-oxidation step is required to convert As(III) to As(V) in order to achieve the removal of As(III). A variety of technologies have been used to remove arsenic from water including conventional co-precipitation, lime softening, filtration, ion exchange, reverse osmosis and membrane filtration [6-9], but due to the extended use of chemicals, bulky sludge, high cost, these techniques are not feasible at small-scale or household level [5, 10]. The adsorption technique is regarded as very attractive and promising due to its simplicity, ease of operation and handling, sludge free operation, and regeneration capacity [11, 12]. Many types of adsorbents have been used including activated carbon [13-15], calcite [16], natural and synthetic zeolites [17], etc. However because of the selectivity and affinity of Fe(III) toward inorganic arsenic species, zero valent-iron[18], Fe(III) bearing mixed oxide [19-22], iron oxide-coated materials [12, 22-25], iron (hydr)oxide [1, 26, 27], laterite [2, 28] and Fe(III)-loaded resins [29-33] are the most used in arsenic adsorption. Arsenic is strongly attracted to the sorption site of these solids and thus can be effectively removed from solution. The purpose of the work presented within this chapter was to evaluate the adsorption performances of Fe_2O_3 obtained after annealing different iron salts: iron oxalate – $Fe(COO)_2 \cdot 2H_2O$, ammonium ferric alum – $Fe^{3+}(NH_4)(SO_4)_2 \cdot 12H_2O$ and Mohr salt – $Fe^{2+}(NH_4)_2(SO_4)_2 \cdot 6H_2O$ in the removal process of As(III) from aqueous solution. In order to resolve the issue of the resulted sludge after adsorption we proposed the recycling of the exhausted adsorbent with arsenic content in iron glass applications and thus, the presented method generates no secondary contaminated substances at the end of the waste recycling/ depollution process. This approach starts from the fact that in glass applications arsenic is used under As_2O_3 form. Even if this is very toxic, it is practically irreplaceable in these mixtures, due to its acting as an effective O_2 carrier at high temperatures [34-36].

In order to make it more suitable for this purpose we added SiO_2, as Ultrasil VN3 (Degussa), in the obtained Fe_2O_3 adsorbent composition. The focus was on these oxide mixtures because the iron oxide is responsible for arsenic removal and the presence of SiO_2 during adsorbent preparation leads to higher surface area, so we expect to obtain higher adsorption capacities. Although these materials are selective and efficient in removing arsenic, their applicability is limited because of their cost. Hence the goal for our research is to develop an arsenic removal system by using a low cost adsorbent. The material tested in the present chapter is an iron containing waste sludge (IS) resulted from the hot-dip galvanizing process.

The main points described in this chapter are: (a) obtaining the Fe_2O_3 and the $Fe_2O_3:SiO_2$ mixtures; (b) characterization of the obtained synthetic adsorbents and the low cost adsorbent (IS); (c) investigation of As(III) adsorption using the new adsorbents in batch tests; (d) study of the multi-component presence influence in the solution; (e) characterization of As (III) uptake on the columns packed with the new adsorbents under different operating conditions; (f) imobilization of the exhausted adsorbents in vitreous matrices.

1. Adsorbents Preparation and Characterization

The iron containing sludge (IS) formed during hot-dip galvanizing process was tested as a low cost adsorbent. Hot-dip galvanizing is a coating process applied in order to prevent iron-based materials from corroding. Hot-dip galvanizing is one of the most common coating methods and it is particularly being used to coat steel. Before the hot-dip galvanizing, pieces are being subjected to a preparation process, which consists of degreasing, chemical cleaning, rinsing, fluidizer treatment and pre-warming. Waste waters result from these operations and they are neutralized with lime. This method can efficiently remove heavy metals from wastewaters, but it generates an iron containing sludge classified as industrial waste, which often causes disposal problems [37]. Arsenic removal from water using the above mentioned iron sludge has been tested, as an appropriate environmental and economical solution, in accordance with the sustainable development principles. The IS used for arsenic removal, dried at 105°C has the particle size below 0.32 µm. The chemical composition of the IS resulted from energy dispersive X-ray analysis (EDX) is presented in table 1.

Table 1. Chemical composition of the IS

No.	Element	Composition (wt,%)
1.	Fe	33.5
2.	Ca	11.2
3.	Cl	22.8
4.	P	0.45
5.	S	0.71
6.	Zn	2.74
7.	O	28.6

The major component of the IS is iron, which makes it suitable for arsenic removal from water, due to the high affinity of arsenic towards iron. The chlorine ions come from the hydrochloric acid used at the chemical cleaning of the steel parts and the calcium ions come from the neutralization agent used for the residual waters treatment [38]. The other ions evidenced by the analysis in low proportion come either from the degreasing solution of the steel parts, or from the water used in various stages of the technological process. The zinc ions result from the parts, which were poorly coated, subjected to de-zincking and then re-entered in the technological process. The IS was characterized through: XRD, and BET – accelerated surface area analysis.

The adsorbents based on Fe_2O_3 and Fe_2O_3:SiO_2 mixtures were obtained by annealing at 550°C of the iron oxalate – $Fe(COO)_2 \cdot 2H_2O$, and at 800°C of the ammonium ferric alum – $Fe^{3+}(NH_4)(SO_4)_2 \cdot 12H_2O$ and Mohr salt – $Fe^{2+}(NH_4)_2(SO_4)_2 \cdot 6H_2O$. In the case of samples with SiO_2 content, these have been wet homogenized with the iron salts prior to annealing, in the mass ratio 1:1, in a porcelain dish, dried at 110°C and the resulted mixtures have been grinded in a mortar. The iron salts and the mixtures with Ultrasil have been loaded in porcelain crucibles and annealed in an electric kiln at the minimum temperatures necessary for their decomposition with the formation of Fe_2O_3. After annealing the obtained iron oxides have been characterized through XRD phase analysis, and BET – accelerated surface area analysis. The XRD patterns were recorded on a Bruker D8 Advance System with monochromator, Cu

$_{K\alpha}$ radiation. The specific surface area and the pore volume of the studied materials were measured using a Micrometrics ASAP 2020 BET surface area analyzer, by cold nitrogen adsorption. The characteristic of the studied adsorbents are presented in table 2. The XRD pattern of the IS reveals the predominance of the iron oxy-hydroxides as goethite and lepidocrocite. Can be observed that the IS sample dried at 105°C is more amorphous then the sample annealed at 200°C (figure 1). The XRD pattern of the samples obtained by annealing of the iron salts alone or mixed with SiO_2 reveals that the only crystalline phase present in the sample is the hematite (figure 2).

The BET surface area analysis of the IS sample was carried out after degasing at 200°C in order to remove the volatile compounds from the sample. The BET surface area of the IS is higher than the BET surface area of the iron oxide obtained after annealing of the iron salts. This is explained by the fact that IS is more amorphous then the synthetic oxides. The presence of SiO_2 leads to the increase of the surface area of the synthetic iron oxides. From the BET surface area one can aspect that the materials with the highest adsorption capacity are the IS and the mixtures $Fe_2O_3:SiO_2$.

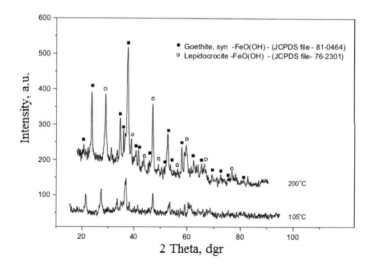

Figure 1. X-Ray diffraction patterns of the IS samples dried at 105°C and annealed at 200°C.

Table 2. Characteristics of the studied adsorbents

Sample no. and symbol	Salt used	SiO_2 content ($Fe_2O_3:SiO_2$ mass ratio)	Anealing temperature °C	BET surface area (m²/g)	Pore volume cm³/g	Average pore size (Å)
1.×	IS	-	200	50.5	0.168	111.5
2.▼	$Fe(COO)_2 \cdot 2H_2O$	-	550	14.6	0.089	255
3.▽	$Fe(COO)_2 \cdot 2H_2O$	1:1	550	78.0	0.556	287
4.■	$Fe^{3+}(NH_4)(SO_4)_2 \cdot 12H_2O$	-	800	11.1	0.035	133
5.□	$Fe^{3+}(NH_4)(SO_4)_2 \cdot 12H_2O$	1:1	800	57.0	0.467	309
6.●	$Fe^{2+}(NH_4)_2(SO_4)_2 \cdot 6H_2O$	-	800	8.22	0.037	178
7.○	$Fe^{2+}(NH_4)_2(SO_4)_2 \cdot 6H_2O$	1:1	800	57.0	0.427	287

In order to understand the characteristics of the surface charge for the studied adsorbents, the pH_{pzc} of the IS and of the iron oxide obtained from the annealing of iron oxalate was determined by mass titration method [13, 14]. For the determination of the pH_{pzc} a definite amount (0.2 g) of sorbent was suspended in 100 mL 0.005 M NaCl solution, used as an inert/background electrolyte with initial pH varying between 2 and 12. The initial pH of the NaCl solution was adjusted to the desired value using 0.1M/2M NaOH or 0.1M/2M HNO_3, so that the volume variation of the solution was as low as possible. The pH of the suspension was measured using a CRISON MultiMeter MM41. The suspensions were stirred for 1 h at 300 rpm, using a magnetic stirring device IKA RTC and then the final pH was read. The pH_{pzc} was determined through the plot of the pH_f versus pH_i (figure 3).

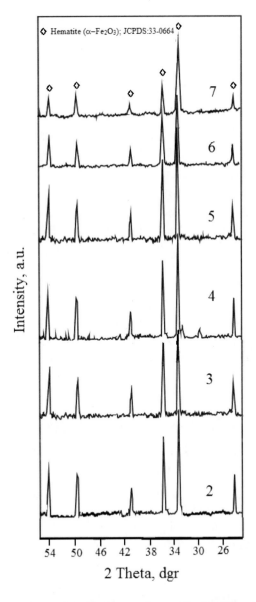

Figure 2. X-Ray diffraction patterns of the Fe_2O_3 samples and Fe_2O_3-SiO_2 mixtures.

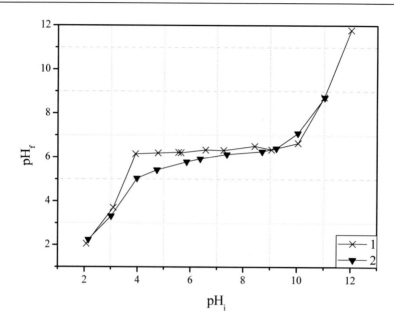

Figure 3. Plot of the final pH (pH_f) versus the initial pH (pH_i) allowing the determination of the value of pH_{pzc} for the IS and iron oxide obtained from iron oxalate.

The acid–base properties of the studied material play an important role in the use of the materials as adsorbents. Figure 5 shows the corresponding plot of pH_i versus pH_f allowing the determination of the point of zero charge (pH_{pzc}) for the IS and for the Fe_2O_3 obtained by annealing of iron oxalate. The pH value of the plateau exhibited in this plot corresponds to $pH_{pzc} = 6.2$ for IS and $pH_{pzc} = 6.1$ for the Fe_2O_3. The presence of such a plateau indicates that the studied materials were ampholytes and behaved as acid–base buffers [13, 14, 39]. These plateaus correspond to the pH range where buffering of the studied materials surface occur, thereby making all values of pH_i between 4 and 9 equal to the corresponding value of pH_f. The pH_{pzc} value suggest that the surface of the material was predominantly positive at pH values lower than 6.2 for IS and 6.1 for Fe_2O_3, respectively and negative at pH values higher than these values. The surface charge density of the materials should increase/decrease as the pH value of the system decreased below pH_{pzc} or increased above these values [38, 40].

2. KINETIC, EQUILIBRIUM AND THERMODYNAMIC STUDIES REGARDING THE AS(III) ADSORPTION ONTO THE STUDIED MATERIALS

In order to establish the adsorption performance of the studied materials in the removal process of As(III) from solution equilibrium, kinetic and thermodynamic studies were performed. The influence of different physicochemical parameters (stirring rate, contact time, arsenic initial concentration and temperature) upon the arsenic adsorption onto the studied materials was investigated.

For all adsorption experiments, the initial pH of the solutions was maintained within the range 6.5–7.0, i.e. at the mid-point of the plateau obtained in the plot of pH_f versus pH_i

mentioned above. This also corresponded to the most common pH value found in natural waters.

Prior to such experiments a stock solution was prepared by diluting an appropriate amount of 0.05 M $NaAsO_2$ solution (Merck TitriPUR). Other solutions of As(III) ions were prepared from the stock solution by appropriated dilution. The adsorption experiments were conducted using agitating device with glass rod to obtain the data for the equilibrium, kinetic and thermodynamic parameters for As(III) adsorption onto the studied materials. The experiments were performed with 0.1 g of adsorbent in 100 mL of As(III) solution of desired concentration and pH.

The samples were stirred at the established temperature and time intervals for the various experimental studies. After stirring, the samples were centrifuged at 1200 rpm for 0.5 h using a Hettich ROTINA 420 centrifuge. The residual concentration of As(III) ions in the resulting solutions was determined using atomic absorption spectrometry with hydride generation [41, 42]. This method uses the selective reduction of As(III) ions to arsine (H_3As) with sodium boron hydride $NaBH_4$ (Merck-Schuchardt; 0.6 w/v% solution) in a NaOH buffer (Chemapol, Prague, Czech Republic; 0.5 w/v%). The arsine gas generated was introduced into the flame of the atomic absorption spectrometer and the absorbance value measured at 193.7 nm was compared with a calibration curve obtained using As(III) ion solutions of various known concentrations prepared from the stock solution. The carrier solution employed in the flow injection system was prepared using HCl (37%, Corozin, Romania; 1:3). A Varian SpectrAA 110 atomic absorption spectrometer with a Varian VGA 77 hydride generation system was used for all measurements.

The extent of adsorption is quantified in terms of the adsorption capacity q_t ($\mu g/g$) of the adsorbent, corresponding to the amount of As(III) ions sorbed per g of adsorbent at a known time, t, as calculated from the following equation [13, 21, 38]:

$$q_t = \frac{(C_0 - C_t)V}{m} \tag{1}$$

where C_0 and C_t are the concentrations of As(III) ions ($\mu g/L$) in the solution initially (t = 0) and after a time t (min), respectively, V is the volume of the solution (L) and m is the mass of adsorbent employed (g).

The effect of stirring rate (50-250 rpm) on As(III) removal was studied with initial As(III) concentration of 100 $\mu g/L$ and adsorbent dose of 1 g/L at a contact time of 90 min. The effect of contact time at ambient temperatures (20-25 °C) was initially studied employing a 100 $\mu g/L$ As(III) ion solution with an adsorbent dosage of 0.1 g, the resulting suspensions being stirred for different contact times (15, 30, 45, 60, 90, 120 and 150 min).

Adsorption isotherm studies were done by varying the initial concentration of As(III) (100, 200, 300, 400, 500, 600 and 700 $\mu g/L$) and keeping the adsorbent dose fixed (1 g/L) at the ambient temperature (20-25 °C), while the influence of temperature was investigated by studying the adsorption process in a 0.1 g of adsorbent/100 $\mu g/L$ As(III) ion solution system at three different temperatures: 20, 25 and 30 ºC, respectively for first sample, 18, 20 and 40°C for second and third samples and 25, 30 and 45 °C for the last four samples. Equilibrium time for the isotherm and thermodynamic studies was kept as 90 minutes.

2.1. EFFECT OF STIRRING RATE

The adsorption of As(III) onto the studied materials was influenced by the rate of stirring. The influence of stirring rate on adsorption of As(III) onto the studied materials was studied by changing the speed of stirring from 50 to 250 rpm and results are given in figure 4.

From the figures it is obvious that the removal of As(III) increases with the increase in stirring rate from 50 to 200 rpm, for all the studied materials. This may be explained by the fact that the increasing stirring rate decreases the boundary layer resistance to mass transfer in the bulk and increases the driving force of As(III) ions. It may be, therefore, assumed that the film diffusion does not dominantly control the overall adsorption process [24]. It is further noted that there was no significant increase in uptake above 200 rpm.

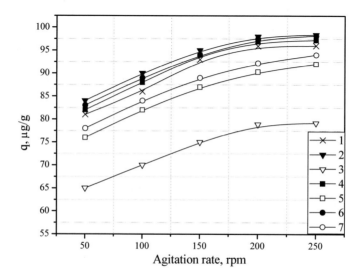

Figure 4. Effect of stirring rate on As(III) uptake by the studied materials. $C_0 = 100\mu g/L$; T=20-25°C; t = 90 min; pH=6.5 – 7; sorbent dose, 1 g/L.

2.2. Effect of Contact Time and Kinetic Studies

In equilibrium sorption experiments, kinetic study is very important to find out the contact time of the adsorbent with adsorbate and for evaluating reaction coefficients, in order to establish the mechanism of adsorption process. The stirring time effect on the adsorption capacity of the studied material in the process of As(III) removal from water are presented in figure 5. Experimental results indicate that the adsorption capacity of all studied materials increased as the contact time was increased up to 90 minutes, and remained constant afterwards. This high initial adsorption rate was due to the local availability of a large number of adsorption sites on the adsorbents. After 90 minutes of contact between adsorbents and adsorbate, the adsorption capacity changed. The rapid removal of As(III) ions coupled with a low equilibration time indicates the existence of highly favorable sorptive interactions. For subsequent experiments, an equilibration time of 90 minutes was chosen.

The kinetics of the adsorption describing the rate of the arsenic removal is one of the important features that define the adsorption efficiency. In order to express the kinetics of arsenic adsorption onto the studied materials the results were analyzed using the models presented as follows.

The pseudo-first-order kinetic model based on the solid capacity and proposed by Lagergren can be used to determine the rate constant for the adsorption process and the integrated form is expressed by the following equation [13, 21, 38-40, 43]:

$$\ln(q_e - q_t) = \ln q_e - k_1 t \tag{2}$$

where: q_t and q_e represent the amounts of the arsenic adsorbed onto the studied materials at t time and at equilibrium time, respectively, $\mu g/g$; k_1 is the specific adsorption rate constant, min^{-1}.

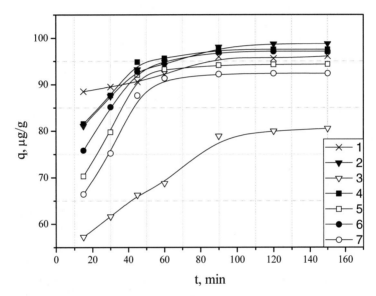

Figure 5. Effect of contact time on the adsorption capacitiy of the studied materials. $C_0 = 100 \mu g/L$; $T=20-25°C$; $pH=6.5 - 7$; sorbent dose, 1 g/L; stirring rate 200 rpm.

The linear form of the pseudo-second-order rate expression of Ho and Mckay, based on the solid phase sorption, is given by [1, 19, 21, 44]:

$$\frac{t}{q_t} = \frac{1}{h} + \frac{t}{q_e} \tag{3}$$

where: $h=k_2 \cdot q_e^2$; k_2 is the pseudo-second-order constant, $min^{-1}(\mu g/g)^{-1}$. Other terms have their usual meanings.

Intra-particle diffusion is an important phenomenon for adsorption processes in porous materials. The functional relationship for this process, which has been used by many authors [11, 13, 21, 44] may be written as:

$$q_t = k_{id} t^{0.5} \tag{4}$$

where k_{id} is the intra-particle diffusion rate constant [$\mu g/(g\ min^{0.5})$], which may be calculated from the slope of the linear plot of q_t versus $t^{0.5}$.

The application of the different kinetic models revealed some interesting features regarding the mechanism and rate-controlling step in the overall sorption process. The pseudo-first-order rate constant may be determined from the linear plot of $\ln(q_e - q_t)$ versus t (figure 6), whilst the pseudo-second-order rate constant can be estimated from the linear plot of t/q_t versus t (figure 7). The values of all these constants determined as indicated, together with the corresponding regression coefficients (R^2), are listed in table 3.

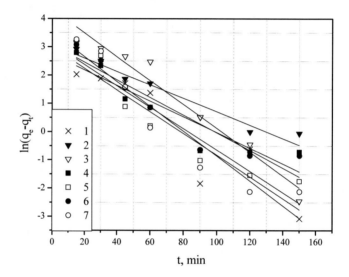

Figure 6. Pseudo-first order kinetic plot.

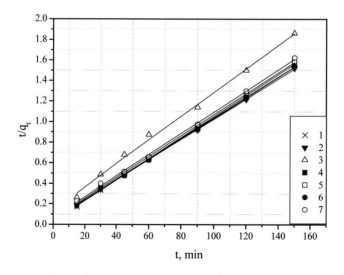

Figure 7. Pseudo-second order kinetic plot.

It was observed that the pseudo-first-order model does not describe in a satisfying manner the kinetics of the sorption process, since the correlation coefficients were much lower than those for the pseudo-second-order rate, for all the studied materials. Furthermore, the calculated equilibrium sorption capacity for the pseudo-first order model, for all the studied materials, $q_{e,calc}$, values differ from the experimental values $q_{e, exp}$, whilst the theoretically predicted equilibrium sorption capacity in the case of the pseudo-second-order model is close to the experimentally determined one, for all studied materials. This shows that the kinetics of As(III) removal using the studied materials with iron oxide content is described by a pseudo-second-order expression, instead of a pseudo-first-order.

Table 3. Kinetic parameters for As(III) sorption onto iron oxide

Sample no.	q_e (exp), $\mu g/g$	Pseudo-first-order			Pseudo-second-order			Intra-particle diffusion
		q_e (calc), $\mu g/g$	k_1, min^{-1}	R^2	q_e (calc), $\mu g/g$	k_2, min^{-1} $(\mu g/g)^{-1}$	R^2	K_{id}, $\mu g/g/ min^{-0.5}$
1.	96.1	20.4	0.0381	0.8513	98.0	0.00387	0.9997	1.31
2.	98.8	21.8	0.0237	0.9925	102	0.00227	0.9999	3.01
3.	80.3	76.7	0.0422	0.9589	86.9	$9.85\cdot10^{-4}$	0.9968	3.77
4.	98	21.3	0.0320	0.9366	100	$2.75\cdot10^{-3}$	0.9998	3.04
5.	93	23.2	0.0216	0.8191	97.1	$1.63\cdot10^{-3}$	0.9990	4.89
6.	97	28.8	0.0411	0.8474	100	$2.18\cdot10^{-3}$	0.9997	3.98
7.	91	26.3	0.0312	0.8713	95.2	$1.65\cdot10^{-3}$	0.9988	4.89

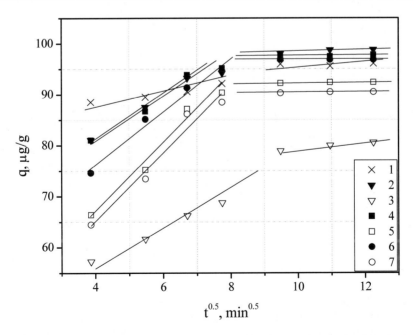

Figure 8. The Weber-Morris plot for intra-particle diffusion of the arsenic sorption kinetic data onto IS and Fe_2O_3 samples.

However, such results provide no information regarding the rate-limiting step in the process. The rate-limiting step (i.e. the slowest step in the process) may either be boundary layer (film) diffusion or intra-particle (pore) diffusion of the solute towards the solid surface. If the rate-limiting step is intra-particle diffusion, the plot of q_t versus the square root of time should be linear and pass through the origin. Any deviation of the plot from the linearity would indicate that the rate-limiting step should be controlled by boundary layer (film) diffusion. The plot of q_t versus $t^{0.5}$ (figure 8) shows two linear sections for all the studied materials. The first may be assigned to intra-particle diffusion, the fact that the line does not pass through the origin indicating that the mechanism of As(III) ion adsorption onto the studied materials is complex, with both surface adsorption and intra-particle diffusion contributing to the rate-limiting step. The second linear section represents the final equilibrium stage [5, 11, 13, 21, 39].

2.3. Effect of Arsenic Initial Concentration and Adsorption Equilibrium Studies

The adsorption isotherms of As(III) removal by the studied materials are presented in figure 9.

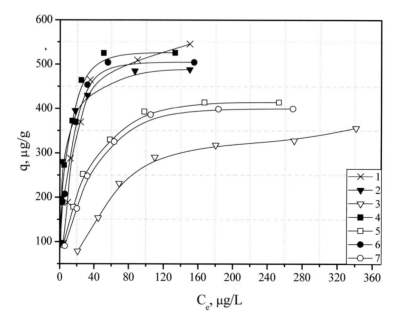

Figure 9. Adsorption isotherms of As(III) onto the studied materials. t=90 min; T=20-25°C; pH=6.5 – 7; sorbent dose, 1 g/L; stirring rate, 200 rpm.

The adsorption capacity increased with the increasing equilibrium concentration of arsenic for all the studied materials. At high equilibrium concentrations, the adsorption capacity approached a limit value. This value represents the experimentally-determined maximum adsorption capacity of As(III) ions onto the studied materials.

Several models are mentioned in the literature to describe experimental data of adsorption isotherms. The Langmuir and Freundlich models are the most frequently employed models. In this chapter, both models were used to describe the relationship between the amount of

As(III) adsorbed by studied materials and its equilibrium concentration in solution for 90 minutes [13, 14, 21, 45, 46].

The linear form of the Freundlich isotherm equation can be written as:

$$\ln q_e = \ln K_F + \frac{1}{n} \ln C_e \qquad (5)$$

and the Langmuir isotherm as the following equation:

$$\frac{C_e}{q_e} = \frac{1}{K_L q_m} + \frac{C_e}{q_m} \qquad (6)$$

where: q_e is the amount of arsenic adsorbed per gram of sorbent, $\mu g/g$; C_e is the equilibrium concentration of arsenic, $\mu g/L$; K_f and $1/n$ are specific constants concerning the relative adsorption capacity of the adsorbent and the intensity of adsorption, respectively; q_m is a measure of monolayer adsorption capacity, $\mu g/g$ and K_L is a constant related to the free energy of adsorption. The curves and parameters, as well as the correlation coefficients (R^2), for As(III) removal trough adsorption onto the studied adsorbent are presented in figures 10, 11 and table 4.

The Freundlich plots (figure 10) have very low regression coefficients suggesting a restricted use of Freundlich isotherm. The constants K_F and $1/n$ computed from the linear plot are presented in table 4. The constant K_F can be defined as an adsorption coefficient, which represents the quantity of adsorbed metal ions for a unit equilibrium concentration. The slope $1/n$ is a measure of the adsorption intensity or surface heterogeneity. For $1/n = 1$, the partition between the two phases does not depend on the concentration.

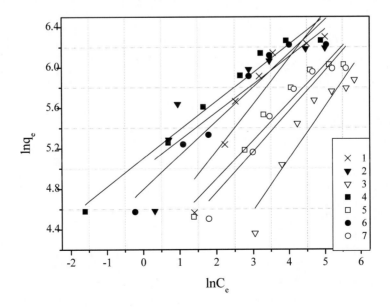

Figure 10. Freundlich plot of As(III) adsorption onto the studied materials.

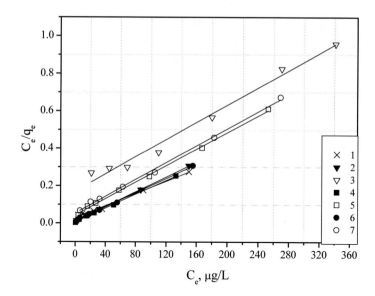

Figure 11. Langmuir plot of As(III) adsorption onto the studied materials.

Table 4. Parameters of Freundlich and Langmuir isotherms for the As(III) adsorption onto the studied materials

Sample no.	$q_m(\text{exp})$, µg/g	Freundlich isotherm			Langmuir isotherm		
		K_F, µg/g	1/n	R^2	$q_m(\text{calc})$, µg/g	K_L, L/µg	R^2
1.	550	70.32	0.4584	0.8550	625	0.0569	0.9952
2.	490	149	0.2744	0.7689	500	0.2470	0.9994
3.	341	20.9	0.5132	0.8865	434.8	0.0133	0.9849
4.	526	164	0.2812	0.9662	555	0.2611	0.9983
5.	415	63.3	0.3751	0.9413	454	0.0511	0.9980
6.	505	121	0.3356	0.9350	526	0.1744	0.9982
7.	400	52.6	0.4007	0.9193	434	0.0427	0.9971

The situation $1/n < 1$ is the most common and correspond to a normal L-type Langmuir isotherm, whilst $1/n > 1$ indicates a cooperative adsorption, which involves strong interactions between the molecules of adsorbate. Values of $1/n < 1$ show favorable adsorption of As(III) ions related to all the studied materials. The Langmuir model effectively describes the sorption data for all the studied materials with correlation coefficient closer to 1. Thus, the isotherm follows the sorption process in the entire concentration range studied for all four materials. Moreover the maximum adsorption capacities of the studied materials obtained from the Langmuir plot are very close to those obtained experimentally. The maximum adsorption capacities of As(III) developed by Fe_2O_3 obtained by annealing of the iron salts are higher than the adsorption capacities of As(III) developed by the mixtures $Fe_2O_3:SiO_2$. This is in contradiction with the conclusion resulted from the BET surface area. If one refers the adsorbent properties only to the Fe_2O_3 content (the active phase responsible for arsenic removal) of the samples 3, 5 and 7, it may be said that the presence of SiO_2 improves the adsorption capacity of the powders. A possible explanation for this situation refers to the hydroxylate character of the Ultrasil particles (even after annealing at 550°C or 800°C), which leads to their hydration and dispersion of the Fe_2O_3 agglomerations in the presence of the aqueous solutions with As(III) content [47]. This behavior is very favorable for the perspective of the waste immobilization in vitreous matrices.

The dimensionless constant, called separation factor (R_L) can be used to describe the essential characteristics of a Langmuir isotherm.

$$R_L = \frac{1}{1 + K_L \cdot C_0} \tag{7}$$

where the terms have their meanings as stated above. In fact, the separation factor is a measure of the adsorbent capacity used. Its value decreases with increasing "K_L" as well as initial concentration. R_L values can be related to the equilibrium isotherm as follows: unfavorable, $R_L > 1$; linear, $R_L = 1$; favorable $0 < R_L < 1$; and irreversible, $R_L = 0$ [13, 14, 16]. The values were calculated for the entire concentration range studied and the results are found to lie in between 0 and 1, in all the cases, demonstrating a favorable sorption process.

Since the IS develop the highest adsorption capacity (amongst the samples tested) in the removal process of As(III) from aqueous solution, its use represent an appropriate environmental and economical solution.

2.4. Effect of Temperature and Thermodynamic Studies

In general, the experimental conditions such as metal ion concentration and temperature have strong effects on the equilibrium distribution coefficient value (K_d), so they can be used to compare the efficiency of various adsorbents. Equilibrium distribution coefficient value (K_d) represents the amount of removed As(III) ions per gram of adsorbent, divided by their concentration in the liquid phase:

$$K_d = \frac{q_e}{C_e} \tag{8}$$

Temperature dependence of the adsorption process is associated with several thermodynamic parameters. Thermodynamic considerations of an adsorption process are necessary to conclude whether the process is spontaneous or not. Thermodynamic parameters such as Gibbs free energy change ($\Delta G°$), enthalpy change ($\Delta H°$) and entropy change ($\Delta S°$) can be estimated using equilibrium constant changing with temperature. The Gibbs free energy change of the adsorption reaction is given by the following equation [1, 13, 21, 45]:

$$\Delta G° = -RT \ln K_d \tag{9}$$

where R represents the universal gas constant (8.314 J/(mol·K)), T represents the absolute temperature (K) and K_d represents the distribution coefficient.

Relation between $\Delta G°$, $\Delta H°$ and $\Delta S°$ can be expressed by the following equations:

$$\Delta G° = \Delta H° - T\Delta S° \tag{10}$$

$$\ln K_d = \frac{\Delta S^\circ}{R} - \frac{\Delta H^\circ}{RT} \qquad (11)$$

The thermodynamic parameters were determined from the slope and intercept of the plot of $\ln K_d$ versus $1/T$ (figure omitted). The values of ΔG°, ΔH° and ΔS° for all the studied temperature are given in table 5.

In case of the obtained Fe_2O_3 and $Fe_2O_3:SiO_2$ mixtures the negative values of ΔG° and positive values of ΔH° indicate that the adsorption of As(III) onto the studied materials is a spontaneous and endothermic process. The more negative value of ΔG° imply a greater driving force to the adsorption process. The values of ΔH° are high enough to ensure the strong interaction between the As(III) and the studied materials. The positive values of ΔS° indicate an increased randomness at the solid-solution interface during the adsorption of As(III) onto the studied materials [1, 13, 21, 38, 45]. The increasing of the adsorption capacities of the studied materials at higher temperatures may be due to the enlargement of pore size and/or activation of the adsorbent surface.

Table 5. Thermodynamic parameters for As(III) adsorption onto the studied materials

Sample no.	Temperature, K	ΔG°, kJ/mol	ΔH°, kJ/mol	ΔS°, kJ/mol·K	R^2
1.	293	-7.68	-161.8	-0.526	0.9802
	298	-5.05			
	303	-2.42			
2.	291	-8.79	131.7	0.483	0.9915
	295	-10.7			
	313	-19.4			
3.	293	-3.16	36.15	0.134	0.9817
	298	-3.83			
	303	-4.50			
4.	298	-15.1	55.8	0.239	0.9963
	303	-16.6			
	308	-20.2			
5.	298	-8.75	58.3	0.225	0.9964
	303	-9.88			
	308	-13.3			
6.	298	-11.9	81.1	0.312	0.9959
	303	-13.4			
	308	-18.1			
7.	298	-7.51	64.9	0.243	0.9964
	303	-8.73			
	308	-12.4			

In case of the IS, the overall free energy change during the adsorption process was also negative for the range of temperatures studied, corresponding to a spontaneous physical process for As(III) ion adsorption. As decreasing the temperature from 303 K to 293 K, the standard free energy change became even more negative, suggesting that the adsorption process was favored at lower temperatures. *This represents an economic advantage, since no heat input would be necessary during the adsorption process.* The negative value of ΔH^0

shows that the adsorption process was exothermal. The negative value of ΔS^0 indicates that the adsorbed species suffers a decrease in its freedom degrees.

3. EFFECTS OF COMPETING ANIONS ONTO As(III) ADSORPTION PROCESS

The existence of multi-component in the solution, as usually found in real-world systems, is of particular importance in practical applications, where competition between different components for the available adsorption sites occurs [3, 48, 49]. Anions directly compete for available surface binding sites and indirectly influence adsorption by alteration of the electrostatic charge at the solid surface. Both direct and indirect effects are caused by solution pH, the relative anions concentrations and intrinsic binding affinities [50, 51].

The present chapter deals with the effect of the presence of some anionic species (nitrate NO_3^-, phosphate PO_4^{3-}, sulphate SO_4^{2-}, chloride Cl^- and carbonate CO_3^{2-}) in the arsenic containing water on the adsorption of As(III) onto the first two studied materials. These two materials were chosen in order to compare the effects of the anions on the arsenic adsorption when a waste material and a synthetic material are used as adsorbents.

For the studies 100 µg/L As(III) synthetic solutions with/without anionic species (NO_3^-, PO_4^{3-}, SO_4^{2-}, Cl^- and CO_3^{2-}) have been used. It has been separately studied the influence of each anionic species and the influence of all mixed anionic species at two concentrations: 10 mg/L and 100 mg/L, respectively. All reagents used to prepare the solutions were in analytical reagent grade. The adsorbents were kept in contact with the arsenic solution in the solid:liquid ratio of 1g/L at laboratory temperature (23°C) for different contact periods (1, 2, 3, 4, 8 and 24 h). After the contact time elapsed the suspensions were filtered and the residual concentration of arsenic in the filtrate was determined. During the adsorption experiments the pH of the suspensions was maintained at a constant value (range between: 6.5-7), due to the natural buffer capacity of the sorbent materials [52].

Experimental data regarding the dependence of the adsorption capacities of the studied materials versus time for two anion concentration are presented in figure 12 for IS and in figure 13 for Fe_2O_3 obtained by annealing the iron oxalate. One may notice that in all situations the adsorption capacity of the studied materials increase as contact time increases. In most cases the increase rate is lower after 6 h of contact time and one may consider that the equilibrium was reached. The figures also show that the adsorption capacities of the studied materials in the removal process of As(III) from water is influenced by the solutions composition.

Figure 12 shows the adsorption of As(III) onto the IS is mainly influenced by the solution composition, according to the following sequence: $PO_4^{3-}>NO_3^->Cl^->CO_3^{2-}>SO_4^{2-}$, regardless the concentration of anionic species (10 mg/L and 100 mg/L).

One may notice that the presence of phosphate anions has a negative influence on the As(III) adsorption in the case of IS, leading to a reduced adsorption capacity. This observation is consistent with literature data, which indicate that arsenic adsorption is seriously reduced by the presence of phosphate, because of the competition for the binding sites of the sorbent material, which appears between arsenic and phosphate [49, 53].

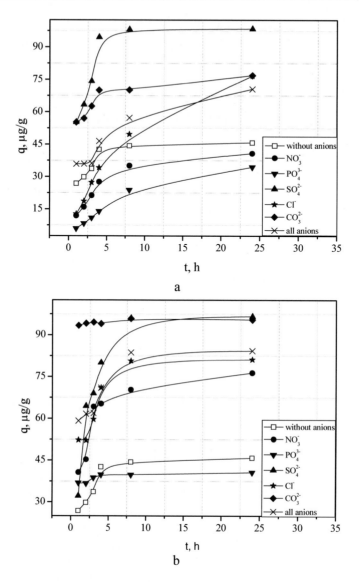

Figure 12. Influence of contact time on the IS adsorption capacity. anionic species concentrations: (a) 10mg/L; (b) 100mg/L.

The individual presence of SO_4^{2-} and CO_3^{2-} has a positive effect on As(III) adsorption, leading to adsorption capacities higher than those obtained in their absence. Other researchers also concluded that the sulphate and carbonate do not affect the As(III) adsorption [49]. The NO_3^-, Cl^- were found to have a negligible effect or even positive effect (when they are in a higher concentration) on the removal of As(III) through adsorption onto IS. The adsorption capacities of the IS is higher in the individual presence of these anions when their concentration is higher (100 mg/L). The simultaneous presence of all anions develops an overall positive effect on arsenic adsorption, but the adsorption capacities were somehow lower that in the presence of SO_4^{2-} and CO_3^{2-} alone, due to the negative influence of PO_4^{3-}.

From figure 13 one may notice that when anions are individually present in the sample matrix at concentrations of 10 mg/L, it's only Cl^- and SO_4^{2-} that have no obvious influence on

As(III) adsorption on Fe_2O_3. All other anions have negative influence according to the following sequence: $PO_4^{3-}>NO_3^->CO_3^{2-}>SO_4^{2-}>Cl^-$. The negative competitive effect of all the studied anions is more visible at higher concentration (100 mg/L). When all anions are present together in solutions, their negative effect on arsenic adsorption remains and adsorption capacities are slightly reduced than compared with the individual anions presence.

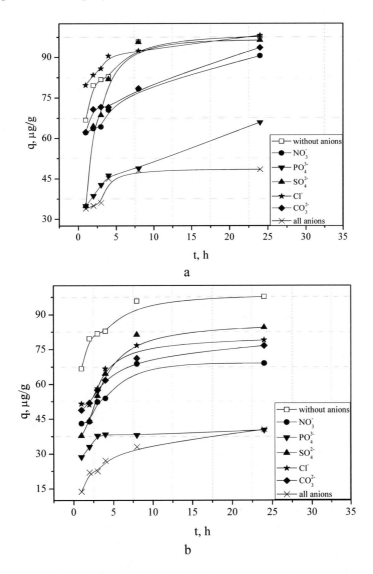

Figure 13. Influence of contact time on the Fe_2O_3 adsorption capacity. Anionic species concentrations: (a) 10mg/L; (b) 100mg/L.

If we compare the efficiency of the two studied materials we can notice that in the absence of the studied anions Fe_2O_3 develop a higher adsorption capacity than IS. The same observation is true for the individual presence of the studied anions in the samples, in a concentration of 10 mg/L. But when the anions are individually present in the solutions in concentrations 10 times higher (100mg/L) the adsorption capacity developed by the Fe_2O_3 is lower than in the case of IS. At the same time, if the As(III) solution contains simultaneously all anions, regardless their concentration, the IS is a more efficient adsorbent for As(III) than Fe_2O_3.

For this reasons we recommend the IS to be used as adsorbent material in the removal process of arsenic from water, since there is highly unlikely to find the studied anions individually in natural underground water. Using the sludge is also advantageous from the economic and environmental protection point of view, since this utilization turns into account a waste resulting from another technological process.

4. ARSENIC ADSORPTION FROM NATURAL UNDERGROUND WATERS

To evaluate the potential use of the studied material as adsorbents for As(III) removal from natural underground waters, we treated 5 sample collected from wells with a known high As(III) concentrations situated in the western area of Romania. The samples of water (100 mL) were treated with the necessary amount of adsorbent (0.1 g) under the optimal conditions established employing synthetic As(III) ion solutions (T=20-25°C, t=90 min, stirring rate=200 rpm, pH=6.5-7).

The initial and residual concentrations of As(III) ions, as well as other metal ions, were determined by atomic absorption spectrometric methods. The concentration of chloride has been determined by $AgNO_3$ titration, using K_2CrO_4 as indicator. Nitrite, nitrate, ammonoum, and phosphate concentrations have been determined by UV-VIS spectrophotometric method using a carry 50 spectrophotometer. The sulphate anion has been determined by direct titration with barium chloride in etilic alchool medium and in presence of barium sulphate using thorin as indicator.

Table 6. The initial compositions of the underground water samples

Parameter	Sample 1	Sample 2	Sample 3	Sample 4	Sample 5
pH	6.7	6.8	6.64	6.74	7
Turbidity (NTU)	9	8.7	8.5	8.7	9
Conductivity (µS)	398	385	378	395	383
P_2O_5 (ppm)	31.6	46.7	32.5	41.2	36.2
NO_2^- (ppm)	0.5	0.3	0.6	0.7	0.4
NO_3^- (ppm)	BDL[*]	22	BDL[*]	BDL[*]	20.2
NH_4^+ (ppm)	1.2	6.6	0.5	0.46	1.1
Cl^- (ppm)	13.8	31.1	10.4	20.7	17.3
SO_4^{2-} (ppm)	48.1	11.1	10.5	12.4	22.8
Ca (ppm)	65	30.8	35	27.4	7.5
Mg (ppm)	47	18.2	11.6	7.39	23.5
Na (ppm)	105	118	98.1	103	81
K (ppm)	1.65	1.67	0.89	1.44	3.78
Fe (ppm)	5	2	1.28	0.75	2
Mn (ppm)	0.5	0.51	0.17	0.2	0.35
As III (ppb)	60.4	80	45.4	62.9	73.5

*BDL= below detection limit.

The compositions of natural underground water samples are presented in table 6. After treatment of the underground water samples with the studied materials the water quality parameters were determined. It was observed that in all the underground water samples for all the studied adsorbents As(III) was practically totally removed from the waters together with iron and manganese ions (residual concentrations were under the detection limit).

In all the cases the pH of the treated waters remained rather unchanged, suggesting that no post treatment is necessary. Treatment with IS did not lead to any increase in the concentration of other metallic ions as a result of possible leaching. All these facts lead to the conclusion that the studied materials could be used as potential adsorbents for As(III) removal from natural underground water.

5. COLUMN STUDIES REGARDING AS(III) ADSORPTION

Batch operation is very easy to be used in the laboratory studies, but less convenient for field applications. Moreover, accurate scale-up data for fixed bed systems cannot be obtained from the adsorption isotherms of batch results, so practical applicability of the adsorbent should be ascertained in column operations. Adsorption on fixed bed columns presents numerous advantages. It is simple to operate, gives high yields and can be easily scaled up from laboratory process [23].

Generally, an arsenic adsorbent should meet several requirements: (1) low cost; (2) granular type; high capacity, selectivity, and rate of adsorption; (3) high physical strength (not disintegrating) in water and (4) able to be regenerated if required [20]. In order to accomplish the first requirement we used the IS in the column test as adsorbent for arsenic removal from water. Since the adsorbent is a powder, in order to avoid the clogging of the column we mixed the adsorbent with quartz sand. From previous studies we concluded the best IS:sand ratio is 1:1, when the highest adsorption capacity was reached [54]. Due to very small size of particles, IS penetrated between the grains of sand, and thus a high contact surface was created, the adsorption process occurred in good conditions. For a larger amount of IS the column was clogged after a short time. For a larger amount of sand, the mixture contains a too small amount of IS active phase responsible for arsenic adsorption, and consequently the adsorption capacity lowers.

Fixed bed adsorption studies were conducted in a 1.5 cm i.d., 20 cm length vertical down flow glass column packed with IS:sand adsorbent for different bed height (5, 7.5 and 10 cm). the experiments were conducted with a 100 μg/L As(III) solution. The column was charged with arsenic solution in the down-flow mode with an influent flow rate, which was supplied and maintained throughout the experiments by the use of variable flow peristaltic pump (Heidolph 6201). The samples were collected at certain time intervals and were analysed for the remaining arsenic concentrations. The shape of the breakthrough curve and the time for the breakthrough appearance are the predominant factors for determining the operation and the dynamic response of an adsorption column. Effects of adsorbent bed height (5-10 cm) and inlet feed flow rate 0.033-0.16 cm^3/s) were investigated on the performance of breakthrough for the adsorption of As(III) by the IS:sand mixture.

Empty bed contact time is a critical parameter that determines the residence time during which the solution being treated is in contact with the adsorbent. Hence, EBCT may strongly affect adsorption, especially if the adsorption mainly depends on the contact time between the adsorbent and adsorbate [55]. The corresponding EBCT in the bed was in the range 55-535 s.

5.1. Effect of Bed Depth

The breakthrough curves obtained for As(III) adsorption onto IS:sand at different bed depth, at a constant linear flow rate of 10 cm^3/min are shown in figure 14.

Sorbent bed height strongly affects the volume of the solution treated or throughput volume. The curves followed characteristic S-shape profile, which is associated with adsorbate of smaller molecular diameter and more simple structure. As it is evident from figure 14, an increase in column depth increased the treated volume due to high contact time. The exhaust time (corresponding to an effluent concentration=0.9C_0) increased with increasing bed height, as more binding sites were available for sorption. The increase in adsorption with bed depth was due to the increase in adsorbent doses in larger beds which provided greater adsorption sites for As(III). The breakthrough time also increased with increasing bed depth, suggesting that it is the determing parameter of the process (table 7). The larger the breakthrough time, better is the intra-particulate phenomena.

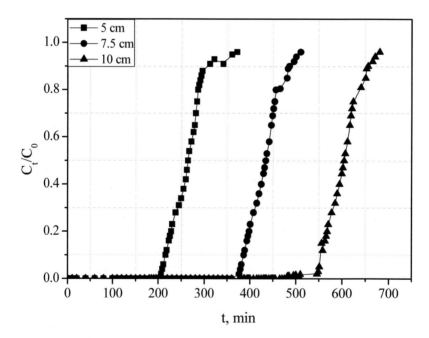

Figure 14. Effect of bed height on breakthrough curves for As(III) adsorption onto IS:sand. C_0=100µg/L, flow rate=10 mL/min, pH=6.8.

5.2. Effect of Flow Rate

The effect of flow rate on As(III) adsorption by IS:sand was investigated by varying the flow rate from 2 to 10 mL/min and keeping the initial As(III) concentration (100 µg/L), bed depth (10 cm) and column diameter (1.5 cm) constant. Breakthrough plots between C_t/C_0 versus time at different flow rate are given in figure 15.

At low flow rate, relatively high volume was observed whilst much sharper breakthrough was obtained at higher flow rates. The time required to reach the breakthrough decreased with an increased flow rate (table 7). This is certainly because of the reduced contact time causes a poor distribution of the liquid inside the column, which leads to a lower diffusivity of the solute through the adsorbent particles [23]. As seen from the results presented in table 7, an increased flow rate reduces the volume treated efficiently until breakthrough and thereby decreases the service time of the bed. This is due to the decrease in the residence time of the As(III) ions within the bed at higher flow rates.

One may notice that the best performance of the IS:sand mixture in the removal process of As(III) on column studies occurs under the process parameters of 10 cm bed depth and 2 mL/min flow rate.

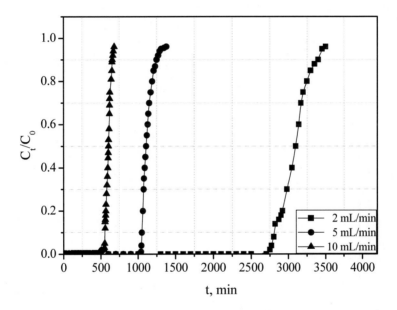

Figure 15. Effect of flow rate on breakthrough curves for As(III) adsorption onto IS:sand. $C_0=100\mu g/L$, h=10 cm, pH=6.8.

Table 7. Adsorption data for fixed bed IS: sand column for As(III) adsorption at different process parameters

Process parameters	Breakthrough time, min	Exhaust time, min	Volume of arsenic solution treated, L
Bed depth, h (cm)			
5	200	371	2/3.71
7.5	376	510	3.76/5.1
10	482	680	4.82/6.8
Flow rate, Q(mL/min)			
2	2750	3500	5.5/7
5	1040	1380	5.2/6.9
10	482	680	4.82/6.8

5.3. Modelling of Breakthrough Curve

The fixed bed column was designed by logit method [2]. The logit equation can be written as:

$$\ln\left[\frac{C}{C_0 - C}\right] = -\frac{KNX}{V} + KC_0 t \tag{15}$$

where C is the solute concentration at any time t, C_0 the initial solute concentration (100 µg/L), V the approach velocity (~ 0.033 mL/h), X the bed depth (10 cm), K the adsorption rate constant (L/(mg·h)), and N is the adsorption capacity coefficient (mg/L). Plot of $\ln[C\cdot(C_0 - C)]$ versus t was shown in figure 16. The value of adsorption rate coefficient (K) and adsorption capacity coefficient (N) was obtained as 0.084,L/(mg·h) and 1.03, mg/L. These values could be used for the design of adsorption column.

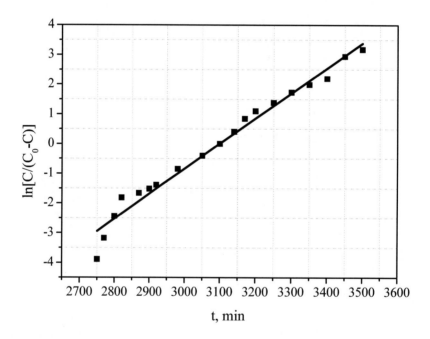

Figure 16. Linearized form of logit model.

6. IMMOBILISATION OF THE EXHAUSTED IS ADSORBENT IN VITREOUS MATRICES

In order to test the possibility of the exhausted IS adsorbent immobilization in vitreous matrices, this has been introduced in the composition of some frits, for further storage in appropriate conditions or reuse in decorative glass applications. 20g of arsenic containing IS were used for 100g frit. The oxide compounds were homogenized and fired at 1250°C, for 30 minutes. The molten glass obtained was poured in cold water, than grinded and classified in different grain size classes. The oxide composition of the obtained frit is presented in table 8:

Table 8. The oxide composition of the obtained frit

Component	Content, % wt.
SiO_2	39,5
B_2O_3	23,6
Na_2O	10,5
PbO	6,7
CaO	3,83
ZnO	0,38
Fe_2O_3	15,3
Al_2O_3	0,14
As_2O_3	0,03

The frit was tested for leaching, as follows: 5g of frit, grain size between 1 and 3 mm were treated with different acids at pH 4÷5 simultaing the effect of acid rains, for 60 days. The pH of the solutions was monitored, and the results are presented in table 9.

One may notice that the pH of the solutions raises up to 7 in the first week and remains constant until week 4. At the end of the test period the pH of the solutions is between 8 and 9, which may be assigned to the sodium leaching from the frits enhanced by the increased contact with the solutions, due to previous grinding. This assumption was confirmed by the chemical analysis of the solution performed by atomic absorption spectrometry (VARIAN Spectr AA 280FS) aimed to establish the metal content. The results are presented in table 10.

Table 9. pH variation of the solutions

Sample	Initial pH	pH variation							
		Week 1	Week 2	Week 3	Week 4	Week 5	Week 6	Week 7	Week 8
Frit + H_2O	5.21	7.10	7.01	7.06	7.13	8.12	8.40	8.47	8.55
Frit + HNO_3	4.64	7.04	7.01	7.06	7.15	8.21	8.34	8.43	8.53
Frit + HCl	4.30	7.03	6.95	7.09	7.19	8.13	8.35	8.57	8.56
Frit + H_2SO_4	4.00	7.09	6.99	7.12	7.23	8.24	8.36	8.46	8.56
Frit + CH_3COOH	4.19	7.10	7.04	7.21	7.56	8.22	8.41	8.43	8.54
Frit + carbonated water	5.47	7.09	6.97	7.12	7.47	8.56	8.99	8.96	9.02

Table 10. The metal content in the leaching solutions

Sample	Metal content, mg/L				
	Ca^{2+}	Na^+	Pb^{2+}	Fe^{n+}	As^{3+}
Frit + H_2O	0.45	63.2	BDL*	BDL	BDL
Frit + HNO_3	0.52	68.2	BDL	BDL	BDL
Frit + HCl	0.63	65.1	BDL	BDL	BDL
Frit + H_2SO_4	0.45	64.2	BDL	BDL	BDL
Frit + CH_3COOH	0.53	68.4	BDL	BDL	BDL
Frit + carbonated water	0.25	88.6	BDL	BDL	BDL

*BDL = below detection limit.

No dangerous species were noticed in the leaching solutions, so one may conclude the exhausted adsorbent IS containig arsenic was succesfully immobilized in the glassy matrix of the frit and it doesn't represent an environmental threat anylonger. Yet, further tests were performed to explore the possibility of actually reclaiming the materials envolved in arsenic adsorption and than immobilized in glass, meaning to find some applications for the resulted frits. For this purpose, another waste was used. CRT waste glass from old TV/ computer screens is available in large amounts and raises disposal issues. Since the CRT glass has high melting temperature (around 1350°C), the need of a fluxing agent occurred. Sodium silicate ($Na_2O\cdot1.5SiO_2$) has been used as fluxing agent. The CRT glass oxide composition is presented in table 11.

Table 11. CRT glass oxide composition

Oxide	Composition, % wt.
SiO_2	62
Na_2O	6.3
K_2O	9
SrO	10
BaO	7
PbO	1
Al_2O_3	2
$Al_2O_3 + Sb_2O_3$	0.5
$TiO_2 + ZrO_2$	2
CeO_2	0.2

The raw materials mixture ratio for obtaining the frit was IS exhausted adsorbent: CRT glass: fluxing agent ($Na_2O\cdot1.5SiO_2$) = 1:1:1. After melting at 1200°C for one hour, the glass melt was poured in cold water and the frit obtained consisted of smooth glass wires or grains, with no incomplete molten solid inclusions or air bubbles, which all indicate a proper melting. An image of the as-obtained frit is presented in figure 14.

Figure 14. Image of the obtained frit.

In order to assess the arsenic and other dangerous elements leaching, the frit was treated with 4% wt. acetic acid solution, according to STAS 708/2-83 [56] for 24h at room temperature. The result of the chemical analysis performed by atomic absorption spectrometry with hydride generation, including the Na^+ content (a good indicator of glass solubility) is presented in table 12:

Table 12. Chemical analysis result of the leaching solution

Leached elements (mg/g frit)			
As	Pb	Fe	Na
BDL *	BDL *	0,35	0,23

*BDL = below detection limit.

The result shows a slight solubility of the frit, but no possible dangerous elements (As or Pb) were noticed in the solution, so the arsenic containing IS immobilization was considered successful. At the same time, using CRT waste glass for this purpose proved to be a viable solution. The frit was further processed in order to obtain a ceramic decorative glaze, since the overall iron content of the frit makes it suitable for obtaining aventurine glazes (generally 10-30% Fe_2O_3 in glaze). After grinding, it was mixed with 5% wt. kaolin and subjected to wet homogenization in a Fritsch Pulverisette ball mill. The slip obtained was applied on a tableware faience ceramic support and after drying, it was fired at 1150°C for 1 hour, followed by a 30 minutes thermal treatment at 780°C, designed to develop the hematite crystals, which impart a certain decorative effect due to their flaky golden appearance. The samples were investigated by optical microscopy in reflected light using a Guangzhou L2020A China microscope with digital camera and by X-Ray diffraction, using a DRON 3 instrument, Cu $_{K\alpha}$ radiation.

Figure 15 presents the glazed samples, as well as optical microscopy images of the hematite crystals developed in the samples, with different magnifications.

50x   100x   400x

Figure 15. Images of the glazed samples showing the decorative effect and the hematite crystals.

Figure 16 presents the X-ray diffraction spectrum of the glaze showing hematite as the single crystalline phase.

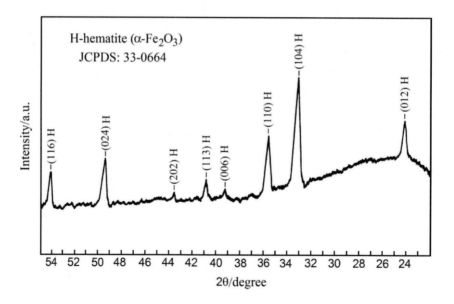

Figure 16. The X-ray diffraction spectrum of the glaze.

Considering the positive results obtained we consider that using the iron containing waste sludge in arsenic adsorption followed by the immobilization of the exhausted adsorbent in vitreous matrices represents a viable solution framed in the direction of closed cycle technologies.

ACKNOWLEDGMENTS

This work was supported by CNCSIS- UEFISCDI, project number PN II-IDEI 927/2008, "Integrated Concept about Depollution of Waters with Arsenic Content, through Adsorption on Oxide Materials, followed by Immobilization of the Resulted Waste in Crystalline Matrices". This work was partially supported by the strategic grant POSDRU/89/1.5/S/57649, Project ID 57649 (PERFORM-ERA), co-financed by the European Social Fund – Investing in People, within the Sectoral Operational Programme Human Resources Development 2007-2013.

REFERENCES

[1]   K. Banerjee, G.L. Amy, M. Prevost, S. Nour, M. Jekel, P.M. Gallagher and C.D. Blumenshein, *Wat. Res.* 42, 3371 (2008).
[2]   S.K. Maji, A. Pal and T. Pal, *J. Hazard. Mater.* 151, 811 (2008).
[3]   H. Guo, D. Stuben and Z. Berner, *Sci. Total Environ.* 377, 142 (2007).
[4]   WHO, Guidelines for drinking-water quality, third ed. Vol. 1: Recommendations, *World Health Organization* (2004).
[5]   T. S. Singh and K.K. Pant, *Sep. Purif. Technol.* 48, 288 (2006).

[6]   V.T. Nguyen, S. Vigneswaran, H.H. Ngo, H.K. Shon and J. Kandasamy, *Desalination* 236, 363 (2009).

[7]   J.R. Parga, D.L. Cocke, J.L. Valenzuela, J.A. Gomes, M. Kesmez, G. Irwin, H. Moreno and M.Weir, *J. Hazard. Mater.* B124, 247 (2005).

[8]   R.Y. Ning, *Desalination* 143, 237 (2002).

[9]   J.Kim and M.M. Benjamin, *Wat. Res.* 38, 2053 (2004).

[10] T. Tuuutijarvi, J. Lu, M. Sillanpaa and G. Chen, *J. Env. Eng.* 136, (2010).

[11] C.Y. Chen, T.H. Chang, J.T. Kuo, Y.F. Chen and Y.C. Chung, *Bioresource Technol.* 99, 7487 (2008).

[12] S. Kundu and A.K. Gupta, *J. Hazard. Mater.* 142, 97 (2007).

[13] D. Borah, S. Satokawa, S. Kato and T. Kojima, *J. Hazard. Mater.* 162, 1269 (2009).

[14] D. Borah, S. Satokawa, S. Kato and K. Kojima, *J. Colloid Interface Sci.* 319, 53 (2008).

[15] P. Mondal, C.B. Majumder and B.Mohanty, *J. Hazard. Mater.* 150, 695 (2008).

[16] H.U. So, D. Postma, R. Jakobsen and F. Larsen, *Geochim. Cosmochim. Ac.* 72, 5871 (2008).

[17] P. Chutia, S. Kato, T. Kojima, S. Satokawa, *J. Hazard. Mater.* 162, 204 (2009).

[18] H.L. Lien and R. T. Wilkin, *Chemosphere* 59, 377 (2005).

[19] Y. Jeong, M. Fan, S. Singh, C.L. Chuang, B. Saha and J. H. Leeuwen, *Chem. Eng. Process.* 46, 1030 (2007).

[20] L. Zeng, *Wat. Res.* 37, 4351 (2003).

[21] K. Gupta and U.C. Ghosh, *J. Hazard. Mater.* 161, 884 (2009).

[22] Z. Ren, G. Zhang and J.P. Chen, *J. Colloid Interf. Sci.* 358, 230 (2011).

[23] S. Kundu and A.K. Gupta, *Chem. Eng. J.* 129, 123 (2007).

[24] S. Kundu and A.K. Gupta, *Sep. Purif. Technol.* 51, 165 (2006).

[25] O.S. Thirunavukkarasu, T. Viraraghavan and K.S. Subramanian, *Water Air Soil Poll.* 142, 95 (2003).

[26] J.C. Hsu, C.J. Lin, C.H. Liao and S.T. Chen, *J. Hazard. Mater.* 153, 817 (2008).

[27] O.S. Thirunavukkarasu, T. Viraraghavan and K.S. Subramanian, *Water SA* 29, 161 (2003).

[28] B.J. Lafferty and R.H. Loeppert, *Environ. Sci. Technol.* 39, 2120 (2005).

[29] M. Vithanage, R. Chandrajith, A. Bandara and R. Weerasooriya, *J. Colloid Interf. Sci.* 294, 265 (2006).

[30] X. Guo and F. Chen, *Environ. Sci. Technol.* 39, 6808 (2005).

[31] B.R. White, B.T. Stackhouse and J.A. Holcombe, *J. Hazard. Mater.* 161, 848 (2009).

[32] T. Moller and P. Sylvester, *Wat. Res.* 42, 1760 (2008).

[33] A. Gupta, V.S. Chauhan and N. Sankararamakrishnan, *Wat. Res.* 43, 3862 (2009).

[34] S. Dakhai, L.A. Orlova and N.Yu. Mikhailenko, *Glass and Ceramics 56*, 177 (1999).

[35] I. A. Levitskii, *Glass and Ceramics 58*, 223 (2001).

[36] I. N. Dvornichenko and S.V. Matsenko, *Glass and Ceramics*, *57*, 67 (2000).

[37] L. Lupa, A. Iovi, P. Negrea, A. Negrea and G. Szabo, *Environ. Eng. Manage. J.* 5, 1099 (2006).

[38] A. Negrea, L. Lupa, M. Ciopec, R. Lazau, C. Muntean and P. Negrea, *Adosrb. Sci. Technol.* 28, 467 (2010).

[39] Lj.S.Cerovic, S.K. Milonjic, M.B. Todorovic, M.I. Trtanj, Y.S. Pogozhev, Y. Blagoveschenskii and E.A. Levashov, *Colloids Surf. A* 297, 1 (2007).

[40] A. Negrea, M. Ciopec, L. Lupa, C. Muntean, R. Lazau, P. Negrea, *Water pollution X, WIT Transaction on Ecology and Environment*, 135, 117 (2010).

[41] P. Negrea, A. Negrea, L. Lupa, and L. Mitoi, *Proc. Int. Symp. "Trace Elements in the Food Chain"*, Budapest, Hungary, May 25–27 (2006).

[42] P. Niedzielski, *Anal. Chim. Acta* 551, 199 (2005).

[43] H.Y. Shan, *Scientometrics* 59, 171 (2004).

[44] Y.S. Ho and G. McKay, *Process Biochem.* 34, 451 (1999).

[45] F. Partey, D. Norman, S. Ndur and R. Nartey, *J. Colloid Interface Sci.* 321, 493 (2008).

[46] A. Negrea, L. Lupa, M. Ciopec, C. Muntean, R. Lazau, M. Motoc, *Rev Chim-Bucharest* 61, 691 (2010).

[47] R. Lazau, L. Lupa, I. Lazau, P. Negrea and C. Pacurariu, *Rom. J. Mater.* 40, 71 (2010).

[48] K. Tyrovola, N.P. Nikolaidis, N. Veranis, N. Kallithrakas-Kontos and P.E. Koulouiridakis, *Wat. Res.* 40, 2375 (2006).

[49] X. Meng, S. Bang and G.P. Korfiatis, *Wat. Res.* 34, 1255 (2000).

[50] X. Guan, H. Dong, J. Ma and L. Jiang, *Wat. Res.* 43, 3891 (2009).

[51] F. Frau, R. Biddau and L. Fanfani, *Appl. Geochem.* 23, 1451 (2008).

[52] A. Negrea, C. Muntean, L. Lupa, R. Lazau, M. Ciopec and P. Negrea, *Chem. Bull. "Politehnica" Univ. (Timisoara)*, 55, 46 (2010).

[53] S. Wang and C.N. Mulligan, *J. Hazard. Mater.* B138, 459 (2006).

[54] A. Negrea, L. Lupa, C. Muntean, M. Ciopec, P. Negrea and R. Istratie, *Chem. Bull. "Politehnica" Univ. (Timisoara)* 55, 123 (2010).

[55] H. Guo, D. Stuben, Z. Berner, U. Kramar, *J. Hazard. Mater.* 151, 628 (2008).

[56] STAS 708/2-83, Porcelain and tableware faience – chemical tests, acid resistence (Romanian language).

INDEX

J

K

L

T